Image & Environment

Image & Environment

Roger M. Downs
David Stea
editors

Cognitive Mapping and Spatial Behavior

With a foreword by Kenneth E. Boulding

Routledge
Taylor & Francis Group
LONDON AND NEW YORK

First published 1973 by Transaction Publishers

2 Park Square, Milton Park, Abingdon, Oxfordshire OX14 4RN
711 Third Avenue, New York, NY 10017

Routledge is an imprint of the Taylor & Francis Group, an informa business

First issued in hardback 2017

Library of Congress Catalog Number: 2004062014

Library of Congress Cataloging-in-Publication Data

Image and environment : cognitive mapping and spatial behavior / Roger M. Downs and David Stea, editors.
p. cm.
Originally published: Chicago : Aldine Pub. Co., 1973.
Includes bibliographical references and index.
ISBN 0-202-30766-2 (pbk. : alk. paper)
1. Space perception. 2. Spatial behavior. 3. Cognition. I. Downs, Roger M. II. Stea, David.

BF469.D68 2005
153.7'52—dc22 2004062014

ISBN 13: 978-0-202-30766-4 (pbk)
ISBN 13: 978-1-138-52570-2 (hbk)

Contents

Foreword

I have read these papers with very great interest. They are interesting in their own right as an expression of an "invisible college" which represents almost a new discipline, cutting across the old disciplines of geography and psychology, with a considerable dash of other social sciences, and even a tantalizing flavor of history, as we move from space into space-time. They represent, furthermore, a significant contribution towards filling the biggest hole in the sciences, the missing science of human learning. In this connection the importance of any contribution can hardly be overestimated, for it may well be that in the present extraordinary period in the history of this planet a science of human learning is the only thing which can save us from disaster. All social problems, whether war, population control, the structure of authority, the distribution of income, or whatever we would like to mention, can only be solved by a dynamics which includes human learning as its most essential element.

The plain fact is, however, that up to now we know very little about human learning in an explicit model-building sense. Education is still a craft industry. Most of us teach the way we do because it is the way we were taught. The production of human beings in the family is also a craft industry. People raise children, on the whole, in the way that they themselves were raised. The few onslaughts from the scientific subculture in the way of learning theory, in the one case, and Drs. Gesell and Spock in the other, have made only marginal changes. Not, of course, that craft industries should be despised; they often produce very elegant and delightful articles. We must be doing something right both in childrearing in the family and in formal education; otherwise we would not have survived as long as we have. In what I have elsewhere called the "crisis of closure," however, in which our terrified spatial imagination perceives the all-too-near wall of the niche into which the human race has been expanding for so long, the craft production of human adults may not be enough, and any move towards a more systematic and specific knowledge of human learning is a big step towards survival.

My little book *The Image* propounded the then somewhat heretical notion that learning essentially consisted of a growth of knowledge, and that behavior was a function of knowledge; that is, the cognitive structure, which I called the "image," including the valuations which were placed over the structure. On this view most behavior could not be explained in terms of a response to any immediate stimulus. This is, of course, the sort of psychological doctrine one would expect from a simple-minded economist, who thinks mainly in terms of the theory of choice, for choice must be between alternative images of the future. There is nothing else to choose, and though immediate stimuli, like seeing something in a shop window, may affect our images of the future, the stimulus clearly is merely a trigger and the act depends on the whole cognitive-valuation structure, that is, on the image. We are not going to find any stimulus-response regularities, therefore, except in extremely simple cases, any more than we should expect to find regularities between a trigger and its consequences if we knew nothing about the system as a whole. The stimulus-response theory, therefore, seems to take us only a very little way into the system which we are really investigating, simply because between the stimulus and the response lies an image which simply cannot be dismissed by calling it an intervening variable. It is not a variable; it is a vast and complex set of parameters to which we also have some kind of access, even though an imperfect one.

It is one thing to postulate a cognitive-valuative structure as a necessity, almost indeed an evolutionary necessity, as Kaplan points out. It is quite another thing to be able to perceive it as a system and to understand its structure. It is indeed a very real question as to whether this is possible, that is, whether the knower can understand not only what he knows, but also what the systematic structure of that knowing is. Whether we can go all the way with this, however, does not concern us. The important thing is that we can go part way, and we should go as far as we can. In this connection, the study of spatial images is of peculiar importance, not only because they are of themselves perhaps the most significant part of the total structure, but also because they seem to be accessible in a way that other parts of the image are not. Indeed, if we think about other elements of the image, we have a strong tendency to structure this in spatial terms. Thus, an economist thinks of valuation in terms of the utility function, which is a kind of "mountain." We think of personal relationships in such terms as near and far, obscure and clear, devious and direct, exalted and commonplace, all of which have strong spatial connotations and are in fact spatial metaphors. It is a challenge indeed to try to think of a metaphor that is not in some sense spatial. We can perhaps rank the senses, in terms of the spatial quality of information which they give us; in the order of vision, hearing, touch, smell, and taste, with taste metaphors, such as bitter and sweet, salty and bland, having the least spatial connotations.

The spatial connotations of language indeed is a subject which does not seem to be included in this volume, perhaps because it does not yet exist, but it is clearly a fascinating field in which the geographer and the linguist might well get together. One thinks, for instance, of the spatial aspects of grammar. The comparative and the superlative are clearly spatial metaphors. Classificatory structures, like gender, have less direct spatial reference, but always involve spatial metaphors like pigeonholes and represent an extension of the spatial metaphor "here" and "there." Word order, which is of overwhelming importance in some languages, is very clearly spatial. Any ordinal ranking has a spatial referent and is indeed a spatial metaphor. It is hard to believe that this overwhelming importance of the spatial metaphor is not related to some as yet unknown spatial pattern of the human nervous system, which enables us to perceive the four-dimensional space-time world in which we live because of some kind of corresponding structure "inside," this structure coloring all our other perceptions. How startling it is, however, to find one's use of the word "color," which is a non-spatial metaphor, to describe the expansion of the internal spatial structure into a more complex image.

Perhaps the reason why the study of "inner space," or internal spatial structures, is so immediately rewarding is that we now have external "maps" which we believe are very accurate representations of the real world, derived from processes of surveying, the application of trigonometry and geometry, and so on, about which there is virtually no dispute, mainly because there is an extraordinarily powerful and effective feedback of mapping error from anyone who uses maps. It is significant that the error here is the failure of correspondence between an external map and an internal map, in which it is the failure of the external map which produces correction as a function of the social system. In the individual learning process, it is correction of error by the comparing of our internal maps with "maps of maps," internal maps of external maps. The map, therefore, in all its possible forms represents a unique contribution towards error detection and its use, not only in the form of geographical maps but in the form of chemical formulas, physical laws, and so on. The map concept is perhaps the real key to the development of science. This is an aspect of the theory of scientific epistemology which has not been sufficiently explored.

I cannot resist the temptation to conclude with a personal note on the curious nature of what might be called scientific space. The organization of science into disciplines sets up a series of ghettos with remarkable distances of artificial social space between them. I have been interested in establishing communications among the disciplines ever since I went to Iowa State College (as it was then called) at Ames, Iowa in 1943, to convert myself into a labor economist. Until that time I had been an extremely pure economist, believing indeed that the other social sciences either did not exist, or at least could make no contribution to economics. Getting into the study of the labor movement convinced me that without sizeable inputs, certainly of sociology and political science, and more doubtfully of psychology, the phenomenon of the labor movement could not possibly be understood. This got me interested in the problem of developing a general social science.

When I went to the University of Michigan in 1949 it was with the intention of developing a seminar in the integration of the social sciences, and this I carried on for a number of years. It attracted a very diverse group of people—engineers, architects, biologists, sociologists, anthropologists, and political scientists. I cannot recall that I was able at any time to find a psychologist interested in it, but I have to confess that my cognitive map of psychology looks remarkably like the interior of Greenland. The seminar, however, did have a curious indirect by-product. As a result of it, I got into correspondence with the late Ludwig von Bertalanffy, who was approaching general systems from the side of biology as I was approaching general systems from the side of economics. We spent a year together in 1954–55 at the Center for Advanced Study in the Behavioral Sciences at Stanford, and out of conversations begun there between Bertalanffy, Anatol Rapoport, Ralph Gerard and myself, we founded what eventually became the Society for General Systems Research. Even in this enterprise, psychologists played an extremely minor role. In that golden year at the Behavioral Sciences Center I recall very few fruitful contacts with psychologists with one exception, Elsa Frenkel-Brunswick. Her husband Egon's famous "lens model" is an important contribution to cognitive theory, and I am a little surprised that no use seems to have been made of it in this volume. The work of my colleague Kenneth Hammond, which follows very closely that of Frenkel-Brunswick, on computer graphic representations of conflictual positions, seems to me very close to the general problems of cognitive mapping, using spatial analogies rather than space itself.

It was a real surprise to me, therefore, to learn from the papers in this volume that my little book *The Image*, which emerged from an intense nine days of solitary dictating at the end of my time at the Center for Advanced Study in the Behavioral Sciences, should have figured in the history of cognitive mapping in the way it seems to have done. In *The Image* I did make a somewhat half-hearted suggestion that perhaps a new science might emerge out of these concepts, which I dubbed "eiconics." However, after writing *The Image* I became interested in other things—in the theory of conflict, general systems, and so on—and my general impression was that *The Image* had fallen on fairly stony intellectual ground. It aroused interest in surprising quarters—among business executives, among the people who teach humanities in engineering schools, and in departments of communications—but as far as I knew it had never penetrated the boundary of the behavioral sciences, for whose benefit it was ostensibly written. These boundaries obstruct traffic both ways. I did not even run across the book by Miller, Galanter and Pribram until long after it was published. I have to confess to my deep embarrassment, especially as I have been a map lover since childhood and incapable of resting content at any new location until I have provided myself with topographical maps of the area, that the exciting development of cognitive mapping, as revealed in this volume, came to me as an almost complete surprise. Cognitive mapping must certainly be Part One of any textbook of "eiconics," so the surprise had a strong mixture of sheer delight along with the personal embarrassment. It is hard, therefore, for me to avoid looking on this volume, as delightful as it is in its own right, as a prelude to something larger, nothing less indeed than a general theory of the cognitive-valuative structure and the learning processes by which it is created.[1] One returns again, however, to the nagging question, how can the intellectual community be organized to reduce the social space between the disciplines? Walls, no doubt, there should be. Otherwise the interdisciplinary all too easily becomes the undisciplined. On the other hand, there surely doesn't have to be a Sahara desert between the economists and the psychologists. Volumes of this kind are an oasis, but one longs for a time when the oasis will expand and emerge in an uninterrupted fertile and traversable republic of the mind.

KENNETH E. BOULDING

1. In this connection the book *Culture and Cognition*, edited by James P. Spradley, should be noted as another step toward "eiconics." This is an expression of what might almost be called the "eiconics movement" in anthropology. But it is remarkable, and perhaps another symbol of the spatial isolation of the disciplines, that I am the only one out of forty-five authors contributing to both this volume and the Spradley volume.

Preface

Aims of the Book

The genesis of this book was an invitation to edit a special issue of the journal, *Environment and Behavior*, in which we brought together some of the emerging work on environmental cognition. After that task was completed, we embarked on the preparation of this book. It would be a fiction to claim that we had fixed aims—we can more easily postdict a set of aims. Primary among these is to introduce and give coherence to a recent approach to studying human environmental behavior. A concern with the link between human behavior and the environment has always been at least an implicit concern of the social sciences, but never before has this concern been manifested so vocally and forcibly as in the recent past—in fact, display of such concern is now demanded by the general public.

As a result, many disciplines have responded to the call, and in true academic fashion, have generated a series of "–ologies," "–istics," and "–isms." Foremost among these is the attempt to relate human behavior to the physical environment, the latter including both "natural" and man-made components. Hastily joining this effort have been a host of disciplines. Consequently, we are faced with a bewildering list of descriptive names for this sub-field: behavioral geography, environmental perception, environmental psychology, ecological psychology, and human ecology are among those proposed. Some clarification has been made in an edited collection of papers on environmental psychology by Proshansky, Ittelson, and Rivlin (1970), and in the reviews by Craik (1970a) and Wohlwill (1970). Even so, in an area where the geometric law of publication applies, we need overviews and critiques of the scattered and often conflicting work on environmental behavior.

In particular, this book represents a collection of papers on the topics of cognitive mapping and cognitive maps. Cognitive mapping is a construct which encompasses those cognitive processes which enable people to acquire, code, store, recall, and manipulate information about the nature of their spatial environment. This information refers to the attributes and relative locations of people and objects in the environment, and is an essential component in the adaptive process of spatial decision making. Thus, for example, cognitive mapping helps us to solve such diverse spatial problems as finding a supermarket, choosing a safe and quick route to and from our workplace, locating potential sites for a new house or a business, and deciding where to travel on a vacation trip. The cognitive processes are not constant but undergo change with age (or development) and use (or learning). Similarly, a cognitive map is an abstraction which refers to a cross-section, at one point in time, of the environment as people believe it to be.

In presenting this structured overview of cognitive mapping, we are *not* trying to force legitimacy on a field which has yet to earn it. The principal aim is the introduction of a new research area to a wider public, with the hope of encouraging further efforts in research and application. Linked with this aim is the presentation of some categories within cognitive mapping based upon what *has* been done rather than what *ought* to have been done. In achieving these goals, many compromises must be made. We had to temper the desire to include as much as possible with the economics of publishing; we balanced the desire to sample a bit of everything with the need to make the sections into meaningful, related collections of ideas and empirical data. Above all, we have allowed the weight and scope of published material to suggest the nature of the sections and define the boundaries. Nothing could be more inimical to the future of this interdisciplinary effort than premature fossilization.

Obviously, with so many pitfalls available, we could not attain perfection; but, given space limitations and our personal biases, we hope that the six sections—Theory, Cognitive Representations, Spatial Preference, the Development of Spatial Cognition, Geographical and Spatial Orientation, and Cognitive Distance represent a fruitful breakdown of cognitive mapping studies. Other breakdowns of the material are undoubtedly possible, given the overlap among the papers in the different categories we have chosen, though any other scheme would have generated as much or more overlap. We might have taken the suggestion of one reviewer and organized the book in accordance with a three-fold scheme consisting of (1) papers taking a theoretical position; (2) papers dealing with a general empirical overview of the field, and (3) papers dealing with specific experimental studies. We rejected this approach because our experience as teachers of this subject matter convinced us that students, at least at this stage, were more concerned with content areas than with the presence or absence of a theoretical statement. The breakdown of the material we have chosen is that which made most sense to us, our colleagues, and our students. No overall theoretical rationale is intended or, at this stage, probably even necessary. Application of the criterion of "aesthetic balance" would have led us to arrange things so that equal numbers of contributions appeared in each section, but would have accomplished little else.

We expect, however—and in fact hope—that readers will question both the number of sections and the overall coverage of material. We recognize the omission of work on environmental learning: the process of the acquisition of spatial information and its effects on existing cognitive structures. There is a lack of material dealing with pragmatics: the use of cognitive mapping concepts in improving the quality and process of environmental design. However, we were constrained by the paucity either of work in certain areas or of workers currently researching such areas. The omissions are a reflection of the current state of cognitive mapping studies.

We tried to depart from the normal format of a book of readings by providing some genuine integration among dispersed ideas from diverse philosophical backgrounds. This was realized in two ways: first, by our summary paper in the Theory section and by the section introductions; second, by the solicitation and editing of papers designed to fill existing gaps in cognitive mapping research. Thus, our objective is the coherent discussion of a research area and *not* merely the presentation of disparate views of academics bound and linked only by the covers of a book.

A book aims not only at meaningful content but also at a potential audience. In the choice of material and the mode of its presentation, we focused on three interest groups. The first was the individual social scientist attracted to the problems of relating behavior to the spatial environment. We have tried to present a comprehensive overview of cognitive mapping allowing him to grasp basic concepts, methods, and existing bodies of empirical data. Thus, the book is intended as an introduction to a field.

However, we wanted to produce a book satisfying a second reader—the would-be or current researcher in cognitive mapping. To this end, we solicited a wide range of new material, locationally and disciplinarily, indicating the current state of the art. Approximately 80 percent of the book consists of original contributions. In some instances, such as the paper by Kaplan (chap. 4), we sought contributions from outstanding researchers to cover the existing field. We also encouraged contributors to reference heavily and collated these references to form an extensive bibliography for research use.

Our third interest group consisted of students. We wanted to produce a book which could be used as a text or source book in courses dealing with man and his spatial environment. To this end, the sections are relatively autonomous, each having its own brief introduction; thus, the book can be read selectively and in an order suggested by a teacher.

Obviously, these three needs conflict and we have recognized this by not satisfying one exclusively, at the expense of the others. We felt that a state of the art book was required at this time to present the reader with the origins, achievements, and future of cognitive mapping research.

Scope of the Book

In selecting contributions for the book, we aimed at a representative coverage of the content areas being studied by researchers in cognitive mapping. Consequently, the selection in no way indicates the relative volume of research either on an author or area basis. Within this guideline, we employed a set of criteria to select from the "classic" works and new contributions available. Obviously, in a new area, recency itself was an important consideration of our aim toward an up-to-date review of cognitive mapping.

One major criterion was the desire to make a sample of historical milestone papers available (such as chaps. 2 and 16), particularly if they were otherwise inaccessible (chaps. 5, 11, and 15). Second, we sought contributions with a theoretical component since we believe that pragmatic solutions depend upon the development of coherent and tested theory. The papers by Kaplan (chap. 4) and Stea and Blaut (chaps. 3 and 12) show how some of the disparate concepts and data can be meaningfully and productively related. Third, heavy empirical content was stressed because we are as much lacking in "hard" data as in developed theory. In this respect, we feel that the papers by Briggs, Lowrey, and Lundberg (chaps. 17–19) provide the foundations out of which speculations, then hypotheses, and eventually theories will emerge.

Fourth, we sought papers with good reviews of the literature relevant to cognitive mapping studies. Selection on this basis was difficult because so many reviews have touched peripherally on cognitive mapping. We avoided overlap with the works of Proshansky, Ittelson, and Rivlin (1970), and Craik (1970a) mentioned earlier. We felt that the present state of work on geographical and spatial orientation could be covered by the reviews in the papers by Griffin, and Lynch (chaps. 15 and 16).

Whenever we detected a gap which could not be filled in accordance with the above criteria, we solicited contributions from experts in the area. Hence, at a later stage in the compilation of the book, the papers by Francescato and Mebane, and Orleans were obtained (chaps. 7 and 8).

Editing is as much a task of exclusion as inclusion, and so we must indicate the decision rules for excluding papers. First, linguistic bias favored works authored in English and representing the U.S., British, and Swedish orbits. The only serious omissions appear to be the Russian studies of spatial orientation (Angyal, 1965 and Shemyakin, 1962), and some writings of the Piagetian School, which are represented in the paper by Hart and Moore (chap. 14).

Second, we gathered little material from outside the spectrum formed by geography, psychology, and planning. This is a reflection both of editorial competence and of the major interest in cognitive mapping within these disciplines. We were aware of studies in cultural anthropology and sociology which touched upon the topic of cognitive mapping, but whose prime objective, however, is different from that of this book; introductions to this other area are to be found in Brookfield (1969) and Campbell (1968). In addition, Lynch's bibliography (chap. 16) is an excellent primer for cultural and anthropological studies.

Three peripherally related types of work are excluded. First, studies of cartographic perception or psychophysics, though dealing with environment, perception, and behavior, have objectives different from our own. Studies such as Wood's (1968) attempt to relate the graphic map presentation mode to the amount and type of information perceived and abstracted by the map reader. They are oriented more to studying perception (for example, the discriminability of various color shading systems) than to a subject's cognitive grasp of his environment.

Second, we excluded studies of spatial mapping which are *not* cognitively oriented except by inference. Although the studies by Chapin and Brail (1969) or Moore and Brown (1970) show the mapped spatial pattern of human activities, the *cognitive* component is present *by inference only*. This comment should not be interpreted as a criticism of this valuable work, since it affords evidence of human spatial behavior regularities which can be investigated via cognitive methodologies.

Finally, we must stress that although we have not explicitly included the work of such cognitive psychologists as Bruner and Werner, we subscribe to the view expressed in chap. 4 by Kaplan. We believe that spatial cognitive mapping processes are part of a general cognitive process whereby an individual copes with information from his total environment. Although spatial cognitive mapping processes are a sub-set of this general process, studies have reached a stage where they are beginning to generate some higher level concepts and hypotheses: this is apparent in part II on cognitive representations. But again, space and our general theme limited our range of choice, and hence we were forced largely to exclude the non-spatial cognitive mapping processes.

The recently expanded interest in cognitive mapping is the result of many complementary factors. Among the more general ones are closer links among geography, planning, environmental design disciplines, and the social sciences. The more specific factors include the apparent bankruptcy of the regional descriptive approach and emergence of the so-called behavioral approach in the spatial sciences, as well as the rejection of economic formulations which relied heavily upon the concept of "rational" or "economic" man. As this book indicates, studies of cognitive mapping encompass a wide range of phenomena at a variety of spatial scales. Our present ability to explain and predict spatial behavior using the approach of cognitive mapping is limited, but the potential for future success is widely recognized. The six sections that follow should assist the reader in understanding what has transpired, and the probable nature of future directions.

Acknowledgments

We have been fortunate in being able to call upon a wide range of help in preparing this book. Hopefully, we have not overlooked anyone, but, if we have, our grateful acknowledgments extend to them, in addition to: the U.S. Office of Education for grant #OE4493 which supported the work reported in the papers by Hart and Moore, and Stea and Blaut (a,b); Ronald Abler, James Blaut, Roger Hart, David Hodge, Greg Knight, Leonard Mark, George McCleary, Gary Moore, David Seamon, Milton Wend, and Anthony Williams who have all read and commented upon various bits and pieces of the book; Peter Bamford and Suzanne Downs for bibliographic labors; David Hodge for cartographic assistance; and finally, but not least, to Anne Bates, Gloria Graves, Bill Loomis, Nina McNeal, Linda McGovern, and Kathy Yakich for typing, and retyping!

Contributors

Donald Appleyard	Departments of City and Regional Planning, and Landscape Architecture, University of California, Berkeley.
James Blaut	Department of Geography, University of Illinois, Chicago Circle.
Ronald Briggs	Department of Geography, University of Texas at Austin.
Kevin R. Cox	Department of Geography, Ohio State University.
Roger M. Downs	Department of Geography, The Pennsylvania State University.
Donata Francescato	Department of Psychology, University of Houston.
Peter R. Gould	Department of Geography, The Pennsylvania State University.
Donald R. Griffin	The Rockefeller University and the New York Zoological Society.
Roger A. Hart	Graduate School of Geography, Clark University.
Stephen Kaplan	Department of Psychology, University of Michigan.

Terence R. Lee	Department of Psychology, University of St. Andrews.
Robert A. Lowrey	CONSAD Research Corporation, Pittsburgh.
Ulf Lundberg	Psychological Laboratories, University of Stockholm.
Kevin Lynch	Department of Urban Studies and Planning, Massachusetts Institute of Technology.
William Mebane	Graduate School of Business, Harvard University.
Gary T. Moore	Department of Psychology, Clark University.
Peter Orleans	School of Architecture and Urban Planning, University of California, Los Angeles.
Thomas F. Saarinen	Department of Geography, University of Arizona.
David Stea	School of Architecture and Urban Planning, University of California, Los Angeles.
Edward C. Tolman (deceased)	Department of Psychology, University of California.
Georgia Zannaras	Department of Geography, Ohio State University.

Image and Environment

I

Theory

Introduction

There is no generally accepted "party line" by which we can understand and explain cognitive maps or mapping. In this interdisciplinary area, writers bring disciplinary biases with them: thus, Orleans relies on insights from sociology, Briggs from geography, and Appleyard from planning. Each writer ultimately turns to psychology for "the answer," yet apart from Tolman's contributions (chap. 2) and the work of Piaget and Werner (summarized in chap. 14), little *theoretical* effort *per se* has been directed specifically to cognitive spatial representations.

The theorists cited in this introduction have contributed only peripherally to the question of cognitive mapping. Most of the research that they originally inspired was concerned with other problems. The recent empirical explorations of cognitive representation draw heavily upon the "classic" psychological theorists. However, even this apparently unified source has led to a divergence of emphasis—Kaplan and Lee both draw inspiration from physiological psychology; Lundberg and the Swedish investigators rely upon psychophysics, a reliance shared by Briggs and Lowrey; Appleyard has employed cognitive, learning, and perceptual psychology. Recent work into how environmental cognition develops has considered theoretical views in developmental psychology and ethnoscience (see chap. 14).

Thus, there is no unified theoretical framework upon which we can base our understanding of cognitive mapping. The initial paper attempts to develop such a framework. Kaplan (chap. 4) also offers the beginnings of a fruitful move in this direction but his argument, although attractive and intuitively plausible, remains an untested speculation.

The lack of a tested theoretical framework is a serious problem with many significant consequences. We find, for example, that there are terminological difficulties affecting not only our formulation of problems but also our methods of tackling them: are we concerned with perception or cognition, attitude or preference? There are obstacles in the context of research design: where are our *a priori* hypotheses, our standardized and validated measuring procedures? The result of these problems can be summarized as follows: work on the cognitive representation of man's spatial environment has identified interesting consistencies among phenomena, but has not developed any theoretical frameworks providing the necessary explanation or prediction.

However, we must temper this gloomy assessment by recognizing that such a stage is inevitable and normal. The difficulties faced in this interdisciplinary work are no greater than those portrayed by Watson in *The Double Helix*: Rome was not built in a day nor DNA explained overnight. Hence, it is neither the fact of nor the need for interdisciplinary work and cross-fertilization which is at issue, but rather the nature of the material under study, the method by which it is treated, and the basic theoretical underpinnings. This introduction cannot solve these problems, but it can discuss some of the significant issues. Accordingly, it touches upon three theoretical and methodological points:

1. *Models of Man.* Any behavioral theory makes assumptions about the nature of "human nature," and several of the major viewpoints are reviewed.
2. *Psychological Theory: Toward a Gestalt.* Because the works of Koffka and Lewin are not represented in the book, and because they provided an impetus for the theoretical and empirical work of others concerned with spatial behavior (Barker, 1963; 1968; Barker and Gump, 1964; Tolman, chap. 2), some of their relevant positions are summarized.
3. *Methodology.* The research methods used in the various studies reported here are as numerous as the contributors; as yet no single "acceptable" method or body of methods has emerged, and so a basic framework is established.

Theory and data are inseparable. Theoretical ideas are sprinkled throughout this book and, similarly, data is liberally sprinkled through the papers that follow. Stea and Blaut's (chap. 3) is a developmental paper, as is that of Hart and Moore. Either might have been appropriate to this section or to part IV, The Development of Spatial Cognition: similarly, the papers by Appleyard and Orleans have theoretical overtones. Thus, for a more complete outlook, the reader is urged to obtain a "theoretical gestalt" by combining the ideas that follow with those in succeeding sections of this book.

Models of Man

Implicit or explicit in the writing of thinkers over the past several hundred years have been models of human behavior, incorporating assumed internal determinants as well as patterns of interpersonal influence, social interaction processes, motivation for group membership, and decision-making (Simon, 1957). Many writers of the seventeenth, eighteenth, and nineteenth Centuries, from Descartes through La Mettzie, Darwin, Spencer, and Huxley, were concerned with models based upon theological and biological determinants of behavior. The early twentieth Century saw the emergence of models for human decision-making incorporating or specifically neglecting the psychological component. Because our concern is with cognitive factors affecting one aspect of decision-making—vis-à-vis the spatial environment—we will review the crucial models, illustrating the conceptual extremes.

Psychoanalytic man, as delineated by Freud and Jung, was totally nonrational. His adult behavior was determined in large part by the (probably unconscious) resolution of psychological conflicts experienced earlier in life, and was influenced by biologically transmitted traces of earlier experiences in human evolution ("collective unconscious" or "racial memory"). External factors were assumed to play a small role in adult patterns of decision-making: social influence was secondary, and environmental influence negligible. The only exception to this latter statement is the work of Searles (1960) who incorporated influences from the physical (and spatial) environment into psychoanalytic thinking.

Classical economic theory proposed an opposite model: man was, for the purposes of economic exchange, considered totally rational and influenced in his decision-making only by objective factors external to himself, of which he had total knowledge. All constraints were completely known and all were considered:

> There are two principal species of economic man: the consumer and the entrepreneur. Classical economics assumes the goals of both to be given: the former wishes to maximize his utility, which is a known function of the goods and services he consumes; the latter wishes to maximize his profit. The theory then assumes both of them to be rational. Confronted with a pair of alternatives, they will select that one which yields the larger utility or profit, respectively. (Simon, 1957, p. 197)

Like the psychoanalytic model, economic man failed to account for much of the variation in human behavior: for an empirical investigation of the degree of this failure in terms of human spatial behavior, see the paper on Swedish farming systems by Wolpert (1964).

Attempts to consider the finiteness of human rationality in a changing and complex world, the empirical limits upon cognition, and the "objective" weighing of alternative decision criteria led to the *principle of bounded rationality*:

> The capacity of the human mind for formulating and solving complex problems is very small compared with the size of the problems whose solution is required for objectively rational behavior in the real world—or even for a reasonable approximation to such objective rationality. (Simon, 1957, p. 198)

Behavior based on bounded rationality may seem "irrational" and may be characterized as such, but the resemblance is only apparent. Rather, the essential characteristics of the cognitive process are its limited ability to cope with and store information and its attempt to form impressions of and tentative decisions about the environment on the basis of limited, fragmentary information under severe time constraints. In the economic model, man maximizes his utility or profit. In Simon's boundedly rational model, he *satisfices*, finding a course of action which is "good enough" for the situation as he comprehends it. Psychological models based upon observation of human and animal learning (e.g. Estes, 1954) that postulate limitations in the complexity of the choice mechanisms and the capacity of organisms for obtaining and processing information account for observed behavior better than do models of rational choice:

> However adaptive the behavior of organisms in learning and choice situations, this adaptiveness falls far short of the ideal of "maximizing" postulated in economic theory. Evidently, organisms adapt well enough to "satisfice"; they do not, in general, "optimize." (Simon, 1957, p. 261)

Simon's interest in building computer models of man was not new as a philosophic ideal. La Mettzie's idea of *L'Homme Machine* had been promulgated two centuries earlier, and basic computer theory had existed for a century and a half. Newall, Shaw, and Simon (see Gruber, Terrell, and Wertheimer, 1962) extended these ideas to the construction of a program for a "thinking machine" called the "logic theorist," which was set the problem of deriving the axioms of mathematics from the axioms of logic à la Russell and Whitehead. The success of this modeling attempt is less well known than that of the many chess-playing programs which have been written. Simon and others have cautioned that the fact that a machine can play chess in much the same way that a man does (albeit more efficiently) does not imply that the internal hookup of the cerebral cortex must be like the innards of the computer—but the warning, in people's frantic search for the "lugram" of rational thought, has often gone unheeded. A parallel problem is discussed in chapter 1 in terms of representations of the spatial environment. Although the input source is identical, a city-dweller, a camera, a cartographer, and a tape-recorder all employ different signatures to arrive at a representation (or form) of a city; yet all of these forms have the same function. Parallel functions do *not* necessarily imply parallelism of form, but our understanding of cognitive representation has been hampered by overlooking this vital point.

Psychological Theory: Toward a Gestalt

Among theorists in psychology, Koffka (1935) may have been the first to distinguish between the geographical environment (or absolute space) and the behavioral environment (or relative space), although he acknowledges borrowing some concepts from Tolman. Koffka held that the geographical environment is not a stimulus or set of stimuli in itself, but is "stimulus-providing," and that the mediation of the behavioral environment clarifies the relationship between the geographical environment and behavior:

> Behavior takes place in a behavioral environment, by which it is regulated. The behavioral environment depends upon two sets of conditions, one inherent in the geographical environment, one in the organism. But it is also meaningful to say that behavior takes place in a geographical environment . . . (1) Since the behavioral environment depends upon the geographical, our proposition connects behavior with a remote instead of an immediate cause . . . (2) the results of the animal's behavior depends not only upon his behavioral but also on his geographical environment. . . . The geographical environment, not only the behavioral, is changed through all behavior. (1935, p. 31)

Lewin, whose association with Tolman was closer, stressed the relationship of and distinctions among mathematical space, physical space, and psychological life space, concepts which resemble those of Koffka. Lewin developed a "topological" or "hodological" psychology, stressing the connection and paths between psychological regions: "There is a certain topological structuring of the environment in nearly all situations with which psychology deals, and no doubt there is always some structuring of the person" (1936, p. 62). For Lewin, the contrast between physical and psychological space stemmed from the laws appropriate to the two, with the determination of spatial relations in psychology dependent upon psychological processes and, hence, upon the nature and laws of psychological dynamics. While psychological life space was considered potentially "metricisable" in the same sense that physical space is metric, Lewin also clarified distinctions between physical and psychological worlds via differing notions of connectedness and closure. The *single* connected space in which all physical reality is included does not exist within topological psychology, each life space being viewed as dynamically unique and equivalent to the totality of the physical world. The notion of "dynamic closure" entered here as well, the physical world being considered as a "dynamically closed" unity and the psychological world a dynamically enclosed unity.

Tolman (1932, p. 158) acknowledges a greater debt to Lewin than to Koffka. But it was Koffka who foreshadowed the spatial nature of the controversy between the yet-to-emerge Hullian and Tolmanian camps in learning:

> We observe three rats in the same maze, each starting at one end and finally emerging at the other. Then in a way we could say the three rats have run through the maze, a geographical statement. But our observation has convinced us that there were obvious differences in their behavior . . . the behavior within the *behavioral* environment. A rat running for food does not do so only from the moment when it is near enough to see or smell it, but from the very beginning. Tolman's book gives ample evidence for this statement. But the first part of the geographical maze does not contain the food, nor any stimulation emanating from the food. . . . Behavior in the geographical environment is (thus) the activity as it really is, in the behavioral as the animal thinks it is. (1935, pp. 36–37)

Methodology

There is no single correct path to understanding and explanation: hence it is impossible to specify an optimal research methodology. Consequently, we will not comment so much upon the specific methodologies (or *tactics*) used in studies of cognitive mapping as upon the overall research *strategies* utilized. Stea and Downs (1970) identified two basic strategies: system identification and system analysis. The former, a holistic approach concerned with the identification and description of the overall system, isolates relationships (and interactions) between segments of the spatial environment, types of people, and cognitive response typologies.

> On this level, a major concern is the establishment of purely functional relationships between, for example, socioeconomic status variables and the cognition of different segments of the environment. One main outcome is a classification or typology of cognitive responses, where such responses are the product of essentially unknown process interactions. (1970, p. 7)

The second strategy, employed once system interactions have been identified, described, and isolated, is a searching analysis of the system:

> The focus is on the interactions between sets of variables, together with a specification of the system parameters. Knowledge of relationships and of parameters can lead directly to casual models indicating, for example, the process by which information is coded to form part of cognitive representations. . . . An additional factor associated with the second strategy . . . is the use of quantitative methods of data analysis. The questions asked in the second strategy cannot be answered satisfactorily without mathematical and statistical analysis. Thus, for example . . . factor analytic, multiple regression, and analysis of variance models [have been used]. This is not to decry the first strategy as being weaker or less powerful that the second—rather their objectives differ and hence so do the means of attaining them. The relation-

> ship between the two strategies represents an overlap between two distinct stages of growth in studies of environmental cognition. The descriptive first strategy identifies general patterns of interaction between variables and serves to generate hypotheses which can be tested by the adoption of the second strategy. . . . [But] quantification and power are not synonymous. The advantages of a quantitative approach to environmental cognition should not obscure the need for care in its use nor the need for an adequate, preexisting conceptual structure. (Stea and Downs, 1970, pp. 7–8)

Within the literature, the distinction between these two strategies is *not* as clear cut as the preceding would suggest. Just as we lack an overall theoretical framework, so too we lack a consensus as to what constitutes a "good" or "bad" research design. The dangers inherent in this lack are obvious. The current *ad hoc* posture towards methodological questions is acceptable and even necessary in the exploratory stage of any research effort. However, it also runs the risk of producing incompatible results and delaying the development of cumulative scientific knowledge. There are disciplinary orientations in our approaches to research design, with geographers tending to favor the search for relationship via variance compounding (i.e. correlational methods), while psychologists favor variance splitting via analysis of variance methods. Linked with the previous observation is the problem of control and inference. We are faced with four sets of variables—the spatial environment itself, the information or stimulus set, the intervening cognitive processes, and the group and individual differences in the operation of these processes—yet we lack the research design capability to cope with this type of complexity. Thus, although we have overlooked methodological questions until now in our efforts to understand cognitive maps, we cannot afford to continue to do so because theory and data (and therefore methodology) are inextricably linked.

1

Cognitive Maps and Spatial Behavior: Process and Products

ROGER M. DOWNS AND DAVID STEA

Introduction

A surprising fact is associated with studies of cognitive mapping: although the emergence of this vigorously developing research area has been recent, we are not discussing something newly discovered such as a subatomic particle or a cell protein structure. Instead we are concerned with phenomena so much part of our everyday lives and normal behavior that we naturally overlook them and take them for granted.

A series of examples indicate the pervasive influence of cognitive maps and mapping processes. *Newsweek* (June 15, 1970) quoted a London cab driver: "It's crazy, . . . How do they expect anyone to find their way around here?" This plea resulted from an ingenious planning experiment in which sidewalks were widened and streets narrowed and turned into a system of mazes, dead-ends, and one-way routes. The objective was to create a confusing obstacle to drivers, forcing them to abandon habitual short cuts in favor of main streets, or, better still, to give up driving and use public transportation. That the drivers have well-developed cognitive maps is implied in one planner's claim: "You can't make it just difficult. You have to make it nearly impossible or you won't win."

As a graphic example of the value of cognitive maps consider the 1970 Apollo 14 moon walk. Astronauts Shepard and Mitchell were within 75 feet of their objective, the rim of Cone Crater, but returned to "Antares" without completing their mission. The reason? They had become confused and disoriented by the lack of distinctive lunar landscape features and the endless sequence of gullies. It was only later that they realized just how close they had been to their objective.

We are all aware of the image evoked when the news media use a locational term such as "the South" in the U.S.A. or "the Midlands" in England. We share common reactions when told that "it" happened in the South:

images of the climate, the prevailing social system, the attitudes of the people, the food they eat, readily "spring to mind." We rely on these images for understanding and explaining the event (it) because "you would expect that sort of thing to happen there." Cognitive maps are convenient sets of shorthand symbols that we all subscribe to, recognize, and employ: these symbols vary from group to group, and individual to individual, resulting from our biases, prejudices, and personal experiences. In the same way, we respond to an advertisement's exhortation to "come to sunny Florida" or, on the other side of the Atlantic, to "come to sunny Brighton." We associate images of beaches, sun-bathing, amusement parks, golf courses with such simple locational terms; our cognitive mapping processes fill in the necessary details. Thus an advertisement offering ice cream store franchises in the *New York Times* made the following appeal:

> Tired of snow? Tired of crowded city life?
> BE A CONTENTED SOUTHWESTERNER!
> Enjoy life the way it should be; among neighborly congenial folks on the balmy Gulf Coast, or the dry Texas "Hill Country;" or the scenic "Piney Woods" of East Texas. (February 21, 1971, Section 5, p. 2)

So we find that planners try to alter cognitive maps, astronauts need them, the news media use them, and advertisers tempt us with them: they *are* part of our everyday lives. But for research designed to understand and explain them, definitions by example are necessary but *not* sufficient. Consequently, we offer a formal definition: *Cognitive mapping is a process composed of a series of psychological transformations by which an individual acquires, codes, stores, recalls, and decodes information about the relative locations and attributes of phenomena in his everyday spatial environment.*

In this paper, we will expand this definition and examine the conceptual frameworks which are subsumed within it.

An Analysis of Cognitive Mapping Processes

COGNITIVE MAPS AND ADAPTIVE BEHAVIOR

Underlying our definition is a view of behavior which, although variously expressed, can be reduced to the statement that *human spatial behavior is dependent on the individual's cognitive map of the spatial environment.* That this formulation is necessary is indicated by a comparison of the characteristics of the individual with those of the spatial environment.

The environment is a large-scale surface, complex in both the categories of information present and in the number of instances of each category. Things are neither uniformly distributed over this surface, nor ubiquitous: they have a "whereness" quality. In contrast, the individual is a relatively small organism with limited mobility, stimulus-sensing capabilities, information processing ability, storage capacity, and available time. The individual receives information from a complex, uncertain, changing, and unpredictable source via a series of imperfect sensory modalities operating over varying time spans and intervals between time spans. From such diversity the individual must aggregate information to form a comprehensive representation of the environment. This process of acquisition, amalgamation, and storage is *cognitive mapping*, and the product of this process at any point in time can be considered as a *cognitive map*.

Given a cognitive map, the individual can formulate the basis for a strategy of environmental behavior. We view cognitive mapping as a basic component in human adaptation, and the cognitive map as a requisite both for human survival and for everyday environmental behavior. It is a coping mechanism through which the individual answers two basic questions quickly and efficiently: (1) Where certain valued things are; (2) How to get to where they are from where he is.

COGNITIVE MAPS AND SPATIAL BEHAVIOR

Although the cognitive map represents a set of processes of unknown physiological and controversial psychological nature, its effect and function are clear. We believe that a cognitive map exists if an individual behaves as if a cognitive map exists (Stea and Downs, 1970). Above we cited anecdotal examples of the relationship between behavior and the cognitive mapping process. Normal everyday behavior such as a journey to work, a trip to a recreation area, or giving directions to a lost stranger would *all* be impossible without some form of cognitive map. These ubiquitous examples are overlooked and relegated to "second nature" status. Admittedly, much spatial behavior is repetitious and habitual—in travelling, you get the feeling that "you could do the trip blindfolded" or "do it with your eyes shut." But even this apparent "stimulus-response" sequence is not so simple: you must be *ready* for the cue that tells you to "turn here" or *prepare* for the traffic light that tells you to "stop now" or *evaluate* the rush hour traffic that tells you to "take the other way home tonight." Even in these situations you are *thinking ahead* (in both a literal and metaphorical sense) and using your cognitive map. In human spatial behavior, we consider even a series of stimulus-response connections as a "simple" (or "impoverished") form of cognitive map, in which the general aspects of spatial relationship implicit in cognitive mapping play a minimal role. In terms of the two basic questions raised earlier, the person *knows* that an object is valued and one way of getting to it, but knowledge of the "whereness" in relation to the *location* of other objects is absent. The goal is always a part of the cognitive map, however primitive the map might be, even when the degrees of closeness of approach to the goal cannot be articulated. Thus, someone "who knows only one route" knows more about that route than just the appropriate responses at certain choice points, and because he "thinks ahead," is also engaging in cognitive mapping. We are postulating the cognitive map as the basis for deciding upon and implementing *any* strategy of spatial behavior.

However, we must make it perfectly clear that a cognitive map is not necessarily a "map." This apparently paradoxical statement focuses on a misconception which has emerged in research in this area over the past ten years and which our definition might exacerbate. We are using the term *map* to designate a functional analogue. The focus of attention is on a cognitive representation which has the *functions* of the familiar *cartographic* map but not necessarily the physical properties of such a pictorial graphic model (Blaut, McCleary, and Blaut, 1970). Consequently, it is an analogy to be used, not believed. The problem of the map analogy is particularly acute for geographers, a group with a distinctive viewpoint or "spatial style" (Beck, 1967). In fact, if we are to believe Beck, geographers are strangely ambivalent as to whether they prefer features to be "up" versus "down" or "horizontal" versus "vertical". This ambivalence may result from their way of approaching the world, based on concepts of relative location, proximity, and distance, and especially geared to the use of cartographic maps. Boulding (1956) argued that "the map itself . . . has a profound effect on our spatial images (p. 65)." More particularly, drawing upon Lynch's (1960) attractive and appealing series of cognitive maps of U.S. cities portrayed as cartographic maps, we might paraphrase Boulding's statement to read: "The cartographic map has had a profound effect on our concept of a cognitive map."

Spatial information can be represented in a variety of ways. Consider, for example, a street directory in which streets are ordered alphabetically and people ordered spatially (by residences *and* apartments) and contrast it with a telephone directory listing exchange areas spatially and people alphabetically. Further representations include tape-recorded walking tours for museums or European cities, rail and bus route schedules, and electronic media such as radar and laser holograms. All of these media share the same function, not structure; and thus cognitive maps are derived from analogies of process, not product.

COGNITIVE MAPPING SIGNATURES AND COGNITIVE REPRESENTATIONS

As Blaut, McCleary, and Blaut (1970) indicate, the basic functional identity among the media exemplified above can be subsumed under a general "black box" model. All of the media rely upon the same sort of spatial information, and all are employed in the same sorts of spatial behavior: thus, the inputs and outputs are specified while the intervening storage system (the black box) is not. The way in which spatial information is *encoded* (map making) and *decoded* (map reading or interpreting) gives rise to a set of operations called the *signature* of a given mapping code. Thus, a cartographic map signature is dependent upon three operations: rotation of point of view to a vertical perspective, change in scale, and abstraction to a set of symbols (for example, red dots for towns, blue lines for rivers). These operations are more general than the specific signatures, however. Thus, many other signatures are feasible; we have no reason to anticipate that cognitive maps should necessarily have the same form of signature as cartographic maps. Above all, we should avoid getting "locked" into a form of thinking through which we, as investigators, force a subject to "produce" a *cartographic* cognitive map and which we then "verify" against an objective cartographic map. It is significant, therefore, that Lynch (1960) used several input signatures (verbal and graphic) in his original study and a single, graphic output signature to produce his now famous maps of Boston, Jersey City, and Los Angeles.

The issue of mapping signatures involves some fundamental theoretical and methodological issues in the study of cognitive mapping processes. Underlying the whole approach is the basic question: How is information, derived from the absolute space of the environment in which we live, transformed into the relative spaces that determine our behavior? The transformation can be viewed as a general mapping process involving any or all three fundamental operations: change in scale, rotation of perspective, and a two stage operation of abstraction and symbolization, all of which result in a representation in relative space.

We are interested in the class of cognitive representations which result from the transformation of information about spatial phenomena from one set of absolute space relationships into a set which is adaptive or useful in terms of human spatial behavior.

Thus, we should be interested in developing theoretical statements about the *cognitive signatures* that are employed in dealing with information from the spatial environment. We have given these signatures a seemingly bewildering series of labels (cognitive maps, mental maps, images, and schemata) without applying the necessary critical scrutiny. For example, the only differences between Lynch's "images" (1960) and the city maps of cartographers lie in the degree of abstraction employed and in the type of symbols chosen to depict information. The research procedure is the result of a series of transformations: each individual constructs his own relative space based upon approximately the same absolute space. Lynch aggregates and summarizes these relative spaces reconverting the information by using another signature—conventional cartography with associated scale change and rotation to a vertical perspective. Such representations may be heavily *content-loaded*—that is, they may stress *what* is being represented and not the *way* in which it is being represented. Instead, we should be concerned with the nature or signature of relative space as it is construed and constructed by the individual. Only if we do this can we ask how relative and absolute spaces compare and differ. Speculatively, it seems likely that cognitive representations may employ a variety of signatures simultaneously; some aspects of our composite cognitive maps may resemble a cartographic map; others will depend upon linguistic signatures (in which scale and rotation operations are irrelevant), and still others upon visual imagery signatures derived from eye-level viewpoints (in which the scale transformations may be disjointed or convoluted). These remain speculations because we have not yet fully understood *what* we should be looking for—in our view we must regard cognitive maps as the result of these mapping signatures and try to understand the nature of such signatures.

We have defined cognitive mapping as the process of acquiring, amalgamating, and storing spatial information. We have tried to specify more clearly the meaning of a cognitive map. However, before considering the nature and functions of cognitive maps in more detail, we must discuss some basic definitions and attempt to clarify a few misconceptions which currently prevail.

The Concepts of Perception, Cognition, Attitude, and Preference

PERCEPTION AND COGNITION: DISTINCTIONS

Unfortunately, perception and cognition have been employed in a confusing variety of contexts by psychologists and other social scientists. Frequently, the context itself is ambiguous, since the word falls into the "process-product" category (Rudner, 1966, p. 7). When such a word is used it is difficult to determine whether the *process* of perceiving is being discussed or whether the concern is with the *product* of the perception process. Both aspects are of importance in our studies of cognitive mapping, but if we are to develop cumulative scientific knowledge, an explicit statement of the focus of interest must be included. Such a statement is notably lacking in geographic studies. In addition, perception has been used in a variety of ways: to experimental psychologists, it involves the awareness of stimuli through the physiological excitation of sensory receptors; to some social psychologists, it implies both the recognition of social objects present in one's immediate sensory field *and* the impressions formed of persons or groups experienced at an earlier time. To many geographers, perception is an all-encompassing term for the sum total of perceptions, memories, attitudes, preferences, and other psychological factors which contribute to the formation of what might better be called environmental *cognition*.

The complex interrelationship between perception and cognition is illustrated by the definition of a perceptual-conceptual repertory as:

> The stable, recognized patterns of perceptions into which sensory complexes are organized. Species differ radically in their capacity to organize sensory complexes into such patterns. The confirmed city dweller does not see the browning wheatfield as ready for harvest; the country dweller may find the directions in the subway confusing rather than patterned. (English and English, 1958, p. 379)

Given the varied uses of the terms, it is difficult to distinguish between perception and cognition, but we will make a "distinction of convenience," the necessity for which is indicated not so much by the responses of individuals to "stimuli," but by their responses to "labels." Environmental designers and geographers have identified some of the issues in human response to natural and designed environments as "perceptual problems," but, to their dismay, often found that studies of single-neuron preparations and autokinetic effects have little to say about human "perception" of the landscape.

Thus we reserve the term *perception* for the process that occurs because of the presence of an object, and that results in the immediate apprehension of that object by one or more of the senses. Temporally, it is closely connected with events in the immediate surroundings and is (in general) linked with immediate behavior. This accords with the view of perception delineated by experimental psychology. *Environmental cognition* is thus the subject matter of interest to geographers, physical planners, and environmental designers working on behavior issues. Cognition need not be linked with immediate behavior and therefore need not be directly related to anything occurring in the proximate environment. Consequently, it may be connected with what has passed (or is past) or what is going to happen in the future.

However, this distinction falls short of establishing a clear dichotomy. We agree with Levy that the difference between perception and cognition is one of degree and focus (1970, p. 251). Both refer to inferred processes responsible for the organization and interpretation of information, but perception has a more direct sensory referent than cognition. Cognition is the more general term and includes perception as well as thinking, problem solving, and the organization of information and ideas. A more useful distinction from a spatial point of view is offered by Stea (1969). He suggests that cognition occurs in a spatial context when the spaces of interest are so extensive that they cannot be perceived or apprehended either at once or in a series of brief glances. These large-scale spaces must be cognitively organized and committed to memory, and contain objects and events which are outside of the immediate sensory field of the individual. This scale-dependent distinction, intuitively acceptable to a geographer, also suggests that we are concerned with the nature and formation of environmental cognitions rather than with briefer spatial perceptions.

ATTITUDES, PREDICTIONS, PREFERENCES, AND COGNITIVE MAPS

The processes of perception and cognition that lead to predispositions to behave in certain ways toward object classes *as they are conceived to be* are termed *attitudes*. The parallels between the concepts of cognitive map and attitude are marked. For example, we assume that knowledge of an individual's cognitive map is necessary to predict his spatial behavior: a similar claim has been made in psychology with respect to attitudes. Yet as Fishbein says:

> After more than seventy-five years of attitude research, there is still little, if any, consistent evidence supporting the hypothesis that knowledge of an

> individual's attitude towards some object will allow one to predict the way he will behave with respect to the object. (1967b, p. 477)

He rejects the argument that the lack of confirmation of the hypothesis is due to an incorrect operational definition of the terms involved, since continual revisions of measuring instruments have brought no success. Instead, he suggests that the conceptualization of an attitude and its hypothesized links with behavior are faulty, and replaces the holistic concept of an attitude with a formulation containing three components: *cognitions* or beliefs, *affect* or attitude, and *conations* or behavioral intentions.

Fishbein claims that the fact that affect, cognition, and action are not always highly correlated necessitates this more complex typology (1967a, p. 257). The belief component of Fishbein's model is relevant to our definition of a cognitive map. He distinguishes between beliefs concerning the *existence* of an object and about the *nature* of an object, both of which are expressed in probability-improbability dimensions. Significantly, Boulding refers to the *image* (or cognitive map) as being subjective knowledge which "largely governs my behavior (1956, pp. 5–6)."

However, this governing relationship may be both indirect and highly complex. In such a light, work on the perception of environmental hazard and individual locational behavior must be reevaluated. For example, the questions that Kates (1967, pp. 72–73) developed in his study of storm hazard on the Eastern seaboard of the U.S.A. measure the structure and content of belief systems. Through the verbal content of people's responses, Kates attempts to infer the reasons for people choosing to locate in potentially hazardous areas. However, Fishbein points out that attitudes, beliefs, and expressed behavioral intentions are frequently brought into line with actual behavior. Consequently, Kates' approach contains problems of causal relations and inference, since the perception of the hazard may have been adjusted, or *rationalized*, so that it conforms with past behavior (i.e., the decision to locate). In other words, if a behavior can be specified, an attitude can usually be postdicted.

Finally, we must distinguish among attitudes, preferences, and traits. In comparison with attitudes, preferences are usually considered to be: (1) less global—often directed to a specific object rather than a class of objects; and (2) less enduring over time—more subject to change than relatively stable, permanent attitudes. When a given attitude pervades a wide variety of objects over a considerable period of time, it becomes a *personality trait*. Craik has suggested the existence of environmental traits:

> Individuals not only exhibit characteristic styles of relating to other persons, such as "dominant," "assertive," "deferent," and so on, they also display enduring orientation toward the physical environment. Designed to identify the individual's conception of himself in reference to the natural and man-made physical environment, an inventory of environmental traits would

> permit the declaration: "I am the sort of person who reacts in these ways to the molar (large scale) physical environment." (1970a, p. 86)

Hypothetically, one could construct a scale from preference through attitude to trait, increasing in both inclusiveness and duration of the cognitive, conative, and affective components.

These discussions indicate the depth of confusion that exists concerning the key concepts of perception, cognition, and attitude. Part of the confusion is due to obvious interrelationships; for example, cognition is assumed by many to be the major component of perception (Langer, 1969) although affective and conative characteristics are present as well. Similarly, there is interplay between an attitude and the way an object is perceived. Boulding argues that "for any individual organism . . ., there are no such things as 'facts.' There are only messages filtered through a changeable value system (1956, p. 14)." This lack of conceptual clarity is a major problem in an area already overburdened with tentative and unrelated conceptual infrastructures.

The Nature and Functions of Cognitive Maps

To understand more fully what cognitive maps are, how they are formed, and how they work, we need answers to three basic questions: (1) What do people *need* to know? (2) What do people *know?* (3) How do people *get* their knowledge?

WHAT DO PEOPLE NEED TO KNOW?

Given an individual with the limitations specified earlier and a spatial environment with complex characteristics, there are two basic and complementary types of information that he must have for survival and everyday spatial behavior: the *locations* and the *attributes* of phenomena. Cognitive maps consist of a mixture of both. Since location and attribution are properties of objects as well as of phenomena, we must also know what an "object" is.

Locational information is designed to answer the question, Where are these phenomena? and leads to a subjective geometry of space. There are two major components to this geometry, *distance* and *direction.* Distance can be measured in a variety of ways, and we are surprisingly sensitive to distance in our everyday behavior. The claim that "it takes you only half an hour to go and get it" will perhaps receive the reply that "it's too far to go." We think of distance in terms of time cost, money cost, and the more traditional measures, kilometers and miles. Knowledge of distance—the amount of separation between pairs of places and pairs of phenomena—is essential for planning any strategy of spatial behavior. Geography, for example, has developed a series of models of human spatial behavior which depend upon the individual's sensitivity to distance variations and upon his assumed goal of minimizing the distance traveled either by himself or by his products.

Direction is no less important in the geometry of space, although we are less conscious of directional information. We take direction more for granted than we perhaps should. It is only when we cannot find a map in the glove compartment of the car and become lost that the need for directional information becomes acute. The person who "gives" directions by pointing vaguely and saying "it's over there" is no more helpful than one who says "it's on the left"—we need to know *whose* left.

By combining distance and direction we can arrive at locational information about phenomena, but not necessarily the same as that provided by the Cartesian coordinates of cartographic maps. For example, suppose that we wish to visit a drive-in movie theater—what do we need to know? First, we need to know where we are—this means "keying" our cognitive map to our current location. We need to know where the movie theater is, which, at this spatial scale can be accomplished in two ways. Either we know where the theater is *in relation to* where we are now and consequently can select the easiest route to get there, or we know its location *relative* to some other place whose location is known—thus it may be "about five minutes' drive past the Suburbsville Shopping Center." Second, we need to know how far away it is, how to get there, and how long it will take to get there.

Thus, locational information is not as simple as it might appear. We must store many bits of distance and direction data to operate efficiently in a spatial environment, a process involving relatively accurate encoding, storage, and decoding. Use of locational information in formulating a strategy of spatial behavior, however, requires a second type of information: that concerning the *attributes* of phenomena.

Attributive information tells us what *kinds* of phenomena are "out there," and is complementary to locational information, indicating *what* is at a particular location and *why* anybody would want to go there.

An attribute is derived from a characteristic pattern of stimulation regularly associated with a particular phenomenon which, in combination with other attributes, signals the presence of the phenomenon. A concrete example will clarify this definition. Imagine that at the end of the search process specified in the drive-in theater example you are confronted with something that you "recognize" consisting of a large open space surrounded by a wall with an enormous screen at its far end, a small building at a break in the wall, and lots of teenagers driving in and out in cars. Obviously, it is the drive-in movie theater that you were searching for, and the screen and teenagers can be considered attributes of the phenomenon "movie theater." You can interpret the pattern of stimulation (visual in this instance) as indicating a series of attributes that, in this combination, signal the presence of a drive-in theater.

We can divide attributes of phenomena into two major classes: (1) descriptive, quasi-objective, or denotative; and (2) evaluative or connotative. The attributes listed as signaling the presence of the drive-in all belong to the first type, while attributes such as "reasonable prices," "good shows," or "easy to get to" are evaluative or connotative. Here, we are separating attributes which are affectively neutral (descriptive) from those which are affectively charged (evaluative). This process of evaluation involves a relationship between a phenomenon and its potential role in the behavior of the experiencing individual.

What is the relationship between an attribute and an object? An object is identified and defined by a set of attributes and bits of locational information. However, what is an object at one spatial scale can become an attribute at another. Consider the following sequence: at an interurban scale we might view cities as objects with population density, built-up area, and level of industrial growth as examples of attributes; at an intraurban scale, we could consider shopping centers as objects with number of stores and number of different types of retail functions as attributes; at an intracenter scale, the stores become the objects; and finally, at the intrastore scale, the offerings of the store become attributes. The scale of analysis of the problem at hand defines what is an object and what is attributive and locational information.

WHAT DO PEOPLE KNOW?

If we compare a cognitive map with a base map of the real world (whether it be an aerial photograph, a cartographic map, or a scale model), we find that *cognitive* mapping does not lead to a duplicative photographic process with three-dimensional color pictures somehow "tucked away in the mind's eye," nor does it give us an elaborately filed series of conventional cartographic maps at varying spatial scales. Instead, cognitive maps are complex, highly selective, abstract, generalized representations in various forms. As Kates and Wohlwill (1966) argue, we must realize that "the individual does not passively react or adapt to the environmental forces impinging on him, but brings a variety of cognitive activities to bear—expectancies, attitudes, even symbolic elaboration and transformation of the world of reality—which come to mediate and modulate the impact of the environment on him (pp. 17–18)." We can characterize cognitive maps as *incomplete, distorted, schematized,* and *augmented,* and we find that both *group similarities* and *idiosyncratic individual differences* exist.

The Incompleteness of Cognitive Maps. The physical space of the real world is a continuous surface which we have come to understand through a classic geometrical framework, that of Euclid. Even though the amount of the earth's surface within our immediate visual range is limited, we are told that the surface is one of approximately continuous curvature. There are no gaps or bottomless voids, and the Flat Earth Society to the contrary, we cannot fall off the edge. There is always something at the "back of the beyond." Yet all cognitive maps depict discontinuous surfaces. Seemingly, some areas of the earth's surface do not "exist" when existence is defined by the presence of phenomena in the subject's cognitive representation. Carr and Schissler (1969) show how the knowledge of approach routes to a city extends only as far as the visual range attainable from the highway in the journey to work, a finding also apparent in the work of Appleyard, Lynch, and Myer (1964). On a smaller scale, Ladd (1970, pp. 90–94) indicates that black children do not represent certain sections of the neighborhood environment in their cognitive maps.

However, before we accept the discontinuous nature of cognitive maps, we must question the nature of our evidence, particularly negative evidence. As Crane (1961) observes in a provocative review, Lynch's (1960) maps of Boston omitted the city's then most prominent feature, the John Hancock Building. Yet it is difficult to accept that residents did not know of its existence; they chose not to represent it externally on their *drawn* cognitive maps. The reason for this omission may be related to the distinction between denotative and connotative meaning. Although the phenomenon may *denote* something to the individual, it may have no *connotative* meaning; that is, it may play no significant or valued role in the person's behavior. In addition, we frequently find that cognitive maps are distorted so that the size (scale) of represented phenomena, especially in the drawings of young children, indicates relative connotative significance. Therefore, we must be careful in interpreting the absence of phenomena from cognitive maps as reflecting cognitive discontinuity of space.

Distortion and Schematization. By the *distortion* of cognitive maps, we mean the cognitive transformations of both distance and direction, such that an individual's subjective geometry deviates from the Euclidean view of the real world. Such deviations can have major effects upon the patterns of spatial use of the environment. In terms of distance distortions, Lee (1962; 1970) has indicated that, given two urban facilities equidistant from an urban resident, one located on the downtown side is considered closer than the one which is away from the city center. If people are sensitive to distance, consequent spatial behavior patterns will be dependent upon such distance distortions.

Far more significant, and as yet little understood, are the results of *schematization.* By schematization we mean the use of cognitive categories into which we code environmental information and by which we interpret such information. We are, as Carr (1970, p. 518) suggests, victims of conventionality. This conventionality may be expressed in two ways. The first involves the use of those spatial symbols to which we all subscribe and which we use both as denotative and connotative shorthand ways of coping with the spatial environment. Thus, we all understand (or *think* we understand) the intended, value-loaded meanings of "Africa the Dark Continent," "Europe the Center of Culture," "Behind the Iron Curtain," and "The Midwest as the Heartland." Symbols (often mythological), such as the Western route to India and the search for the Northwest Passage, have had major effects upon the course of history. In general, such symbols deal with large spatial areas and are subscribed to by a large part of the population.

However, there are other symbols dealing with geographic entities at many scales; geographic entities which owe their cogency and importance to their mere existence—even to rumored existence. In the aggregate, such entities have been termed the "invisible landscape" (Stea, 1967). As images, these elements are perhaps the most purely symbolic. Included are certain National Parks and Wilderness Areas (of importance to many people who never have and perhaps never will see them); national landmarks such as the Statue of Liberty, once (and perhaps still) *the* symbol of the United States; New York's theater district, for those inhabitants of "The City" who never have gone there and never will go; and even, for some, New York City itself. It could be argued that even though many people do not want to go into New York, it is still important for New York to maintain its image because the same people want to locate around it. "Suburb of *what?*" is perhaps not an insignificant question.

A second aspect of schematization or conventionality involves the very limited set of cognitive categories or concepts that we have developed in order to cope with information derived from the spatial environment. As we were recently told, "Once you've seen one slum, you've seen them all." Are all older center-city areas "slums" to middle-class whites or do they have more sophisticated cognitive categories? Our understanding of the semantics (or the vocabulary) of cognitive maps is remarkably limited.

The controversy over linguistic relativity suggests that there are cross-cultural differences in the ways in which spatial information is coded. Such barriers are not only cross-cultural. Burrill (1968), a geographer studying an Atlantic coast swamp area, found that "swamp" meant a complex, multi-attribute feature to local residents; to Burrill it was a simple, single attribute feature. Communication using the term "swamp" was impossible because of this difference in meaning. Similarly, Downs (1970) assumed that a neighborhood shopping center would be a clearly defined and commonly agreed upon spatial unit, with the edge of the commercial area defining the shopping center boundary. However, residents of the area recognized four distinct subcenters.

A modern counterpart of Dick Whittington's belief that the streets of London were paved with gold may have been the belief of Blacks, Puerto Ricans, Appalachian Whites, and other disadvantaged people that they could find "a piece of the action" in the cities of the Northeastern United States, their desire to share in "the good things of modern society," and their trust in the willingness of those who had "made it" to help them. The gap between such beliefs and reality is almost painful (Brody, 1970). Yet another example of the ways in which people cope with spatial information comes from studies of environmental hazard evaluation: Burton and Kates (1964) indicate the unrealistic nature of people's estimates of the probabilities of hazards. Thus, the Los Angeles earthquake of 1971 has not served as a warning to Californians. Building continues, house sales continue—apparently the "lightning will not strike twice" mechanism is operating in spite of warnings that even more severe earthquakes can be expected in the near future.

Augmentation. Another characteristic of cognitive maps is *augmentation.* There is some indication that cognitive maps have nonexistent phenomena added as embroidery. Ancient cartographers abhorred voids and filled blank spaces with fictitious rivers, mountain ranges, sea monsters, and possible locations of Atlantis. A respondent in Appleyard's (1970) study of Ciudad Guayana drew in a railway line on his map of the city because he felt that one must exist between a particular pair of points. Such distortions may be highly significant, but we know little about their causes, and nothing about their eradication.

Intergroup and Individual Differences in Cognitive Maps and Mapping. Superimposed upon the overall relationship between cognitive maps and the real world are significant intergroup differences in the specific ways in which identical or similar spatial environments are construed. The underlying group perspectives are the result of a combination of three factors. First, the spatial environment contains many regular and recurrent features. Second, people share common information-processing capabilities and strategies. The capabilities are associated with the innate, physiological parameters of human information processing while the common strategies are learned methods of coping with the environment. Third, spatial behavior patterns display similar origins, destinations, and frequencies. These factors in combination yield intergroup differences in cognitive maps. Lucas (1963) indicated that the perceived spatial extent of a wilderness recreation area in the Northeastern United States was defined differently by various subgroups of users and by those who were responsible for its administration. Tindal (1971) studied the spatial extents of home ranges among black children in urban and suburban contexts, and found the spatial extent to be correlated with age and sex.

The individual differences among cognitive maps emerge primarily from subtle variations in spatial activity patterns, variations which can have striking effects on such maps. Ladd (1970) cites the case of two brothers who produced surprisingly different cognitive maps of their neighborhood. Such idiosyncracies are particularly notable in verbal descriptions of cognitive maps—the choice of visual details shows tremendous variation from subject to subject.

In answer, therefore, to the question "What do we know?" we can conclude that we see the world in the way that we do because it pays us to see it in that way. Our view accords with our plans for use of the environment. In other words, differences between the "real world" and cognitive maps based on it serve a useful purpose in spatial behavior. Koffka (1935, pp. 28; 33) expressed this idea well:

> Let us therefore distinguish between a *geographical* and a *behavioral* environment. Do we all live in the same town? Yes, when we mean the geographical, no, when we mean the behavioral "in". . . . Our difference between the geographical and the behavioral environments coincides with the difference between things as they "really" are and things as they look to us, between reality and appearance. And we see also that appearances may deceive, that behavior well adapted to the behavioral environment may be united to the geographical.

People behave in a world "as they see it"—whatever the flaws and imperfections of cognitive maps, they are the basis for spatial behavior.

HOW DO PEOPLE GET THEIR KNOWLEDGE?

We have postulated a set of basic characteristics that our knowledge of the spatial environment should possess, and we have indicated the characteristics that our knowledge (or cognitive map) actually possesses. Some of the differences between these two sets of characteristics can be attributed to the ways in which we *acquire* spatial information. What are the various information processing (or sensory) modalities? What are the basic sources and types of spatial information? How does our knowledge (or stored information) change through time? How do we know that cognitive maps exist?

Sensory Modalities. In our studies of cognitive maps, we have overlooked the *range* and *number* of sensory modalities through which spatial information is acquired, and have ignored the integrative nature of cognitive processes related to spatial information. The visual, tactile, olfactory, and kinaesthetic sense modalities combine to give an *integrated* representation of any spatial environment. The modalities are complementary despite our intuitive belief (and linguistic bias) that visual information is predominant. For example, Manhattan tower-dwellers often know the local Horn and Hardart or Chock-Full-O'Nuts restaurants through the smell of their famous coffee being brewed. Dock areas of cities are memorable because of the distinctive sounds they emit; the sea has a distinctive smell; certain streets, because of cobblestones or frequent railway line crossings, have a certain texture. Thus, the quality of distinctiveness or memorableness is not solely the result of the way the environment looks. Some blind people (Shemyakin, 1962) remember the various paths they traverse through the city by the different feel of each path. Held and Rekosh (1963) have demonstrated that sensory-motor interaction with the spatial environment is necessary for correct perception, for experiencing the world "as it really is."

Direct and Vicarious Sources of Information. Sources have differing degrees of validity, reliability, utility, and flexibility. Direct sources involve face-to-face contact between the individual and, for example, a city and information literally floods the person from all of his sensory modes. He must be selective in what he attends to, but can choose to repeat certain experiences, with such variations as making a given trip in the opposite direction or at a different time of day. Above all, he is "learning by doing" via a trial and error process. Reinforcement and checking are continuous: erroneous beliefs about locational and attribute information are rapidly corrected by feedback from spatial behavior.

Vicarious information about the city is by definition secondhand. It is literally and metaphorically "seen through someone else's eyes." This is true of a verbal description, a cartographic street map, a T.V. film, a written description, a color photograph, or a painting. In the mapping context these modes of representation, though similar in function, are different in form because they display different signatures. In each case, the information is selected by and transmitted through a set of filters that necessarily distort the information, generally in a way useful to the individual in his present context—for example, a travel brochure for a potential vacationer or a street map for a newcomer to an area. The result of this filtering is an incomplete representation which varies with both the individual and his group membership, especially since the individual is also comparing this new data with more familiar information and with a set of expectations.

We are all aware of the difference between image and reality when the travel brochure and the vacation resort do not match, or when a friend's color slides and the place he has photographed appear to bear little resemblance to each other. Many paintings are notably "impressionistic," selecting and highlighting some characteristics of a scene over others. A street map may be useful to a local resident or to someone who can "read" maps; but map reading is a learned skill, and we may not be able to translate from the signature of the street map back to the spatial environment.

However, although we can distinguish between active and passive information processing, they are only typological descriptions of a continuum. The two information-processing strategies operate simultaneously and continuously. Steinitz (1968), in discussing how an individual derives meaning from his spatial environment, makes it obvious that meaning develops as a result of both active and passive modes of information processing.

Both modes share a characteristic which distinguishes them from *inferential* information. Both active and passive information processing are tied to stimuli coming from the spatial environment. Inferential information is indirectly tied to the spatial environment, and results from symbolic elaboration, embroidery, and augmentation. Two examples will indicate how an initial stimulus or set of stimuli can trigger a chain of prior associative processes in the individual's memory, resulting in inferential information. Consider a person driving past a city block of poorly-maintained houses, their paint peeling and streets littered with paper and garbage. Frequently, these stimuli are sufficient to set off a stereotyped chain of thought that conjures up lower class people, probably black, on welfare, with a high crime rate. These inferred characteristics are transferred or generalized to encompass not just the particular block but the whole surrounding area. A second example of the same inferential process is the advertisement advising us "to come to sunny Florida." At this point our cognitive representation (or stored and processed information) comes into play, and we associate (or infer) a whole set of characteristics about vacations (or living) in Florida.

Thus, we have three types of information available to us at any point in time. Each has distinct characteristics, validity, and utility. For example, first impressions based upon what "hits you between the eyes" are notoriously incorrect, especially if they are accentuated by invalid chains of inferences. We all know that "things are not what they seem to be" and that we "should always look twice." We recognize the roles of the foregoing three information types in our everyday language and wisdom—they are also crucial in understanding the bases of cognitive maps.

A Terminology for Change

To this point, our whole discussion of cognitive mapping has been static —concepts of learning, time, and change have been omitted. In our approach to the question "How do we get our knowledge?" we can no longer avoid a thorny philosophical and theoretical issue. First, we must clarify the terminology and concepts necessary to tackle the issue, and second, suggest a typology of change.

We acquire the ability to know things about the environment through the process of development. *Development* must be distinguished from (1) *change*, which represents any alteration in structure, process, or events; (2) *simple accretion*, or growth by addition (typical of nonorganic structures); and (3) *progress*, implying change directed toward a given goal or set of goals, usually positively valued, such that the resultant change is regarded as an improvement. Development clearly includes change taking place over a considerable period of time; such change is assumed to be irreversible in the normally functioning individual, and, to the extent that it results in increased differentiation and complexity, is also regarded as progressive. Development encompasses both *growth* (the organic equivalent of accretion) and *maturation*. Maturation is sometimes used to refer to developmental changes due to hereditary factors, or to those changes that inevitably occur in normal individuals living in a suitable environment. Maturation and learning are to some degree interrelated in behavioral development; some maturation appears to be a necessary though not sufficient prerequisite for learning to commence, while further maturation, in certain respects, is dependent upon the successful acquisition of capability for learning.

What effects (or learned changes) can spatial information induce? Boulding (1956) suggests three possibilities: no effect, simple accretion, and complete reorganization. The "no effect" case is the most frequent in the normal adult: the information simply confirms what he already knows (i.e., the cognitive map). Thus, for example, the necessary turns on the way to the Suburbsville Shopping Center have no effect on his cognitive map because he knows the route. The successful shopping trip to the Center has no effect on his attribute information because he already knows that it is a good place for shopping. Most of the spatial information that we receive, although essential for the successful use of the environment at any point in time, has no effect on the stored knowledge or cognitive map.

A Typology of Change: Accretion, Diminution, Reorganization

The simple accretion case relates to minor changes in the cognitive map. Thus, for example, the *route* to the drive-in movie theater is learned during the trip there and back. Also the individual begins to form an *evaluation* of the movie-theater—is it good or bad, cheap or expensive, easy or difficult to get to? Both locational and attribute information are added to the cognitive map: a simple additive change has occurred through learning. As an example of the alternate change, deletion, consider what happens when a street previously used becomes one way: an alteration in route selection is required. If a store previously used in the Suburbsville Shopping Center closes down, an alternative outlet for purchasing those goods once supplied by that store becomes necessary.

Diminution develops directly from deletion. There is no need to assume that cognitive maps undergo only *progressive* change such that we increasingly approximate the asymptote of economic man's perfect knowledge (or accurate representation). Either through the passage of time or through maturation, we forget—the amount of information available through the cognitive mapping process diminishes. If a long period of time elapses before we try to drive to the movie theater again, we may have forgotten the route and where to make the appropriate turns. All stored knowledge is subject to this time decay: we need to repeat a spatial experience in order to "remember" the route in the future. This is in line with our earlier argument that "learning by doing," with its associated processes of feedback and reinforcement, is vitally important.

Diminution may also be an adaptive process. Appleyard (1969a and b;

1970), in his study of adjustment to the new Venezuelan city of Ciudad Guayana, found that people initially enriched their cognitive maps with a mass of detailed information about the city. However, over time their cognitive maps lost detail (became "improvised") as they required less information to use accustomed paths and to "live" in the city. We must not lose sight of the limitations upon the human capacity to cope with information, as discussed by Miller (1956) in his now famous "magic number 7 ± 2" paper. Given our limited capacity to store and handle information, diminution may be an adaptive process ensuring that "excess" information is lost but important information retained.

Maturation can also lead to diminution and forgetting: as the person ages, the capacity to remember and perform certain tasks diminishes (Pastalan and Carson, 1969). Whether this is an inability to retrieve the stored information or decay of the storage mechanism is irrelevant to our argument: the effect is the same. Thus, diminution is a parallel but opposite process to simple accretion.

The most dramatic changes in cognitive maps are the result of total reorganization. Boulding (1956) suggests that images are relatively resistant to change in their overall nature. It requires an accumulation of contrary evidence before a complete reorganization can occur. For example, the realization that the Earth is spherical among those people who initially considered it to be flat came slowly and only as the result of a massive accumulation of evidence. The most frequent spatial example of such a complete reorganization is to be found in long-distance human migration and subsequent residential site selection, largely the result of vicariously received information. Even the latter is often of doubtful utility, since having been once installed in a new environment, one's expectations and hopes can be markedly altered by new spatial information inputs (see Brody, 1970).

We have examined some aspects of our cognitive maps and how they came to be. We know that they are modes of structuring the physical environment, that "blooming, buzzing confusion" which surrounds us at birth and that we must later sort out in order to survive. Much of the support in contentions concerning their existence is behavioral, stemming from introspection and anecdotal evidence, but the "harder" experimental data is beginning to emerge, even, quite recently, within neurophysiology. Thus, the face of cognitive mapping is growing clearer—only the features have yet to be fully filled in.

2

Cognitive Maps in Rats and Men

EDWARD C. TOLMAN

I shall devote the body of this paper to a description of experiments with rats. But I shall also attempt in a few words at the close to indicate the significance of these findings on rats for the clinical behavior of men. Most of the rat investigations, which I shall report, were carried out in the Berkeley laboratory. But I shall also include, occasionally, accounts of the behavior of non-Berkeley rats who obviously have misspent their lives in out-of-State laboratories. Furthermore, in reporting our Berkeley experiments I shall have to omit a very great many. The ones I *shall* talk about were carried out by graduate students (or underpaid research assistants) who, supposedly, got some of their ideas from me. And a few, though a very few, were even carried out by me myself.

Let me begin by presenting diagrams for a couple of typical mazes, an alley maze and an elevated maze. In the typical experiment a hungry rat is put at the entrance of the maze (alley or elevated), and wanders about through the various true path segments and blind alleys until he finally comes to the food box and eats. This is repeated (again in the typical experiment) one trial every 24 hours and the animal tends to make fewer and fewer errors (that is, blind-alley entrances) and to take less and less time between start and goal-box until finally he is entering no blinds at all and running in a very few seconds from start to goal. The results are usually presented in the form of average curves of blind-entrances, or of seconds from start to finish, for groups of rats.

All students agree as to the facts. They disagree, however, on theory and explanation.

Edward C. Tolman, "Cognitive Maps in Rats and Men," *Psychological Review*, 55, 1948, 189–208.

(1) First, there is a school of animal psychologists which believes that the maze behavior of rats is a matter of mere simple stimulus-response connections. Learning, according to them, consists in the strengthening of some of these connections and in the weakening of others. According to this 'stimulus-response' school the rat in progressing down the maze is helplessly responding to a succession of external stimuli—sights, sounds, smells, pressures, etc. impinging upon his external sense organs—plus internal stimuli coming from the viscera and from the skeletal muscles. These external and internal stimuli call out the walkings, runnings, turnings, retracings, smellings, rearings, and the like which appear. The rat's central nervous system, according to this view, may be likened to a complicated telephone switchboard. There are the incoming calls from sense-organs and there are the outgoing messages to muscles. Before the learning of a specific maze, the connecting switches (synapses according to the physiologist) are

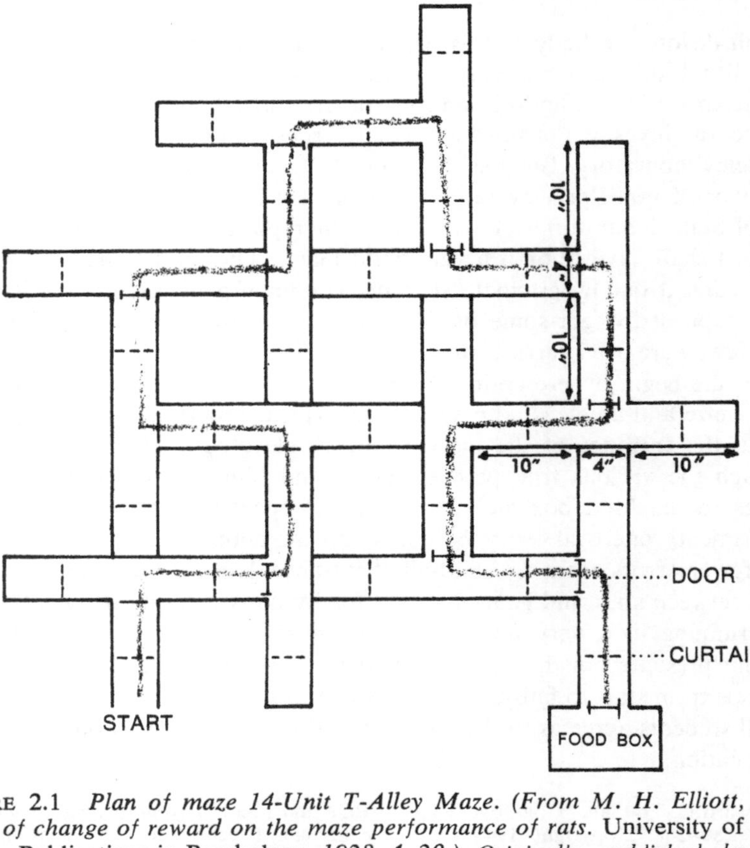

FIGURE 2.1 *Plan of maze 14-Unit T-Alley Maze. (From M. H. Elliott, The effect of change of reward on the maze performance of rats.* University of California Publications in Psychology, *1928, 4, 20.) Originally published by the University of California Press; reprinted by permission of The Regents of the University of California.*

closed in one set of ways and produce the primarily exploratory responses which appear in the early trials. *Learning*, according to this view, consists in the respective strengthening and weakening of various of these connections; those connections which result in the animal's going down the true path become relatively more open to the passage of nervous impulses, whereas those which lead him into the blinds become relatively less open.

It must be noted in addition, however, that this stimulus-response school divides further into two subgroups.

(a) There is a subgroup which holds that the mere mechanics involved in the running of a maze is such that the crucial stimuli from the maze get presented simultaneously with the correct responses more frequently than they do with any of the incorrect responses. Hence, just on a basis of this greater frequency, the neural connections between the crucial stimuli and

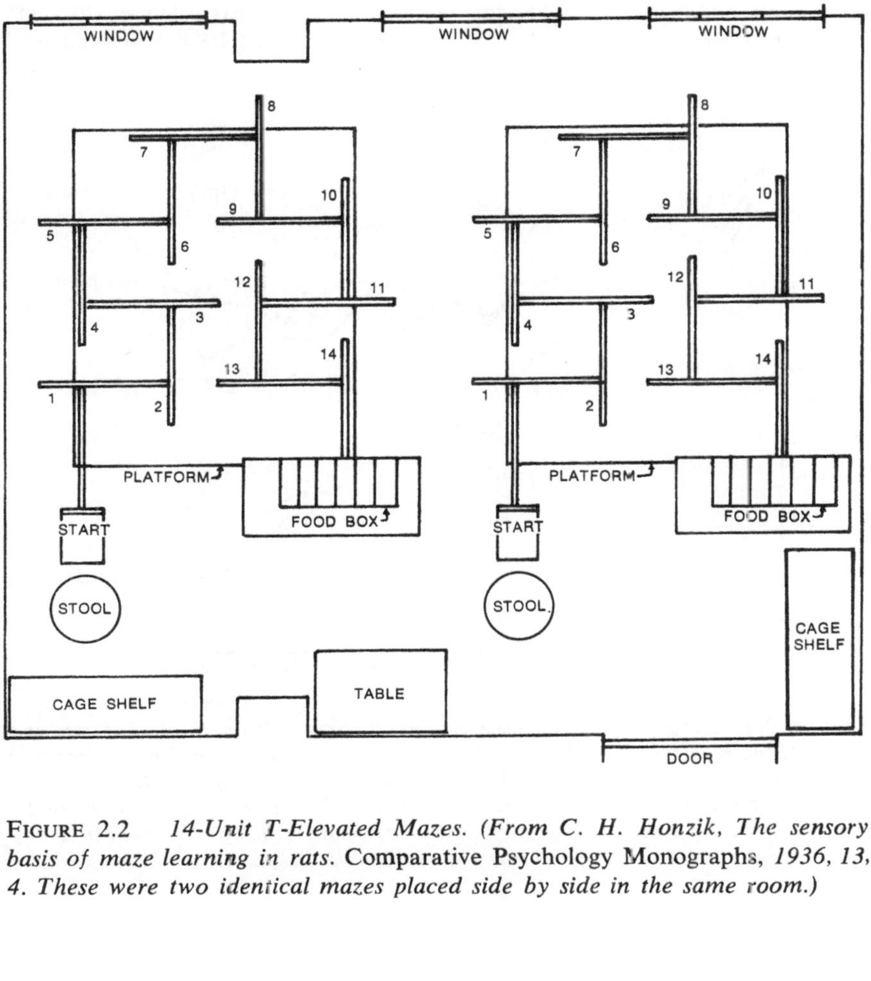

FIGURE 2.2 *14-Unit T-Elevated Mazes. (From C. H. Honzik, The sensory basis of maze learning in rats.* Comparative Psychology Monographs, *1936, 13, 4. These were two identical mazes placed side by side in the same room.)*

the correct responses will tend, it is said, to get strengthened at the expense of the incorrect connections.

(b) There is a second subgroup in this stimulus-response school which holds that the reason the appropriate connections get strengthened relatively to the inappropriate ones is, rather, the fact that the responses resulting from the correct connections are followed more closely in time by need-reductions. Thus a hungry rat in a maze tends to get to food and have his hunger reduced *sooner* as a result of the true path responses than as a result of the blind alley responses. And such immediately following need-reductions or, to use another term, such "positive reinforcements" tend somehow, it is said, to strengthen the connections which have most closely preceded them. Thus it is as if—although this is certainly not the way this subgroup would themselves state it—the satisfaction-receiving part of the rat telephoned back to Central and said to the girl: "Hold that connection; it was good; and see to it that you blankety-blank well use it again the next time these same stimuli come in." These theorists also assume (at least some of them do some of the time) that, if bad results—"annoyances," "negative reinforcements"—follow, then this same satisfaction-and-annoy-

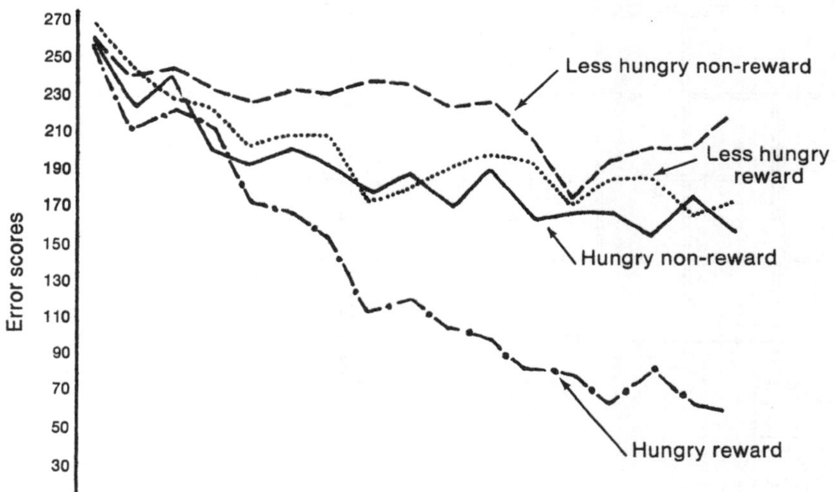

FIGURE 2.3 *Error curves for four groups, 36 rats. (From E. C. Tolman and C. H. Honzik, Degrees of hunger, reward and non-reward, and maze learning in rats.* University of California Publications in Psychology, *1930, 4, No. 16, 246. A maze identical with the alley maze shown in Fig. 2.1 was used.) Originally published by the University of California Press; reprinted by permission of The Regents of the University of California.*

ance-receiving part of the rat will telephone back and say, "Break that connection and don't you dare use it next time either."

So much for a brief summary of the two subvarieties of the "stimulus-response", or telephone switchboard school.

(2) Let us turn now to the second main school. This group (and I belong to them) may be called the field theorists. We believe that in the course of learning something like a field map of the environment gets established in the rat's brain. We agree with the other school that the rat in running a maze is exposed to stimuli and is finally led as a result of these stimuli to the responses which actually occur. We feel, however, that the intervening brain processes are more complicated, more patterned and often, pragmatically speaking, more autonomous than do the stimulus-response psychologists. Although we admit that the rat is bombarded by stimuli, we hold that his nervous system is surprisingly selective as to which of these stimuli it will let in at any given time.

Secondly, we assert that the central office itself is far more like a map control room than it is like an old-fashioned telephone exchange. The stimuli, which are allowed in, are not connected by just simple one-to-one switches to the outgoing responses. Rather, the incoming impulses are usually worked over and elaborated in the central control room into a tentative, cognitive-like map of the environment. And it is this tentative map, indicating routes and paths and environmental relationships, which finally determines what responses, if any, the animal will finally release.

Finally, I, personally, would hold further that it is also important to discover in how far these maps are relatively narrow and strip-like or relatively broad and comprehensive. Both strip-maps and comprehensive-maps may be either correct or incorrect in the sense that they may (or may not), when acted upon, lead successfully to the animal's goal. The differences between such strip maps and such comprehensive maps will appear only when the rat is later presented with some change within the given environment. Then, the narrower and more strip-like the original map, the less will it carry over successfully to the new problem; whereas, the wider and the more comprehensive it was, the more adequately it will serve in the new set-up. In a strip map the given position of the animal is connected by only a relatively simple and single path to the position of the goal. In a comprehensive-map a wider arc of the environment is represented, so that, if the starting position of the animal be changed or variations in the specific routes be introduced, this wider map will allow the animal still to behave relatively correctly and to choose the appropriate new route.

But let us turn, now, to the actual experiments. The ones, out of many, which I have selected to report are simply ones which seem especially important in reinforcing the theoretical position I have been presenting. This position, I repeat, contains two assumptions: First, that learning consists not in stimulus-response connections but in the building up in the nervous system of sets which function like cognitive maps, and second, that such cognitive maps may be usefully characterized as varying from a narrow strip variety to a broader comprehensive variety.

The experiments fall under five heads: (1) "latent learning", (2) "vicarious trial and error" or "VTE", (3) "searching for the stimulus", (4) "hypotheses" and (5) "spatial orientation."

1. "Latent Learning" Experiments

The first of the latent learning experiments was performed at Berkeley by Blodgett. It was published in 1929. Blodgett not only performed the experiments, he also orginated the concept. He ran three groups of rats through a six-unit alley maze, shown in Fig. 2.4. He had a control group and two experimental groups. The error curves for these groups appear in Fig. 2.5. The solid line shows the error curve for Group I, the control group. These animals were run in orthodox fashion. That is, they were run one trial a day and found food in the goal-box at the end of each trial. Groups II and III were the experimental groups. The animals of Group II, the dash line, were not fed in the maze for the first six days but only in their home cages some two hours later. On the seventh day (indicated by the small cross) the rats found food at the end of the maze for the first time and continued to find it on subsequent days. The animals of Group III were treated similarly except that they first found food at the end of the maze on

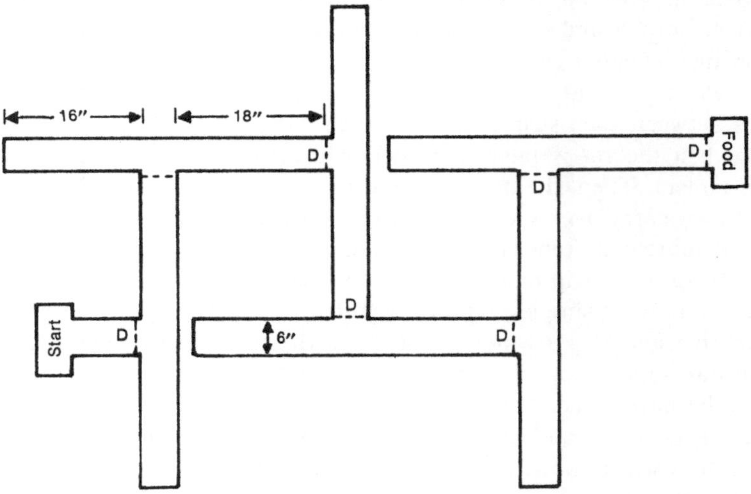

FIGURE 2.4 *6-Unit Alley T-Maze. (From H. C. Blodgett, The effect of the introduction of reward upon the maze performance of rats.* University of California Publications in Psychology, *1929, 4, No. 8, 117.) Originally published by the University of California Press; reprinted by permission of The Regents of the University of California.*

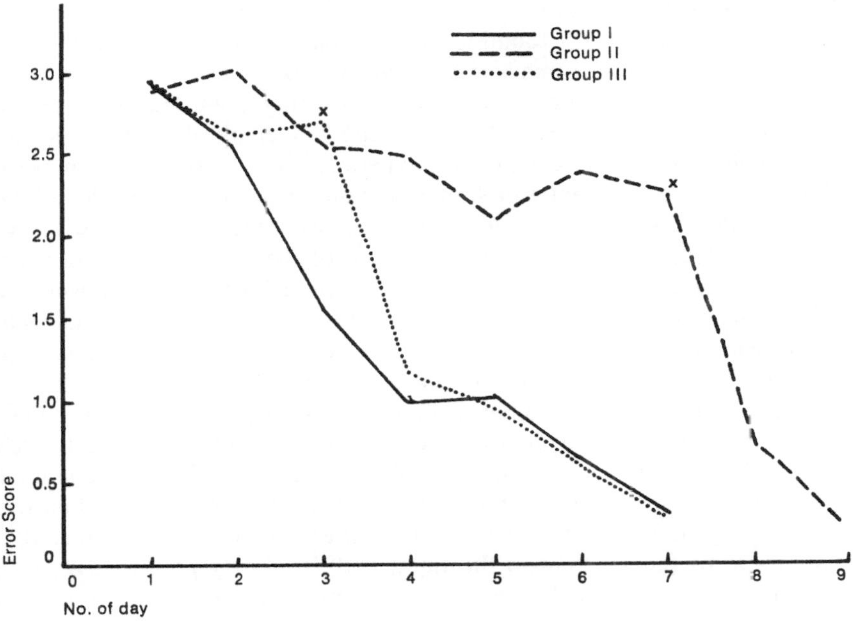

FIGURE 2.5 *(From H. C. Blodgett, The effect of the introduction of reward upon the maze performance of rats.* University of California Publications in Psychology, *1929, 4, No. 8, 120.) Originally published by the University of California Press; reprinted by permission of The Regents of the University of California.*

the third day and continued to find it there on subsequent days. It will be observed that the experimental groups as long as they were not finding food did not appear to learn much. (Their error curves did not drop.) But on the days immediately succeeding their first finding of the food their error curves did drop astoundingly. It appeared, in short, that during the non-rewarded trials these animals had been learning much more than they had exhibited. This learning, which did not manifest itself until after the food had been introduced, Blodgett called "latent learning." Interpreting these results anthropomorphically, we would say that as long as the animals were not getting any food at the end of the maze they continued to take their time in going through it—they continued to enter many blinds. Once, however, they knew they were to get food, they demonstrated that during these preceding non-rewarded trials they had learned where many of the blinds were. They had been building up a "map", and could utilize the latter as soon as they were motivated to do so.

Honzik and myself repeated the experiments (or rather he did and I got some of the credit) with the 14-unit T-mazes shown in Fig. 2.1, and with larger groups of animals, and got similar results. The resulting curves are shown in Fig. 2.6. We used two control groups—one that never found food in the maze (HNR) and one that found it throughout (HR). The experimental group (HNR–R) found food at the end of the maze from the 11th day on and showed the same sort of a sudden drop.

But probably the best experiment demonstrating latent learning was, unfortunately, done not in Berkeley but at the University of Iowa, by Spence and Lippitt. Only an abstract of this experiment has as yet been published. However, Spence has sent a preliminary manuscript from which the following account is summarized. A simple Y-maze (see Fig. 2.7) with two goal-boxes was used. Water was at the end of the right arm of the Y and food at the end of the left arm. During the training period the rats were run neither hungry nor thirsty. They were satiated for both food and water before each day's trials. However, they were willing to run because after each run they were taken out of whichever end box they had got to and put into a living cage, with other animals in it. They were given four trials a day in this fashion for seven days, two trials to the right and two to the left.

In the crucial test the animals were divided into two subgroups one made solely hungry and one solely thirsty. It was then found that on the first trial the hungry group went at once to the left, where the food had been, statistically more frequently than to the right; and the thirsty group went to the right, where the water had been, statistically more frequently than to the left.

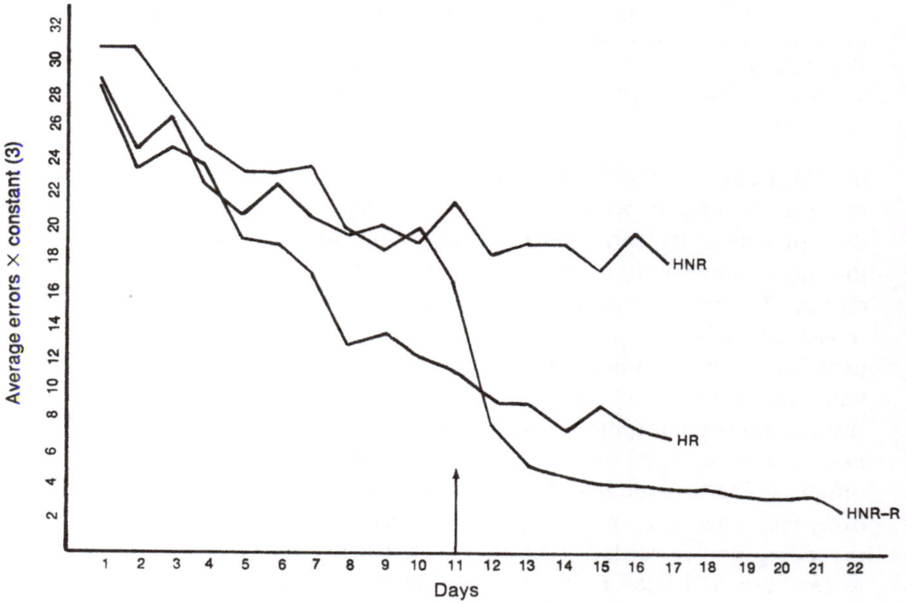

FIGURE 2.6 *Error curves for HR, HNR, and HNR-R. (From E. C. Tolman and C. H. Honzik, Introduction and removal of reward, and maze performance in rats.* University of California Publications in Psychology, *1930, 4, No. 19, 267.) Originally published by the University of California Press; reprinted by permission of The Regents of the University of California.*

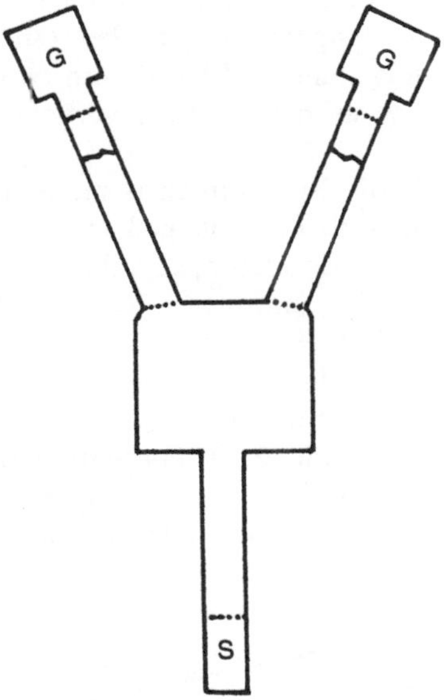

FIGURE 2.7. *Ground plan of the apparatus. (Taken from K. W. Spence and R. Lippitt, An experimental test of the sign-gestalt theory of trial and error learning.* Journal of Experimental Psychology, *1946, 36, 494. In this article they were describing another experiment but used the same maze.)*

These results indicated that under the previous non-differential and very mild rewarding conditions of merely being returned to the home cages the animals had nevertheless been learning where the water was and where the food was. In short, they had acquired a cognitive map to the effect that food was to the left and water to the right, although during the acquisition of this map they had not exhibited any stimulus-response propensities to go more to the side which became later the side of the appropriate goal.

There have been numerous other latent learning experiments done in the Berkeley laboratory and elsewhere. In general, they have for the most part all confirmed the above sort of findings.

Let us turn now to the second group of experiments.

2. *"Vicarious Trial and Error" or "VTE"*

The term Vicarious Trial and Error (abbreviated as VTE) was invented by Prof. Muenzinger (1938) at Colorado to designate the hesitating, looking-back-and-forth, sort of behavior which rats can often be observed to indulge in at a choice-point before actually going one way or the other.

Quite a number of experiments upon VTEing have been carried out in our laboratory. I shall report only a few. In most of them what is called a discrimination set-up has been used. In one characteristic type of visual discrimination apparatus designed by Lashley (shown in Fig. 2.8) the animal is put on a jumping stand and faced with two doors which differ in some visual property say, as here shown, vertical stripes vs. horizontal stripes.

One of each such pair of visual stimuli is made always correct and the other wrong; and the two are interchanged from side to side in random fashion. The animal is required to learn, say, that the vertically striped door is always the correct one. If he jumps to it, the door falls open and he gets to food on a platform behind. If, on the other hand, he jumps incorrectly, he finds the door locked and falls into a net some two feet below from which he is picked up and started over again.

Using a similar set-up (see Fig. 2.9), but with landing platforms in front of the doors so that if the rat chose incorrectly he could jump back again and start over, I found that when the choice was an easy one, say between

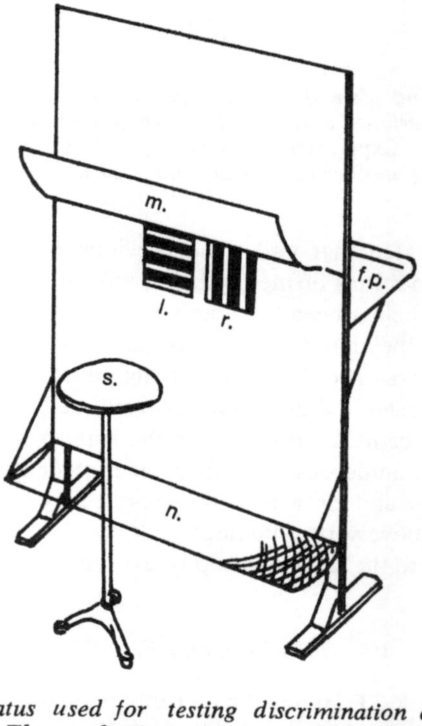

FIGURE 2.8. *Apparatus used for testing discrimination of visual patterns. (From K. S. Lashley, The mechanism of vision. I. A method for rapid analyses of pattern-vision in the rat.* Journal of Genetic Psychology, *1930, 37, 454.)*

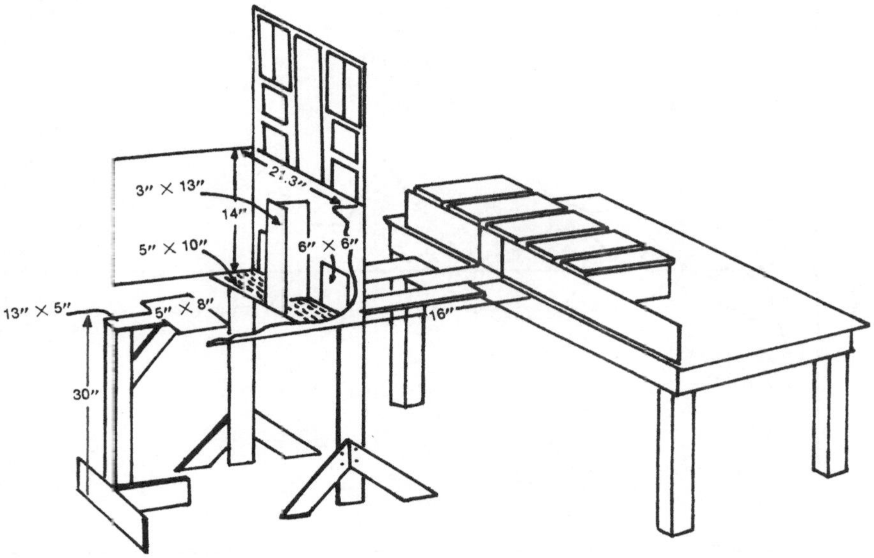

FIGURE 2.9. *(From E. C. Tolman, Predictions of vicarious trial and error by means of the schematic sowbug.* Psychological Review, *1939, 46, 319.)*

a white door and a black door, the animals not only learned sooner but also did more VTEing than when the choice was difficult, say between a white door and a gray door (see Fig. 2.10). It appeared further (see Fig. 2.11) that the VTEing began to appear just as (or just before) the rats began to learn. After the learning had become established, however, the VTE's began to go down. Further, in a study of individual differences by myself, Geier, and Levin (actually done by Geier and Levin) using this same visual discrimination apparatus, it was found that with one and the same difficulty of problem the smarter animal did the more VTEing.

To sum up, in *visual discrimination* experiments the better the learning, the more the VTE's. But this seems contrary to what we would perhaps have expected. We ourselves would expect to do more VTEing, more sampling of the two stimuli, when it is difficult to choose between them than when it is easy.

What is the explanation? The answer lies, I believe, in the fact that the manner in which we set the visual discrimination problems for the rats and the manner in which we set similar problems for ourselves are different. *We* already have our "instructions." We know beforehand what it is we are to

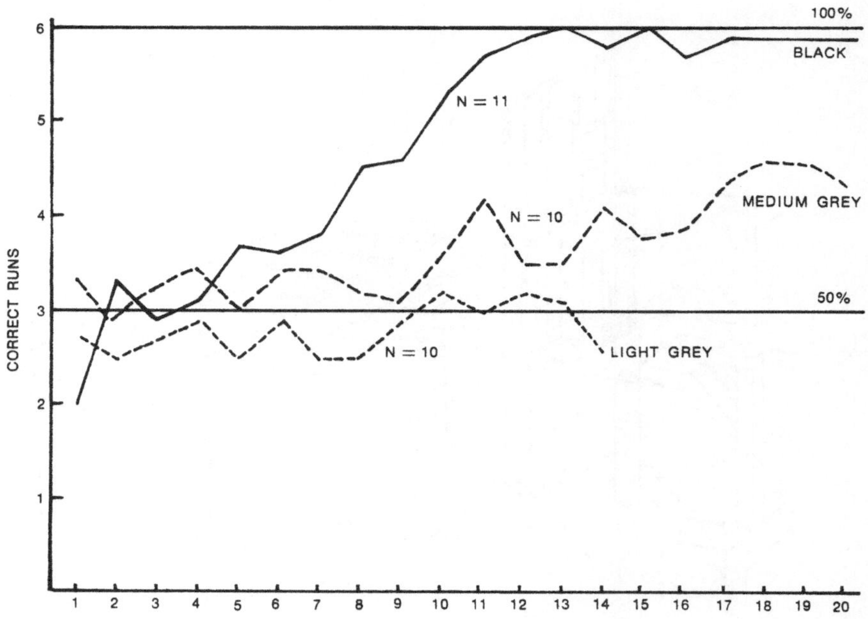

FIGURE 2.10. *Learning curves: average number of correct runs per day (From E. C. Tolman, Prediction of vicarious trial and error by means of the schematic sowbug.* Psychological Review, *1939, 46, 319.)*

do. We are told, or we tell ourselves, that it is the lighter of the two grays, the heavier of the two weights, or the like, which is to be chosen. In such a setting we do more sampling, more VTEing, when the stimulus-difference is small. But for the rats the usual problem in a discrimination apparatus is quite different. They do not know what is wanted of them. The major part of their learning in most such experiments seems to consist in their discovering the instructions. The rats have to discover that it is the differences in visual brightness, not the differences between left and right, which they are to pay attention to. Their VTEing appears when they begin to "catch on." The greater the difference between the two stimuli the more the animals are attracted by this difference. Hence the sooner they catch on, and during this catching on, the more they VTE.

That this is a reasonable interpretation appeared further, from an experiment by myself and Minium (the actual work done, of course, by Minium) in which a group of six rats was first taught a white vs. black discrimination, then two successively more difficult gray vs. black discriminations. For each difficulty the rats were given a long series of further trials beyond the points

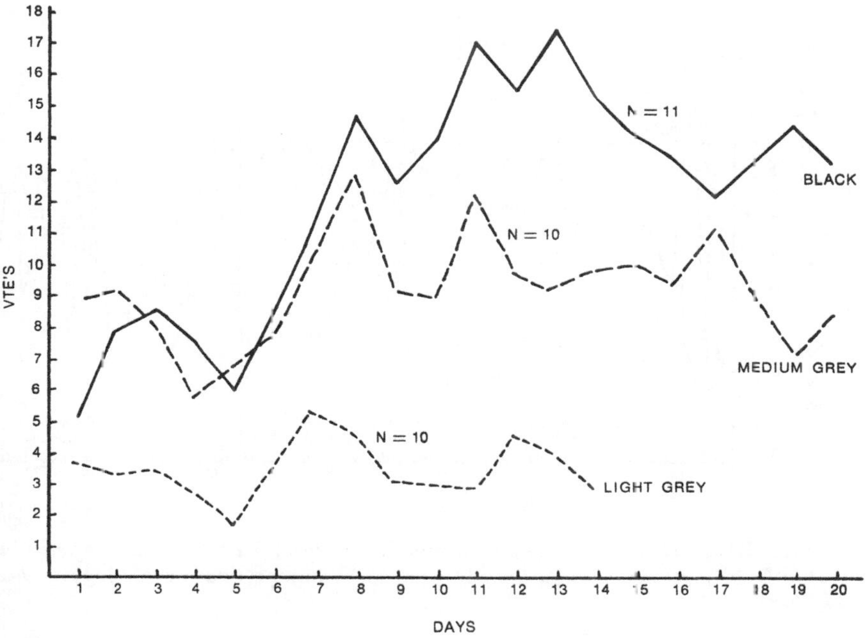

FIGURE 2.11. *Average number of VTE's per day. (From E. C. Tolman, Prediction of vicarious trial and error by means of the schematic sowbug.* Psychological Review, *1939, 46, 320.)*

at which they had learned. Comparing the beginning of each of these three difficulties the results were that the rats did more VTEing for the easy discriminations than for the more difficult ones. When, however, it came to a comparison of amounts of VTEing during the final performance after each learning had reached a plateau, the opposite results were obtained. In other words, after the rats had finally divined their instructions, then they, like human beings, did more VTEing, more sampling, the more difficult the discrimination.

Finally, now let us note that it was also found at Berkeley by Jackson (Jackson, 1943) that in a maze the difficult maze units produce more VTEing and also that the more stupid rats do the more VTEing. The explanation, as I see it, is that, in the case of mazes, rats know their instructions. For them it is natural to expect that the same spatial path will always lead to the same outcome. Rats in mazes don't have to be told.

But what, now, is the final significance of all this VTEing? How do these facts about VTEing affect our theoretical argument? My answer is that these

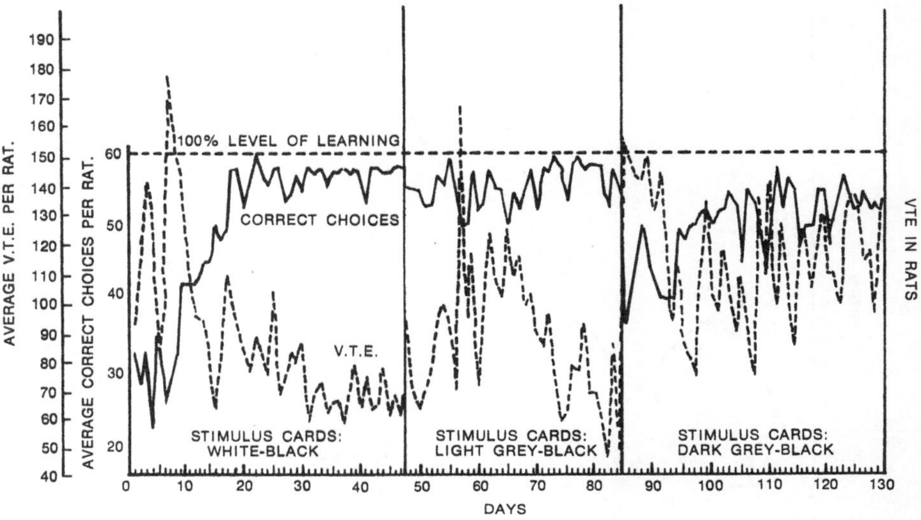

FIGURE 2.12. *(From E. C. Tolman and E. Minium, VTE in rats: overlearning and difficulty of discrimination.* Journal of Comparative Psychology, *1942, 34, 303.)*

facts lend further support to the doctrine of a building up of maps. VTEing, as I see it, is evidence that in the critical stages—whether in the first picking up of the instructions or in the later making sure of which stimulus is which—the animal's activity is not just one of responding passively to discrete stimuli, but rather one of the active selecting and comparing of stimuli. This brings me then to the third type of experiment.

3. "Searching for the Stimulus"

I refer to a recent, and it seems to me extremely important experiment, done for a Ph.D. dissertation by Hudson. Hudson was first interested in the question of whether or not rats could learn an avoidance reaction in one trial. His animals were tested one at a time in a living cage (see Fig. 2.13) with a small striped visual pattern at the end, on which was mounted a food cup. The hungry rat approached this food cup and ate. An electrical arrangement was provided so that when the rat touched the cup he could be given an electric shock. And one such shock did appear to be enough. For when the rat was replaced in this same cage days or even weeks afterwards, he usually demonstrated immediately strong avoidance reactions to the visual pattern. The animal withdrew from that end of the cage, or piled up sawdust and covered the pattern, or showed various other amusing re-

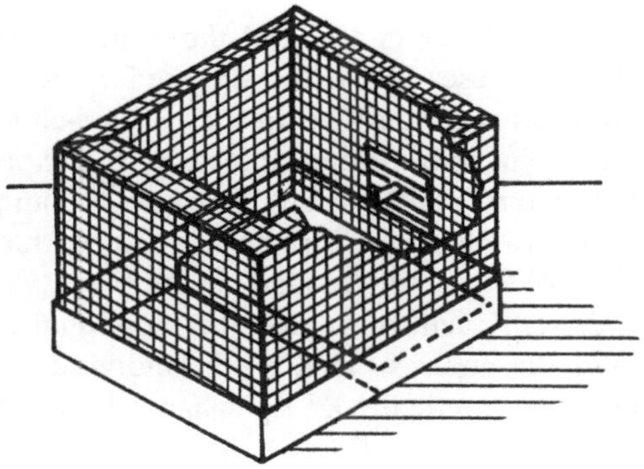

FIGURE 2.13. *(From Bradford Hudson. Ph.D. Thesis: "One trial learning: a study of the avoidance behavior of the rat." On deposit in the library of the University of California, Berkeley, California.)*

sponses all of which were in the nature of withdrawing from the pattern or making it disappear.

But the particular finding which I am interested in now appeared as a result of a modification of this standard procedure. Hudson noticed that the animals, anthropomorphically speaking, often seemed to look around *after* the shock to see what it was that had hit them. Hence it occurred to him that, if the pattern were made to disappear the instant the shock occurred, the rats might not establish the association. And this indeed is what happened in the case of many individuals. Hudson added further electrical connections so that when the shock was received during the eating, the lights went out, the pattern and the food cup dropped out of sight, and the lights came on again all within the matter of a second. When such animals were again put in the cage 24 hours later, a large percentage showed no avoidance of the pattern. Or to quote Hudson's own words:
"Learning what object to avoid . . . may occur exclusively during the period *after* the shock. For if the object from which the shock was actually received is removed at the moment of the shock, a significant number of animals fail to learn to avoid it, some selecting other features in the environment for avoidance, and others avoiding nothing."

In other words, I feel that this experiment reinforces the notion of the largely active selective character in the rat's building up of his cognitive map. He often has to look actively for the significant stimuli in order to form his map and does not merely passively receive and react to all the stimuli which are physically present.

Turn now to the fourth type of experiment.

4. The "Hypothesis" Experiments

Both the notion of hypotheses in rats and the design of the experiments to demonstrate such hypotheses are to be credited to Krech. Krech used a four-compartment discrimination-box. In such a four-choice box the correct door at each choice-point may be determined by the experimenter in terms of its being lighted or dark, left or right, or various combinations of these. If all possibilities are randomized for the 40 choices made in 10 runs of each day's test, the problem could be made insoluble.

When this was done, Krech found that the individual rat went through a succession of systematic choices. That is, the individual animal might perhaps begin by choosing practically all right-hand doors, then he might give this up for choosing practically all left-hand doors, and then, for choosing all dark doors, and so on. These relatively persistent, and well-above-chance systematic types of choice Krech called "hypotheses." In using this term he obviously did not mean to imply verbal processes in the rat but merely referred to what I have been calling cognitive maps which, it appears from his experiments, get set up in a tentative fashion to be tried out first one and then another until, if possible, one is found which works.

Finally, it is to be noted that these hypothesis experiments, like the latent learning, VTE, and "looking for the stimulus" experiments, do not, as such, throw light upon the widths of the maps which are picked up but do indicate the generally map-like and self-initiated character of learning.

For the beginning of an attack upon the problem of the width of the maps let me turn to the last group of experiments.

5. "Spatial Orientation" Experiments

As early as 1929, Lashley reported incidentally the case of a couple of his rats who, after having learned an alley maze, pushed back the cover near the starting box, climbed out and ran directly across the top to the goal-box

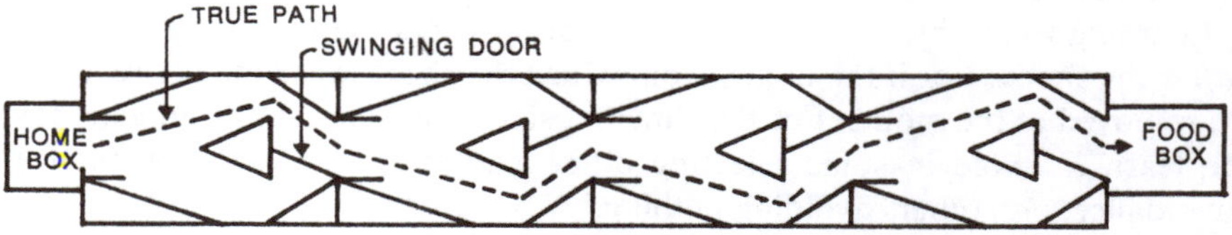

FIGURE 2.14. *(From I. Krechevsky (now D. Krech), The genesis of "hypotheses" in rats.* University of California Publications in Psychology, *1932, 6, No. 4, 46.) Originally published by the University of California Press; reprinted by permission of The Regents of the University of California.*

where they climbed down in again and ate. Other investigators have reported related findings. All such observations suggest that rats really develop wider spatial maps which include more than the mere trained-on specific paths. In the experiments now to be reported this possibility has been subjected to further examination.

In the first experiment, Tolman, Ritchie and Kalish (actually Ritchie and Kalish) used the set-up shown in Fig. 2.15.

This was an elevated maze. The animals ran from A across the open circular table through CD (which had alley walls) and finally to G, the food box. H was a light which shone directly down the path from G to F. After four nights, three trials per night, in which the rats learned to run

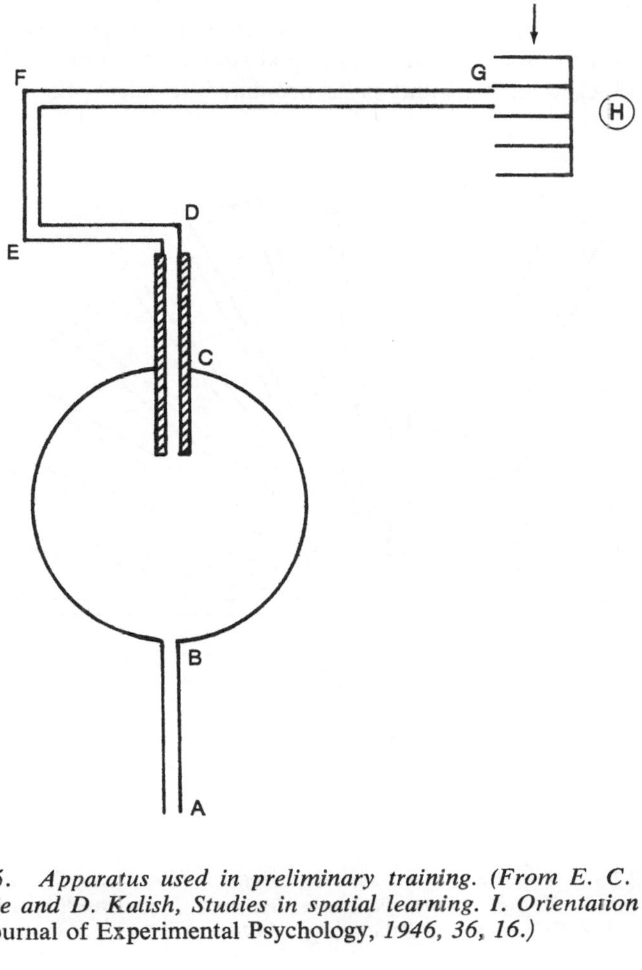

FIGURE 2.15. *Apparatus used in preliminary training. (From E. C. Tolman, B. F. Ritchie and D. Kalish, Studies in spatial learning. I. Orientation and the short-cut.* Journal of Experimental Psychology, *1946, 36, 16.)*

directly and without hesitation from A to G, the apparatus was changed to the sun-burst shown in Fig. 2.16. The starting path and the table remained the same but a series of radiating paths was added.

The animals were again started at A and ran across the circular table into the alley and found themselves blocked. They then returned onto the table and began exploring practically all the radiating paths. After going out a few inches only on any one path, each rat finally chose to run all the way out on one. The percentages of rats finally choosing each of the long paths from 1 to 12 are shown in Fig. 2.17. It appears that there was a preponderant tendency to choose path No. 6 which ran to a point some four inches in front of where the entrance to the food-box had been. The only other path chosen with any appreciable frequency was No. 1—that is, the path which pointed perpendicularly to the food-side of the room.

These results seem to indicate that the rats in this experiment had learned not only to run rapidly down the original roundabout route but also, when this was blocked and radiating paths presented, to select one pointing rather

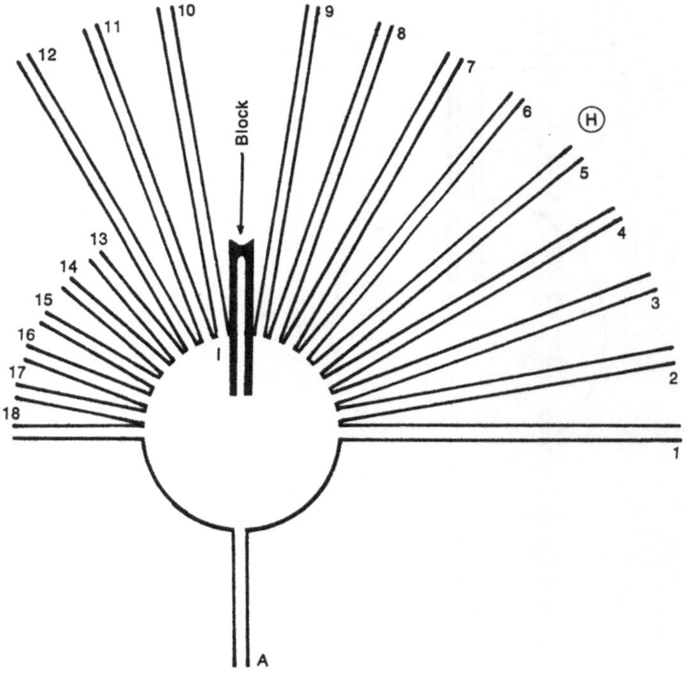

FIGURE 2.16. *Apparatus used in the test trial. (From E. C. Tolman, B. F. Ritchie and D. Kalish, Studies in spatial learning. I. Orientation and the short-cut.* Journal of Experimental Psychology, *1946, 36, 17.)*

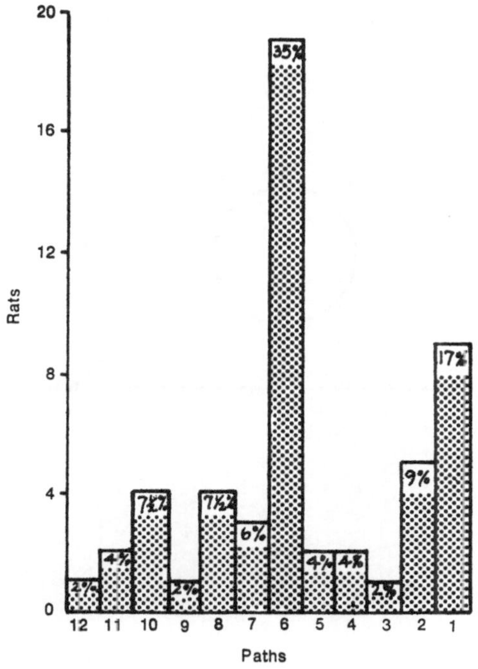

FIGURE 2.17. *Numbers of rats which chose each of the paths. (From E. C. Tolman, B. F. Ritchie and D. Kalish, Studies in spatial learning. I. Orientation and the short-cut.* Journal of Experimental Psychology, *1946, 36, 19.)*

directly towards the point where the food had been or else at least to select a path running perpendicularly to the food-side of the room.

As a result of their original training, the rats had, it would seem, acquired not merely a strip-map to the effect that the original specifically trained-on path led to food but, rather, a wider comprehensive map to the effect that food was located in such and such a direction in the room.

Consider now a further experiment done by Ritchie alone. This experiment tested still further the breadth of the spatial map which is acquired. In this further experiment the rats were again run across the table—this time to the arms of a simple T. (See Fig. 2.18.)

Twenty-five animals were trained for seven days, 20 trials in all, to find food at F_1; and twenty-five animals were trained to find it at F_2. The L's in the diagram indicate lights. On the eighth day the starting path and table top were rotated through 180 degrees so that they were now in the position shown in Fig. 2.19. The dotted lines represent the old position. And a series of radiating paths was added. What happened? Again the rats ran across the table into the central alley. When, however, they found themselves

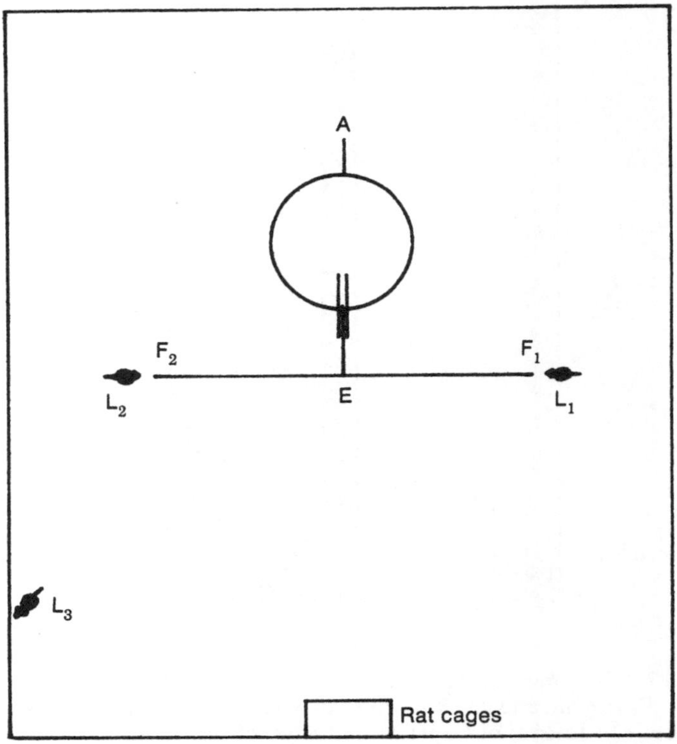

FIGURE 2.18. *(From B. F. Ritchie. Ph.D. Thesis: "Spatial learning in rats." On deposit in the Library of the University of California, Berkeley, California.)*

blocked, they turned back onto the table and this time also spent many seconds touching and trying out for only a few steps practically all the paths. Finally, however, within seven minutes, 42 of the 50 rats chose one path and ran all the way out on it. The paths finally chosen by the 19 of these animals that had been fed at F_1 and by the 23 that had been fed at F_2 are shown in Fig. 2.20.

This time the rats tended to choose, not the paths which pointed directly to the spots where the food had been, but rather paths which ran perpendicularly to the corresponding sides of the room. The spatial maps of these rats, when the animals were started from the opposite side of the room, were thus not completely adequate to the precise goal positions but were adequate as to the correct sides of the room. The maps of these animals were, in short, not altogether strip-like and narrow.

This completes my report of experiments. There were the *latent learning experiments,* the *VTE experiments,* the *searching for the stimulus experi-*

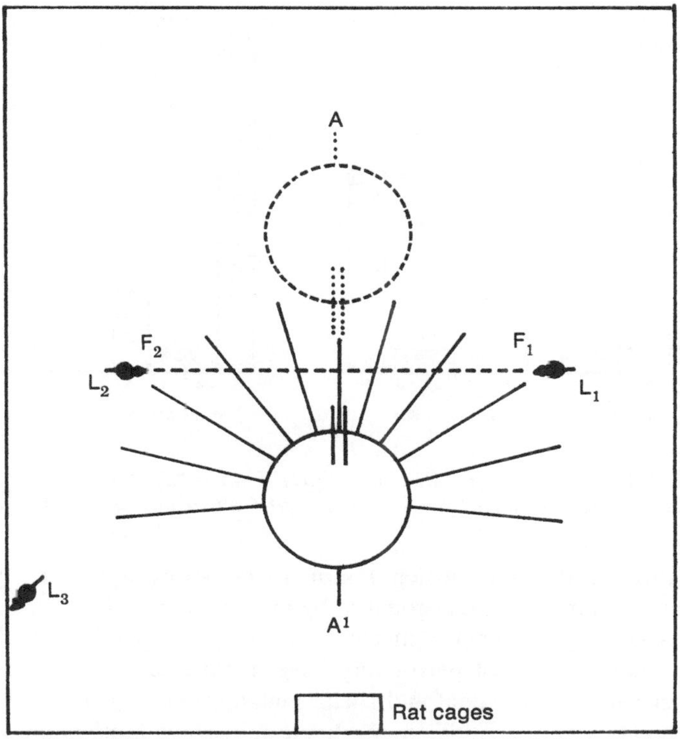

FIGURE 2.19. *(From B. F. Ritchie. Ph.D. Thesis: "Spatial learning in rats.' On deposit in the Library of the University of California, Berkeley, California.)*

ment, the *hypothesis experiments*, and these last *spatial orientation experiments*.

And now, at last, I come to the humanly significant and exciting problem: namely, what are the conditions which favor narrow strip-maps and what are those which tend to favor broad comprehensive maps not only in rats but also in men?

There is considerable evidence scattered throughout the literature bearing on this question both for rats and for men. Some of this evidence was obtained in Berkeley and some of it elsewhere. I have not time to present it in any detail. I can merely summarize it by saying that narrow strip maps rather than broad comprehensive maps seem to be induced: (1) by a damaged brain, (2) by an inadequate array of environmentally presented cues, (3) by an overdose of repetitions on the original trained-on path and (4) by the presence of too strongly motivational or of too strongly frustrating conditions.

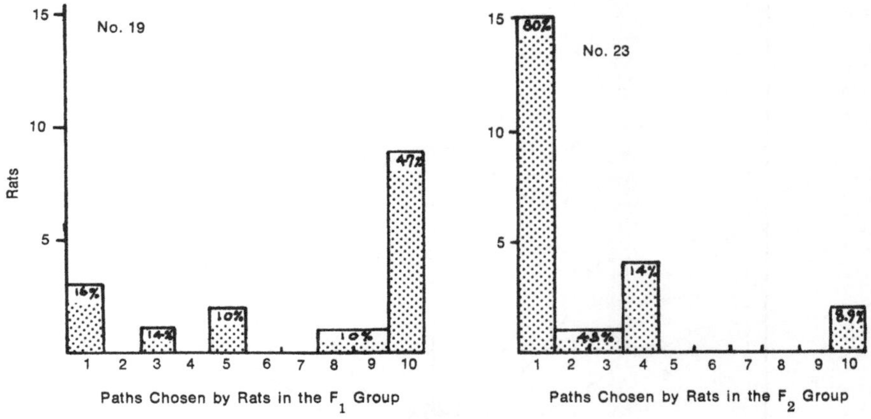

FIGURE 2.20. *(From B. F. Ritchie. Ph.D. Thesis: "Spatial learning in rats." On deposit in the Library of the University of California, Berkeley, California.)*

It is this fourth factor which I wish to elaborate upon briefly in my concluding remarks. For it is going to be my contention that some, at least, of the so-called "psychological mechanisms" which the clinical psychologists and the other students of personality have uncovered as the devils underlying many of our individual and social maladjustments can be interpreted as narrowings of our cognitive maps due to too strong motivations or to too intense frustration.

My argument will be brief, cavalier, and dogmatic. For I am not myself a clinician or a social psychologist. What I am going to say must be considered, therefore, simply as in the nature of a *rat* psychologist's *rat*iocinations offered free.

By way of illustration, let me suggest that at least the three dynamisms called, respectively, "regression", "fixation", and "displacement of aggression onto outgroups" are expressions of cognitive maps which are too narrow and which get built up in us as a result of too violent motivation or of too intense frustration.

(a) Consider *regression*. This is the term used for those cases in which an individual, in the face of too difficult a problem, returns to earlier more childish ways of behaving. Thus, to take an example, the overprotected middle-aged woman (reported a couple of years ago in *Time Magazine*) who, after losing her husband, regressed (much to the distress of her growing daughters) into dressing in too youthful a fashion and into competing for their beaux and then finally into behaving like a child requiring continuous care, would be an illustration of regression. I would not wish you to put too much confidence in the reportorial accuracy of *Time*, but such an extreme case is not too different from many actually to be found in our mental hospitals or even sometimes in ourselves. In all such instances my argument would be (1) that such regression results from too strong a present emotional situation and (2) that it consists in going back to too narrow an earlier map, itself due to too much frustration or motivation in early childhood. *Time*'s middle-aged woman was presented by too frustrating an emotional situation at her husband's death and she regressed, I would wager, to too narrow adolescent and childhood maps since these latter had been originally excessively impressed because of overstressful experiences at the time she was growing up.

(b) Consider *fixation*. Regression and fixation tend to go hand in hand. For another way of stating the fact of the undue persistence of early maps is to say that they were fixated. This has even been demonstrated in rats. If rats are too strongly motivated in their original learning, they find it very difficult to relearn when the original path is no longer correct. Also after they have relearned, if they are given an electric shock they, like *Time*'s woman, tend to regress back again to choosing the earlier path.

(c) Finally, consider the *"displacement of aggressions onto outgroups."* Adherence to one's own group is an ever-present tendency among primates. It is found in chimpanzees and monkeys as strongly as in men. We primates operate in groups. And each individual in such a group tends to identify with his whole group in the sense that the group's goals become his goals, the group's life and immortality, his life and immortality. Furthermore, each individual soon learns that, when as an individual he is frustrated, he must not take out his aggressions on the other members of his own group. He learns instead to displace his aggressions onto outgroups. Such a displacement of aggression I would claim is also a narrowing of the cognitive map. The individual comes no longer to distinguish the true locus of the cause of his frustration. The poor Southern whites, who take it out on the Negroes, are displacing their aggressions from the landlords, the southern economic system, the northern capitalists, or wherever the true cause of their frustration may lie, onto a mere convenient outgroup. The physicists on the Faculty who criticize the humanities, or we psychologists who criticize all the other departments, or the University as a whole which criticizes the Secondary School system or, vice versa, the Secondary School system which criticizes the University—or, on a still larger and far more dangerous scene—we Americans who criticize the Russians and the Russians who criticize us, are also engaging, at least in part, in nothing more than such irrational displacements of our aggressions onto outgroups.

I do not mean to imply that there may not be some true interferences by the one group with the goals of the other and hence that the aggressions of the members of the one group against the members of the other are necessarily *wholly* and *merely* displaced aggressions. But I do assert that often and in large part they are such mere displacements.

Over and over again men are blinded by too violent motivations and too intense frustrations into blind and unintelligent and in the end desperately dangerous hates of outsiders. And the expression of these their displaced hates ranges all the way from discrimination against minorities to world conflagrations.

What in the name of Heaven and Psychology can we do about it? My only answer is to preach again the virtues of reason—of, that is, broad cognitive maps. And to suggest that the child-trainers and the world-planners of the future can only, if at all, bring about the presence of the required rationality (i.e., comprehensive maps) if they see to it that nobody's children are too over-motivated or too frustrated. Only then can these children learn to look before and after, learn to see that there are often round-about and safer paths to their quite proper goals—learn, that is, to realize that the well-beings of White and of Negro, of Catholic and of Protestant, of Christian and of Jew, of American and of Russian (and even of males and females) are mutually interdependent.

We dare not let ourselves or others become so over-emotional, so hungry, so ill-clad, so over-motivated that only narrow strip-maps will be developed. All of us in Europe as well as in America, in the Orient as well as in the Occident, must be made calm enough and well-fed enough to be able to develop truly comprehensive maps, or, as Freud would have put it, to be able to learn to live according to the Reality Principle rather than according to the too narrow and too immediate Pleasure Principle.

We must, in short, subject our children and ourselves (as the kindly experimenter would his rats) to the optimal conditions of moderate motivation and of an absence of unnecessary frustrations, whenever we put them and ourselves before that great God-given maze which is our human world. I cannot predict whether or not we will be able, or be allowed, to do this; but I *can* say that, only insofar as we *are* able and *are* allowed, have we cause for hope.

3

Toward a Developmental Theory of Spatial Learning

DAVID STEA AND JAMES M. BLAUT

Introduction

The general area of learning has occupied the attention of psychological theorists over the past several decades; geographers, over almost as long a period, have been concerned with land utilization or geographical behavior; and those who deal in questions of visual aesthetics and environmental design have, during a similar period, pondered the problem of visual thinking. But the intersection of these interests, the subject matter of the field which may be termed "environmental learning," has received relatively little attention. Such learning deals with comprehension of a different environment at a different scale than that usually dealt with by psychologists involved in research on perception. It appears that this form of learning must incorporate the cognitive representation of geographical space, and we have chosen to call such representations by the name apparently coined by Tolman: *cognitive maps*. Recent studies in the field have focused largely on the analysis of behavioral products which presumably reflect the existence of cognitive maps and much less upon the process itself, a process which we shall refer to as *cognitive mapping*. A few rather vague theoretical positions have been proposed for what some have termed "appraisive cognitive mapping" and these draw heavily upon prior work of Tolman, Lewin, and other psychologists of the classic era. The present paper tries to provide certain underpinnings for a general theory of cognitive mapping, drawing upon our own findings and those of a number of other investigators.

We begin by reviewing briefly the evolution of psychological and geographical thought related to this problem area. The next step is an existential

The research reported here was supported in part by a grant (OE 4493) from the Bureau of Research, United States Office of Education. It was first presented at the Environmental Design Research Association Conference held in Pittsburgh, Pennsylvania in October, 1970.

leap to a general framework for cognitive mapping, involving the presentation of a hologram model to conceptualize the process and to clarify the roles played by sensorimotor variables in terms of both deprivation and enrichment. Consideration of the latter leads to a discussion of the development of cognitive mapping abilities, especially the early stages, when direct contact with a variety of large scale environments is severely limited. Thus, we hypothesize that early environmental learning necessitates the use of environmental surrogates, and that these lead to the formation of primal cognitive maps. Evidence from our studies of mapping and toy play behavior in young children is then presented as a basis for developmental hypotheses concerning the role of perceptual surrogates in the process. Finally, we turn to the question of environmental design: What are the implications of our findings for the process of creating environments for children and adults?

Historical Sketch

Interest in cognitive mapping processes has waxed and waned, and waxed once more since the birth in Leipzig in 1879 of scientific psychology. Psychology was born in structuralism; its subject matter was the minimal elements of experience and introspection was its method. *Imagery* (though a different form of imagery than is discussed in this paper) was one of its major concerns, with much speculation centering on such concepts as *eidetic* imagery—the ability to visualize precisely a stimulus once seen—and "imageless thought." Reaction to this introspective approach, reflecting the logical positivist movement in philosophy and a desire to emulate the methods and exactness of physics, gave rise to Behaviorism. In the second decade of the twentieth century, J. B. Watson, regarded by many as the father of Behaviorism, tried to dismiss imagery in all its forms from the proper study of psychology and succeeded in driving it far into the background. Nevertheless, concepts related to spatial imagery were reintroduced by a number of psychologists. Ironically, when Tolman came forward with his explicit concept of cognitive mapping, the thinking behind it—most notably Kurt Lewin's (1936) concept of the psychological field—drew heavily on early twentieth-century introspective psychology.

Back in the more orthodox Behaviorist camp, Clark Hull (1952) was simultaneously developing a theory of learning based upon chains of discrete stimulus-response connections. Because learning built upon such connections seemed diametrically opposed to learning based on cognitive mapping processes, a controversy flamed up during the late 1940's and early '50's, generating a flood of experiments with albino rats designed to demonstrate, that one law (S–R) or the other (cognitive mapping) was a root explanation for all learning. After a time, the controversy simply petered our, never having been satisfactorily resolved in the context of spatial or environmental learning. But while interest in generalized cognitive mapping waxed and waned, interest in human spatial imagery remained nearly dormant within psychology until Boulding (1956) suggested that the long-lost stepchild of nineteenth century perception was due for a revival. Boulding's *The Image* suggested relations between nonvisual imagery and cognitive mapping that inspired two simultaneous but independent contributions, one by Miller, Galanter, and Pribram (1960) and the other by Lynch (1960), thus bringing imagery and the notion of cognitive plans for the utilization of environments back into legitimacy.

Then began the present epoch of rapid growth in research dealing with cognitive mapping and other components of environmental behavior. The main impetus seems to have come from two *client* fields, each of which had a pressing need for predictive theory concerning human interaction with the spatial environment. One of the fields was environmental design, which at that time was desperately searching for a way to adapt physical planning to human needs, but possessed no reliable means of discovering what those needs called for in the way of a redesigned environment. The other field was geography, which had finally recovered from the trauma of environmental determinism and was no longer afraid to search for the causes of environmental behavior, but possessed little theoretical capital of its own with which to initiate the search. To serve these client fields, a new area emerged within psychology (Craik, 1970a; Proshansky, Ittelson, and Rivlin, 1970).

Geography in the nineteenth century had been as much concerned as psychology with the problem of explaining environmental behavior, or "man-land relations", as the problem was labeled by geographers. Unfortunately, geographers had chosen to explain "man" in terms of "land," rather than vice versa—perhaps a pardonable mistake in the days when the physical environment seemed to conform to strict and predictive laws, while social science was still largely social history, and psychology was hardly predictive at all. But the theory of environmentalism collapsed when anthropology and sociology produced a few contrary facts and alternative models in the first decades of this century (Tatham, 1951). The geographers' reaction was to abandon environmental behavior as a field along with environmental determinism as a theory, and to sink into the apathy of nonexplanatory description (Blaut, 1962). Yet one cannot persist very long in describing the man-environment system without developing a thirst for explanation. Pseudo-explanations obtained by defining behavior as a dependent part of a macrogeographic, superordinate whole—the Region—did not satisfy this thirst, and the search for unashamed theories of environmental behavior was finally recommenced in the 1960's. While educational geographers were searching for a basis for geographic education in the work of Piaget and Inhelder (1967), some behavioral geographers turned back to the earlier views of Tolman and Lewin for an environmental learning framework. The subject matter was originally conceptualized as "environmental perception," but it soon became apparent that the basic problem was neither perception as studied by psychologists nor environmental determinism in any of the forms promulgated earlier by geographers, but rather environmental cognition—and notably its most intriguing element, cognitive mapping.

These new problems have forced us to reexamine some old theoretical viewpoints and controversies. In psychology, for example, what may have been most relevant in the Hull-Tolman differences mentioned above is not the question of whether environmental behavior is grounded in stimulus-response or cognitive mapping, but the empirical fact that both seem possible and indeed ought to occur in the specific framework of spatial learning. Rarely has it been explicitly noted that the experiments and thus the learning tasks involved in the controversy were always spatial. Geographers have suffered from an almost opposite difficulty: through an overzealous concern for the forest, they have failed to note the trees. What seems to be needed is the application of a sense of theoretical rigor to the geographer's newly rekindled interest in environmental behavior. Only by means of a broadened approach from one discipline and a more focused attack from the other will it be possible to arrive at useful and predictive theory.

Environmental Stimulation and Psychological Holograms

The environment provides an array of stimuli which the perceiving organism must somehow take in, assess, and act upon. Each organism is endowed with many ways of taking in environmental information, and these are traditionally termed sensory modalities. Such sensory modalities are organized hierarchically, though differently for different organisms. The hierarchical organization of sensory modalities in humans appears to have four bases: first, these modalities differ in the sorts and amounts of information—information useful for weighing environmental evidence—that they convey; second, they differ in their order of "reliability"; third, they differ in their order of "availability"; and fourth, they differ in the order in which their utility is learned by the developing child. Here we are clearly adopting the position that perception is learned.

All the above is already relatively well known. But what is suggested here is that these sense modalities also differ in the number of dimensions of which they can take account in various kinds of environments. As an example of one of these environments—albeit a microenvironmental object—consider a ball, a ball that is small enough to be held in the hand. The ball is placed in front of a person so that he can see it and observe that it is a three dimensional object at rest whose location can be readily changed (displacement in four-space) without changing its shape. Importantly, he can also feel the ball, with eyes closed, and get the same amount of information in almost as little time.

When we move up to larger environments, we find that the differences among modalities become more significant. Suppose a person is facing a meadow: he can look out across the meadow; he can see the grass and the trees and note their coloration; and, if he is developmentally sufficiently advanced, he can utilize sizes of known reference objects to measure such things as relative distances. The perceiver sees objects moving across the meadow and can detect very readily and estimate fairly accurately the distance, direction, and velocity of motion. Four-dimensional visual perception is again almost instantaneous at this environmental scale. But suppose that we somehow restrict this individual to the use of his tactile sense by placing him barefoot in the meadow in such a way that he is forced to get all his environmental information through his feet. If he does not move, his information field is essentially restricted to what is under his feet—really zero-dimensional or "point" information relative to the larger environment. *Vis-à-vis* the almost instantaneous visual experience of such an environment, tactile information is essentially zero-dimensional. Our subject's experience becomes one-dimensional when he walks a line across the landscape and integrates the zero-dimensional inputs over time (time here indicates perceived change rather than duration—no space-time dichotomy is implied).

What is proposed is that the process of moving through the environment increases the dimensionality of the information being provided by the senses a person is using. Sometimes this increase is an expansion by one dimension and sometimes by more than one. This suggests that it ought to be possible to compensate for the loss of four-dimensional vision by the utilization and integration of information accrued by the other sense modalities, combined with motion, and integrated over time. For example, when the person standing in his bare feet in the field, receiving zero-dimensional (point) information, begins to move across the field along a certain line, he has increased his informational input to a one dimensional (linear) system. Memory enables him to integrate the "feel" of various things along the way. More generally, the more propinquitous a perceptual experience is to an experiencing individual, the larger the number of sensory modalities that can be brought into play; the more remote the object or environment experienced, the higher the dimensionality of the modality which must be brought into play, or the larger the number of low-dimensional modalities which must be integrated to provide the experience. This simply restates the idea that there are a number of routes to the goal of perceptual experience, and will later lead us to a discussion of sensory and experimental deprivation.

With this ammunition, we return to the question of cognitive mapping. There is only minimal physiological evidence for the existence of cognitive maps as either "place" or "process" in the central nervous system; yet it requires but a few moments of introspection to realize that such representations must certainly exist. To convey with some precision what we all know on a common-sense level is, in this case, no mean task. The search for a physical analogue to the cognitive map has led us to consider the hologram, a recent product of laser technology, as a model or, perhaps, a metaphor.

Holograms have become familiar to the general public via "tricks" such as the creation of enlarged three-dimensional images of chess boards, each piece being the size of a man, so that chess tournaments can be watched by crowds much as they now watch such spectator sports as football. But the importance of the hologram is much more general. If we are dealing with very small objects such as chessmen, we can enlarge them and see the consequences of manipulating these objects in three dimensions very readily with holograms. Conversely, it seems that it ought to be possible to reduce the imaged sizes of very large objects to make them more amenable to convenient investigation.

Extending this to spatial conceptualization, we might speak of a "psychological hologram", a conceptual three-dimensional projection of the three dimensional object. It is a representation presumably located structurally or functionally somewhere in the central nervous system, the availability of which enables one to walk around a cognitive model, to rotate it, or to imagine the consequences of rotation, expansion, or compression. It is therefore a form of "model thinking" in the sense that it is a "thought-up model."

What sorts of representations are then possible? A sculpture is three-dimensional and often representational, but it does not change with every changing idea about the object represented. A hologram is a "dynamic" representation, and inserting changes occurring over time, whether real or psychological, makes it four-dimensional in scope. A three-dimensional man can thus engage in four-dimensional thinking. In the next section, we shall consider the question of reduction and enrichment of the "normal" dimensionality of environmental experience as reflected by studies of child development.

Forms of Environmental Deprivation

We noted before that it ought to be possible to partially substitute perceptual memory—the integrated memory of things once perceived—for immediate, or instantaneous perceptual experience. This is really what a blind person does, and the reason that a blind person is able to have much the same spatial experience over a somewhat longer period of time as a sighted person. This is also why a blind child is able, when asked, to reproduce a very good drawn image map of the house in which he lives. Details likely to be left out in such a reproduction are windows (Shemyakin, 1962), which have minimal importance to the blind child. What we suggest, then, is that the blind person possesses no strange "sixth sense" but that he simply learns to do two things: first, as is well known, he learns to utilize cues that the average sighted person shuts off because they appear either unimportant, redundant, or contradictory to the cues coming from another (the visual) modality; second, he learns to integrate dimensional information from all his other sense modalities to enable him eventually to create an accurate impression—however imaged—of an experienced environment.

When sight is restored to cataract victims who have been blind from birth, typically several months pass before they are able to deal with visual perception in any really useful way. They can distinguish color immediately, as children can at birth, but they have a difficult time with forms. The cataract victim suffers only a small degree of disability in comparison with sighted people when operating on small-scale environments near at hand because he can utilize the tactile system he has already learned to gather the same information that a visual system would give him about the shapes of objects, and he discriminates fairly rapidly on this basis. When his sight is restored, the new visual cues are not only redundant and lacking in utility; they sometimes actively interfere with perception, and the "newly-sighted" person falls back readily on other cues.

It is when operating on (and in) the larger environment that the real disability of the recently-blind person comes into play. He is initially unable to deal with visual gestalts, and attempts to identify distant objects on the basis of key stimulus elements alone (e.g., the well-known case of the little girl who, after removal of her cataract, called a fish a camel because it had a "hump"—its dorsal fin). There is the breath of a suggestion here that people cured of cataracts might learn to operate more rapidly with gross visual cues dealing with larger environments—"macro-gestalts"—and less rapidly with near cues which they can apprehend non-visually by substituting information from their other sense modes.

The basic question is, of course, one of perceptual learning of large environments via perceptual experience and sensorimotor interaction. Deprivation, via blindness or other means, leads to decreased ability to perceive in what humans call a "normal" fashion, but perhaps increased ability to perceive in other ways as the following example may illustrate. Turnbull (1961), in reporting on the perceptual world of the forest-dwelling Ituri pygmy, states that at least one of the pygmies had developed to only a limited extent the fundamental perceptual attribute of size-constancy and associates this with influences of the pygmy's environment—dense forest, where vistas of more than 100 yards are virtually impossible to obtain. In other words, Turnbull suggests that the pygmy had been "deprived" of those perceptual experiences necessary to develop fully the mechanism of size constancy. He further suggests that the utility of vision is generally reduced in the pygmies' environment; they appear to hunt less by sight than by sound, and while they do not possess a highly developed visual art form, their auditory art form (music) is developed to a high degree.

The work of Held and his associates (e.g., Held and Rekosh, 1963) further strengthens the contention that *motor* experience and sensorimotor interaction are important to the development of normal perception. Such experiments indicate that subjects exposed to equivalent sensory stimulation —e.g., visual—but differential opportunities to interact with an experienced environment, show different degrees of perceptual attainment. Specifically, it appears that the less the motor-environment interaction available to a subject, the less "veridical" his perception becomes. In terms of environmental learning, these findings seem to have implications for selection of transportation modes for both children and adults. "Passive systems" (e.g., buses) may lead to less environmental knowledge (impoverished imagery) than "active" systems (walking, bicycling).

Still another issue is the relation of the way in which environmental information is absorbed by the perceiver to the way in which it is retrieved (input-output modes). Observations (Stea, Douglas, Emerson, and Hart, 1969) that a blind subject can produce a better outline map of the United States than many sighted geography graduate students are startling at first, but seem less surprising when we consider that the blind, who learn maps by Braille, using tactile modality with proprioceptive feedback, are asked to reproduce these maps much in the same way as they are learned. Sighted subjects, who learn visually and reproduce otherwise, might thus be expected to suffer at least some handicap in a task of this sort.

Environmental Enrichment and Surrogates to Direct Experience

Thus far the discussion has centered upon direct environmental experience and its cognitive effects. The average adult is able to partake extensively of such experience; he moves quite freely through the world of large-scale objects and landscapes surrounding him, sometimes under his own power, and sometimes with the aid of transportation systems under his or another's control. The very small child, however, is denied such experience; the world he sees is sized for adults ("human scale" must mean very different things to children and adults, as architects ought to note; in fact there is no unitary human scale) and his movement through the world is severely restricted. As indicated in the following paragraphs, however, the development of the child's knowledge, or his cognitive mapping abilities, appears to antedate extensive firsthand dynamic experience with the larger environment. We will argue that this knowledge is acquired through the observation and manipulation of surrogates for direct experience; through acquisition of the ability to "model" the real world. But sensorimotor interaction *is* a necessary precedent and a most important prior experience.

We have been working with three-to-eleven-year-old children in Massachusetts, Puerto Rico, and St. Vincent testing in various ways their abilities to deal with map-like environmental surrogates: black and white aerial photographs, maps traced from these photographs, standard topographic maps, free-drawn maps, and landscape toys (Blaut, McCleary, and Blaut, 1970; Blaut and Stea, 1969; Muir and Blaut, 1969; Stea and Blaut, chapter 12; Wisner, 1970; Zerner, 1970). One group of studies with five- and six-year-olds in all three areas shows that almost all children of this age, with no prior exposure to aerial photographs, no training, and, in the case of St. Vincent, no prior exposure to television, can interpret a wide range of micro-features on the photographs. They can in most cases discover and name an aggregate feature (e.g., "town"). In a test carried out only in Massachusetts, they were able to prepare traced maps from the photographs, attach noniconic color codes to the tracings, and operate such tracings as maps in the solution of simulated navigational problems. At six, the children in Massachusetts were able to learn formal map-making and map-reading, starting with photo-interpretation, and ending at the point where interpretation becomes a matter of reading in the natural language (Muir and Blaut, 1969); the test was partially replicated in St. Vincent. A longitudinal study in Puerto Rico showed an increase in photo-interpretative ability from ages five through nine and thereafter a leveling off or perhaps slight decline. Finally, in tests of free toy play just completed in Massachusetts, preliminary analysis suggests that five-year-olds possess the ability to assemble toys representing landscape features at the perceptual scale—houses, cars, roads, churches, trains, and the like—into communities and other macro-graphic entities; at least half of the four-year-olds tested can do the same. Data for three-year-olds have not been analysed as yet. However, five children of this age had previously been tested on their ability to interpret micro-features on aerial photographs and all had done so. Hence, with cross-cultural data from nearly 500 children, we can state that children are able to read and use aerial photographs as maps.

These findings show that children are able to solve all essential problems of mapping—rotation from a horizontal to an orthogonal view of the landscape, reduction of scale, and abstraction to semi-iconic signs—before they are exposed to maps. The findings also show that children who have never seen the earth from a vertical perspective can nevertheless recognize a landscape image in this perspective. Finally, the data show that children who have never interacted with large spaces can nevertheless recognize the aerial properties of such spaces and solve mapping problems concerning them. We can explain these abilities only if we assume that a very highly evolved cognitive map has already been formed in many children by the age of five. But how can we explain this developmental attainment?

When we consider the development of cognitive maps in terms of the evolution of spatial experience, "perception" cannot be divorced from movement through the environment. What we regard as the "normal" experience of relative motion in a large environment is movement of the person or experiencing organism relative to a stationary environmental field: traveling around and among cities, for example. However, another experience which is just as reasonable and, in fact, occurs almost as frequently (in watching films and television, for example), is the opposite form of relative motion—the spectator remains stationary and the environment moves about him. This is in part what a child experiences when he engages in some forms of toy play, using houses, cows, and trees to model communities. His toy play involves more active control over the environment than television experience because two manipulations are open to him that are not available in his attempts to assess aspects of the real world. In toy play not only can he move the entire environment about, thus changing his perspective while remaining stationary, but he can choose the positions of the objects relative to each other, and examine the effects of changing these relative positions.

Data from some recent studies of toy play (Blaut and Stea, 1969) indicate that the extent to which children correctly label landscape elements; build model landscapes resembling communities from these elements in undirected play (in place of groupings based simply on similarity of form and color); and supply "gestalt" landscape names for their creations all vary monotonically with age in the 3–5 age range. Thus it seems that toy play activity is both a source of enrichment and an index of development.

We hypothesize that it is partly by these and even earlier manipulations, that the child develops shape constancy, especially the shape constancy implicitly applied to large-scale objects. We assume that some degree of size constancy has previously been established (Bower, 1966) and that the child knows that the model house represents the same thing as the full-size house does. Working with such a representation he can learn what reference objects look like from (to him) impossible vantage points: that is, from above. It seems probable that understanding how such shape constancies operate in the larger environment enables him to pick out "unfamiliar" shapes in aerial photographs, a skill already well developed long before school-entering age.

There is another way of looking at the issue: environmental learning is enhanced through surrogates because, as in experiences of the above sort, more sensory modalities are being brought to bear. In other words, visual experience is *enriched* in toy play through the use of different sensory modalities such as touch and proprioceptive feedback (experienced when an object is moved). The very small child may also need toys he can *taste*. We could probably argue that taste is one of the earliest, primal forms of sensory experience.

What the child experiences is then a three- (or four-) dimensional spatial system, whereas the same components of the real environment, perceived at a distance, would have tended to be like a stage setting or a background: essentially two dimensional. Increase in distance between perceiver and object is correlated with a decrease in the number of senses and possible sensory dimensions brought into play. One might think of surrogate manipulation as a verification process, in effect using one sense or several senses to verify what was previously learned by another. Toy play utilizes touch and other senses as a means of verifying what was purely a visual experience in the real spatial environment. Not only can model environments be manipulated in the sense of being altered, testing different combinations of objects in a model sense (true model thinking), but the child can verify the veridicality of his sensory experience by bringing to bear some modalities generally unused or unavailable in environmental experience.

The child not only scales down space, but also scales down time (again, no dichotomy is implied). A child in the real environment experiences changes in perspective very slowly, while the child engaging in environmental toy play can experience such changes quite rapidly and thereby connect the object viewed from one perspective with the object viewed from another. It thus seems likely to him that these two views are not two separate objects. In large scale spatial perception, environmental experience often appears timeless to children because large elements (other than mobile parts of transportation systems) move very slowly, if at all, and seem to the observer to be stationary. The dimension of time is inserted in toy play, a further enriching power of this surrogate to direct environmental experience.

Thus we see the dependence of environmental learning upon environmental experience: direct and surrogate, "normal" and otherwise. We have not touched upon the critical role played by the earliest experimental stages in development, including manipulation of hanging objects in the crib, and, later, crawling behavior.

Some Implications for Environmental Design

There are a number of lessons here for environmental designers; for example:

1. Spatial learning and thinking are not identical to visual learning and thinking. The training of designers is such that their bias is, quite naturally, visual (Stea, 1967). But spatial representation (cognitive mapping) in the absence of models and boards involves much more than just spatial perception: it involves the utilization of other sensory inputs, the integration of these inputs over time, movement, descriptive modes, the input of value systems and other biases, and—the idea central to this paper—a framework for putting all this together: early spatial learning.

2. Consider the question of why some architectural concepts "go wrong" in the light of "model thinking" as engaged in by architects. Beautiful three-dimensional models of designed environments, created at micro-scale, often

yield unbeautiful environments at macro-scale. There are many reasons why this may be so: the nonrepresented visual environment (e.g., smog) or nonvisual aspects of the full-scale creation may loom larger than expected; and details which seem too small or too insignificant to include in a model (e.g., guy wires, T. V. aerials) may, at a larger scale, completely change the intended effect.

3. The details which may be important to an adult client are likely to be even more important to the child client. Playgrounds, for example, ought to provide the child with more than an opportunity for exercise; they should be settings for three-dimensional sensorimotor environmental experiences at the meso-scale, intermediate between the levels of toy play and the geographical environment. Much more often, however, they are designed primarily to please an ill-defined adult aesthetic sense, and for ease of custodial care, rather than for the enhancement of spatial learning.

We are studying how spatial learning takes place. Our future understanding of this process may say something to designers about how they themselves and their clients experience space, about the language they use (or perhaps ought to use) to convey spatial ideas to others, and about modes of presentation. Some of the more subtle aspects of what is termed "the architectural experience" may never be identified by this means, but architectural experiences themselves, for adults as well as for the very young, may thereby improve.

4

Cognitive Maps in Perception and Thought

STEPHEN KAPLAN

Although the man on the street persists in attributing certain foibles of the species to "human nature," for many years psychology has tended to cast doubt on the hypothesis that man's behavior is influenced by inherited natural tendencies. Rather, it is argued, man is a collection of learned connections or associations. Man is, in these terms, what he has learned.

Recently a counterattack has been launched (e.g., Ardrey, 1961, 1966; Lorenz, 1966). Based on observation of animals in their natural setting and on new evidence on man's origins, a position has emerged that not only asserts that man has a nature, but also that it is a rather aggressive sort of nature. The image of a "killer ape" provides a caricature of what man is assumed to be like.

Undoubtedly there is truth to both these opposing positions. Learning plays a pervasive role in man's behavior, as do innate aspects—his nature. Neither of these positions, nor even both of them together, do justice to the organism in question. It is the central thesis of this paper that man indeed has a nature, and that much of this nature concerns the acquisition and use of knowledge. In other words, man gained his selective advantage in a difficult and dangerous world in large part through the development of quick and efficient mechanisms for handling information. This paper is concerned with a theoretical consideration of how these mechanisms might work. These information-handling mechanisms pertain directly to the issue of cognitive maps, and offer a basis for a more comprehensive view of human nature than any currently available.

Work on this paper was supported by the Forest Service, USDA. I greatly appreciate the help I received from Rachel Kaplan, Kenneth Winter, and W. J. McKeachie. The excellent assistance of the editors, Roger Downs and David Stea, also contributed substantially to the evolution of the paper.

It may seem a bit odd to place so much emphasis on man's nature. The issue may seem rather abstruse and highly academic. Yet in many ways this is our most practical and most pressing question. As man becomes harder to live with, and an ever increasing threat to his (and all other) species, coming to terms with this dangerous organism becomes the ultimate necessity. Yet it is difficult to deal with anything about which we know so little. All programs for reform and all proposals for burning-it-all-down-and-starting-over rest alike on certain implicit conceptions of human nature. Is man basically good but corrupted by the institutions he has created? Is man rational under certain conditions, and if so, which ones, and how can we obtain them fast? Is man a passive recipient of the impinging stimuli? If he does not live by bread alone, what is required? As Kates (1962) has pointed out, neither economic man nor psychoanalytic man is adequate to the challenge (see the introduction to Pt. I.)

Another aspect of this paper may seem slightly strange. In a book concerned with the environment and how we know it, the emphasis on human nature, that is, on man's genetic endowment, may seem out of place. Yet the notion that we somehow add the influences of genetics and environment has not turned out to be a useful conception (Hebb, 1966). The ways we deal with the environment are largely dependent on the sorts of mechanisms that have evolved for this purpose. Undoubtedly, we are profoundly influenced by the environment, but in ways mediated by our sensitivities, our structures, and our inherited initial condition. It is the interaction of these mechanisms and our initial environment that yields a basis for the processing of subsequent environments.

Man's Dependence on Knowledge

Man is an animal, a part of a larger ecological system. Man is a product of evolution, an organism that carries with it biases and tendencies that were required for survival millions of years ago (e.g., Howell, 1965; Pfeiffer, 1969). A central feature of man's prehistory was the necessity of resorting to skill, to wit, to intelligence for survival. As a grounded ape of not particularly formidable proportions, man was forced to survive by anticipating what would happen next in order to capture game and to avoid danger. The arboreal environment had led to the development of excellent vision and uniquely flexible response capability through the combination of the upright posture and the grasping hand. But a cleverly crafted plan and a cleverly crafted tool in the hand were necessary to take advantage of these natural assets. The planning and anticipating required for survival favored the development of a larger and more flexible information-handling capacity.

Some anthropologists argue that early man was a hunter of big game. A sketch of this argument is presented here to provide some concrete and vivid imagery for the assertion that man's survival throughout his evolution has been keyed to information acquisition and handling. Other anthropologists emphasize trapping and gathering as food-getting strategies. The informational requirements in such situations, although slightly less dramatic, would be equally extensive.

The argument for early man as a knowledge-oriented big game hunter is effectively presented in a stimulating paper by Laughlin (1968). He begins by pointing out that man has been, for over 99% of his existence as species, a hunter. He points to two factors that support the idea that early man's major source of food was big game. First, this was the predominant source of on-the-hoof meat in Africa where man evolved. Second, in contemporary African game preserves where both predator and herbivore are protected, the latter are increasing out of control and endangering the plant life in the area. It appears that their control in prior times was in large part the work of the one predator excluded from these preserves. Add to this information the one crucial limiting factor for man as a big game hunter: his weapons were useless at distances of greater than thirty feet. It is difficult to bring down a large animal with such weapons without knowing a great deal about that animal. Hunting in general, and hunting big game in particular, clearly required vast knowledge. It demanded predictive power, planning, and the capacity to communicate.

It is unlikely that an animal so dependent on information for his livelihood processes information in any bland, neutral fashion. In this perspective it would seem that a discussion of informational processes and structures might best begin with an exploration of some kinds of knowledge that might be particularly useful in enhancing survival probabilities. On intuitive grounds, there appear to be four major types of knowledge:

1. *Where one is.* The identification of one's current situation is a critical starting point for adaptive behavior. This requires both perception of the present stimulus array and memory of immediately preceding events.

2. *What is likely to happen next.* The identification of a range of future situations and the estimation of the relative probabilities of occurrence are the essence of prediction.

3. *Whether it will be good or bad.* The question of evaluation of the goodness or badness (or payoff) of predicted situations is also critical in decision making.

4. *Some possible courses of action.* The adaptive organism must not allow himself to become lost in thought. Tolman's cognitive map theory (1932; see also chapter 2) was the target of Guthrie's famous criticism: "In his concern with what goes on in the rat's mind, Tolman has neglected to predict what the rat will do. So far as the theory is concerned the rat is left buried in thought" (1935, p. 172). It should be noted that the model described in this paper is as much indebted to Guthrie as to Tolman. It depends on many small discrete elements to achieve an approximation to

Tolman's continuity while still retaining its associational character. Associational theories rarely leave their organisms lost in thought; their problem is to allow him to think at all. The necessity of considering possible actions demonstrates that the organism's knowledge must include that of himself and his capabilities.

Thus the knowledge man required for survival includes familiarity with the objects and situations characteristic of his environment and with the array of actions he can undertake. In addition he must be able to anticipate what is likely to come next. This requires that man store in his head a great deal of information concerning what leads to what. He must have in his head many possible situations and the relations between them.

Any advocate of the cognitive map concept is, at least implicitly, concerned with the structural basis of thought. Another way of putting this structural concern is in terms of the underlying machinery or mechanism; from this perspective understanding requires a description of the mechanisms involved.

It might seem logical at this point to introduce the basic workings of the proposed cognitive map structure. There is, however, a strategic basis for not doing so at the outset. The purpose of the next two sections is to specify the structure required to handle information in an adaptive fashion. This structure turns out to be a cognitive map. There is thus no need to begin by postulating a cognitive map; it *emerges* out of other (and more basic) considerations.

Where One Is: Perception and Survival

The first of the four kinds of knowledge required for survival, the determination of "Where one is," can be considered as the outcome of the perceptual process. This section is concerned with the importance of efficiency in perception to an organism under time pressure and the structure that would be required for such efficient perception to be possible.

EFFICIENCY IN PERCEPTION AS A REQUIREMENT FOR SURVIVAL

The first kind of knowledge required for adaptive behavior is knowing where one is; that is, the identification of the current situation. We achieve this with apparent ease much of the time, but we should not be misled by this impression. Identification is an enormously difficult problem which only seems simple because of the power of the mechanisms we have available for coping with it. Our environment is rich, diverse, and uncertain. The total pattern of stimulation impinging on the organism at any moment in time is very likely unique. On the other hand treating each moment as unique—that is, starting anew at each successive point in time—would clearly be disastrous. The feeling that one "has never seen anything like that before" must be an infrequent luxury for an animal competing for survival. Past experience is, of course, useless if all events are unique. But the problem is not only that the organism could not learn under such circumstances. Two major constraints make the problem of identification even more acute. The organism has limited time in which to decide and act, and limited storage capacity in his head. Both constraints point to the necessity of being sensitive to the situational regularities that underlie the endless flow of unique events. Since there is neither space for storing nor time for reviewing an indefinitely large number of unique memories, somehow the organism must operate on a more schematic basis. Out of the diversity and uncertainty the organism must be able to extract the essence of an object. In some sense the organism must have a prototype or ideal of an object, the instances of which are often poor shadows. Note, however, that this essence is presumably not an innate preconception of the object; rather it must be some sort of statistical summary based on many individual experiences. Efficient perception therefore depends on objects and situations having recurrent properties; that is, properties that occur together with reasonable frequency in the environment.

Thus from an adaptive point of view, there appears to be a strong argument for efficient perception; that is, for identifying the current situation in a way that capitalizes on past regularities and requires only a relatively small amount of information out of the diverse and uncertain environment. It not only seems reasonable that perception should operate in this way; the evidence indicates that it does. In fact it was some years ago that William James (1892) pointed out that perception is of "the probable and the definite." "Probable" because it reflects past frequencies of occurrence in the environment: one tends to interpret uncertain situations in terms of the familiar; missing details may be "filled in" where required (see chapter 6). "Definite" because the result of commerce with an ambiguous configuration is usually an unambiguous percept: survival requires definite percepts. One cannot be overly "tolerant of ambiguity" and expect to survive in a dangerous world. As Bruner (1957b) has pointed out, organisms often cannot afford the luxury of a second look.

A STRUCTURAL BASIS FOR EFFICIENT PERCEPTION

In this perspective the perceptual mechanism must possess a number of essential properties. (1) It must have a structural basis since it depends on past experience, which is necessarily stored in some structural fashion (Hebb, 1949), (2) The structure must be sensitive to the recurring features of the objects. (3) The mechanism must not require the presence of all of the features, or even of any particular feature, in order to signal the presence of some particular object. The environment is uncertain and the organism may be in a hurry. (4) Survival requires that the percept must in general be definite and complete; the organism must often "go beyond the information given" (Bruner, 1957a). That is, the features available in the stimulus pattern may be incomplete, but the organism perceives the object as possessing all its usual features.

There is a conceptual structure that not only meets these requirements but also has a number of other interesting properties. It is a general purpose structure that provides a hold on a wide range of different problems. While it is described here in rather abstract terms, those who are comfortable with physiological concepts should be aware of the true identity and source of the proposed structure. The neural net hypotheses of D. O. Hebb (1949, 1963, 1966) underlie this discussion. Readers desiring a richer perspective on the concepts presented here should refer to those sources.

Consider a set of threshold elements, where each element can have outputs to and inputs from any other element. In other words, they can send messages to and receive messages from each other.

To enable these elements to behave in an interesting fashion it is necessary to allow them to assume different states, and to provide some rule for determining what state they will be in at any given moment in time. For simplicity, let us allow them only two states, "on" (or active) and "off" (or resting). When an element is "on," it sends a message with a positive nonzero value on all its output lines. When an element is "off," all its output lines have a value of zero. An element will be in the "on" state only when the sum of the values on all of its input lines is greater than some threshold value. Thus an element will be active when the elements that send messages to it are active if their cumulative effect exceeds the element's threshold. Elements can be said to be associated with each other if they receive inputs from and send outputs to each other. Strength of association can be defined in terms of how influential the output of an element is in changing the state of the target element from "off" to "on."

In order to deal with the problems of perception, this abstract system must be placed in the head of some organism and tied in some fashion to experience in the world. The key issue here is the concept of *correspondence*. An element may be said to correspond to some feature in the world when the appearance of the feature in the immediate vicinity of the organism is promptly followed by the element in question going into the "on" state. This correspondence need not be perfect in the sense that the element must be "on" every time the feature is present and never when it is not. In fact, it turns out that the correspondence cannot be perfect and remain useful to an organism trying to make its way in an uncertain world.

The associative pattern within a collection of threshold elements is necessarily rich and relatively unconstrained. Each element can have many inputs and many outputs. Restricted associative patterns (e.g., the "chains" or linear structures of the radical behaviorists (Skinner, 1953; Keller, 1954) and the decision trees of some cognitive theorists (Hunt, 1962)) tend to enforce an order or sequence. The attributes of an object in general do not appear in a unique order, but in many different orders. The associations of any given element must code a multitude of predecessor and successor relationships between the corresponding attribute and other attributes.

A structure that meets the requirements for the efficient perception of some object can be conceptualized as a collection of elements which correspond to attributes of the object. The strength of association within this collection corresponds to the frequencies experienced in the environment. While some attributes will be more strongly or more closely associated with the collection, in general no single attribute will be necessary to signal the presence of the object. This signaling function can be conceived of as activity within this collection of elements. In a collection of this kind the pattern of associations is such that positive feedback occurs once the level of activity exceeds a critical value; that is, the more elements that are active, the more input messages the nonactive elements receive. As a result a great majority of elements will go "on" at once. Thus the collection tends to behave as a unit. (Positive feedback systems require careful controls to prevent instability [Ashby, 1952; Wiener, 1948]. Two useful constructs for this purpose are inhibition [Milner, 1957] and fatigue [Pomerantz, Kaplan and Kaplan, 1969; Kaplan, 1970].)

Representations. This collection of associated elements corresponds to some object in the environment in much the same way as the elements correspond to attributes. In other words, the presence of the object is in general the occasion for activity in the collection of elements.

Again there is not a perfect correspondence. There is no guarantee that the presence of the object in the immediate environment of the organism will lead to activity of the collection of associated elements that in general correspond to that object. The object may not be present sufficiently long as a stimulus pattern, or there may not be a sufficient number of familiar cues available. Likewise under circumstances of strong expectations, very few features or even a lack of contradictory information may be sufficient to activate the collection of associated elements. The very existence of imagination, of dreams, and of hallucination point to the necessity of assuming imperfect correspondence if human behavior and experience is to be modeled adequately.

At this point a few comments on terminology are in order. "The collection of associated elements that in general correspond to an object" is a rather clumsy way to refer to what may well be the basic component of thinking and perceiving. There are, however, more compact alternatives available, each associated with theory and data that make important contributions to our understanding of this mechanism. Bruner's "category" is an important example (Bruner, 1957b). The "schema" concept can be traced from Oldfield (1954) to Attneave (1957) to the work of Evans and his group (e.g., Brown and Evans, 1969; Edmonds and Edmonds, 1969; see chap. 5 for a discussion of a schema as a cognitive map). Despite the substantial quantity of research under the "schema" heading, the field appears to be converging on the term "representation." In a pair of pivotal papers on this concept, Shepard (1968; Shepard and Chipman, 1970) uses the expression "internal representation." Neisser (1968) also seems to favor this term. Reed (1969), who prefers the expression "central representation," provides a helpful discussion of some of the less frequently used terms with essentially the same content. Posner (1969, 1970; Posner and Keele, 1968), whose ingenious research designs have been uniquely valuable in nailing down this concept, varies between "abstract idea" and "internal representation." MacKay (1951) speaks of "acts of representation" in a thoughtful and rather philosophical examination of mind. Stea and Downs (1970) use the term "cognitive representation."

To those well versed in the classic cognition literature the term "representation" may seem to be synonymous with "concept." Both terms undoubtedly refer to an abstraction or equivalence class. However, the distinction lies in the theoretical stance that is implied. A concept is an abstraction in the sense that it is not at the level of an object but of classes of objects (e.g., all small red objects independent of their shape). A representation, by contrast, may stand for an object. In terms of a representational theory, the central correspondence standing for an object is itself an abstraction. In other words, from this perspective all perception is abstractive.

Generic properties of representation. A representation is developed through extended experience with the various stimulus configurations arising from a particular object or situation. The process is one of dropping out irrelevant attributes as much as adding new ones. Thus perception will tend to be generic. Generic perception can be viewed as a categorization process (Bruner, 1957b); as the placing of stimulus patterns into appropriate classifications. This process is often viewed as one that tidies up the organism's view of the world, translating the vagaries of fleeting stimulus patterns into orderly equivalence classes. While this trend does occur in perception, there are other aspects of the process that tend to modify or oppose this trend. First, stimulus information that at first contact is a poor fit for any of the preexisting representations is likely to be distorted in the process of perception. A generic system tends to appear neat and orderly primarily when the material to be categorized is neat and orderly. Second, not all perception is generic. For example a car can come in many colors. The representation for car presumably does not include color, since no particular color could be associated with the stimulus configuration "car" with any degree of reliability. At the same time, the color of a particular car is in general perceived. Color in this case is an attribute that is neither associated with nor contradictory to the representation in question. Its perception may depend on the activity of the element corresponding to that attribute; it thus serves to modify generic perception. When an attribute is contradictory to a portion of a representation, the modification of the perception may be more severe. Thus in certain perceptual experiments there are reports of "a circle but it has a gap in it," "an elephant but with chicken feet," and "an ace of spades but it's red" (e.g., Bruner and Postman, 1949).

These are instances of what has been called "schema-with-correction" (Woodworth, 1938). (In terms of the present model, the mechanism for this phenomenon involves the mutual inhibition of pairs of opposing attributes. The "correction" occurs when the member of an attribute pair that does not belong to the representation successfully dominates the one that does. While the detailed explanation is beyond the scope of this paper, this provides one further way in which a generic approach need not distort the perceptual process by making it seem neater than it actually is.)

Adaptive Decision Making

The previous section concerned perception, the first of the four kinds of survival-related knowledge proposed early in this paper. The other three items on that list, prediction, evaluation, and action, can be subsumed under the concept of decision.

The representation has been proposed as the basis for the organism's impressive capacity to identify the recurrent patterns in his environment. Figuring out where he is, however, is only the first step in adaptive information processing. In order to make suitable decisions he must also be able to make a variety of predictions about possible subsequent developments. Fortunately, it turns out that the representation concept provides a powerful basis for dealing with the problem of prediction.

WHAT IS LIKELY TO HAPPEN NEXT: PREDICTION

Consider an organism who has just identified the current situation. His first concern is to consider what might happen next. Note, however, that each "what-might-happen-next" is itself a situation. The current situation is processed in terms of a representation. Likewise, possible future situations are codable in terms of representations. Thus the problem of prediction becomes one of going from a representation that is active (due to the presence of its corresponding features in the environment) to other representations which are not yet active. In other words, prediction requires that representations be connected or associated with each other. (While the previous discussion of associative structure concerned threshold elements, the argument here is at a more molar level, with representations serving as the basic elements.) Further, these associations must correspond to the environmental frequencies with which one situation follows another. Psychology has been influenced so long by associative models (cf. Voss, 1969) that the immediate temptation is to propose that representations are associated to each other and that the strength of association codes how frequently the one situation follows the other in the environment. This approach has considerable merit. The associative concept has a long and noble history in psychology and its usefulness is undeniable. On the other hand, many different uses have been made of this concept, and some of them are quite unsatisfactory for our purpose.

The first issue that is usually raised in this context is the question of *what* is associated. As Voss (1969, p. 84) has pointed out, "The nature of the events that are said to be associated has varied with the orientation of the writer: ideas, contents of ideas, sensations, images, stimuli, stimuli and responses, and neurons have all been thought to be associated." This issue can be settled promptly since, based on the discussion in the preceding section, representations are what are associated. They are in the same family with ideas and images and are closely related to stimuli and sensations. Representations are, however, considerably more powerful and flexible as far as information processing requirements are concerned.

The second issue concerns the permissible *kinds* of associations. In particular it deals with how restricted associative patterns are to be. Some associative models permit each element only a single output, while others allow only a single input per element. Restrictions of this kind are inappropriate to the present model. Representations stand for generic situations, and their associative patterns code the sequencing of these possible situations. For any given generic situation there can be a variety of different situations both preceding and following it. For example, finding oneself splashing around in the water can have been preceded by a variety of different situations. One can have dived, have fallen, or been pushed. To take a more spatial example, there are many different routes to Rome, or to the corner drug store. Translated into associational terms, in general a representation must receive many inputs. It must also have many outputs. The basis for this is simply the converse of the previous argument. Any given situation can lead to various next situations. Thus, for one standing on the shore overlooking the water, there is no unique next situation that inevitably follows.

In order to utilize knowledge of a common object or event across a wide range of situations, the associations cannot be limited on either the input or the output side. Generic information processing requires a network type of structure (or, more precisely, a nonplanar semi-lattice; Berkhoff, 1940; Ore, 1962). These considerations are rather obvious and perhaps unnecessary. Yet there have been sufficient arguments for alternative structural arrangements to make the point of clarification necessary. While linear or chainlike structures and hierarchical (or branching or treelike) structures can occur as subparts of a network, they are not sufficient in themselves.

WHETHER IT WILL BE GOOD OR BAD: EVALUATION

Prediction is not in itself a sufficient basis for decision. The possible next situations and their probabilities of occurrence do not in themselves indicate to the organism what he ought to do. Decision requires in addition an evaluation of these potential occurrences, an indication of which are good and which bad.

The proposed solution is to provide just such motivational coding. Those situations that are associated with favorable outcomes will come to take on a positive quality of their own; likewise for negative situations. This hypothesis at one time might have seemed rather vague or even unlikely. Physiological evidence collected over the last ten or fifteen years makes this proposal considerably more reasonable. The discovery of the pleasure and pain systems in the brain by Olds and Milner (1954, cf. Olds, 1969, for a more recent account) provides a concrete mechanism. In the light of these data, pleasure and pain are defined very much like sensory properties. Just as the color red is signaled by the activity of some particlar neural unit (or set of units), pain and pleasure have their corresponding units.

Pleasurable and painful situations presumably result in the stimulation of these units. There is an additional hypothesis necessary for evaluation of possible situations, namely that through experience, pain and pleasure units become associated with particular representations. If this is the case, then certain situations will be experienced as to some degree painful (or pleasurable) even when physical injury (or physical delight) is not present. Further, the contemplation of a possible next situation that has been associated with pain will in itself be experienced as to some degree painful.

All three of the concepts developed so far—the generic properties of perception, the network of associations, and the motivational coding—combine to give the organism a highly efficient decision-making capability. When the organism identifies the current situation and predicts possible next situations and actions, these can possess attractiveness or unpleasantness as determined through a wide range of different experiences. Just as a representation is not problem-specific, but may have been active in a variety of different circumstances, so it will have had many opportunities to become associated with outcomes, good and bad. In this way preferences and biases can arise that serve as guidelines for dealing with new problems. While not necessarily appropriate to the new circumstances, they at the very least reduce the indecision that arises out of a lack of preferences. If the Faqueroux Discount Shoppe has a good lunch counter and a good selection of pots and pans, it constitutes a good choice in a search for imitation woogle rugs, given no other information. The emphasis here is not on optimization (cf. Simon, 1957, for a discussion of "satisficing" as an alternative to optimizing). The organism has incomplete knowledge, limited information-processing capacity, and no time to linger. The proposed structure appears appropriate, not to aid in finding the best solution, but to get the organism in the direction of a reasonable solution with minimal delay.

SOME POSSIBLE NEXT STEPS: ACTION

To this point the discussion has dealt with two aspects of the decision process, that is, the prediction and evaluation of next *situations*. A closely related problem is the consideration of possible next *actions*. This issue is fascinating and complex, involving strength comparisons, criteria, and strategy. While a discussion of this problem is beyond the scope of this paper, it is interesting to note that the framework developed here may provide a useful conceptual basis for action as well. It seems reasonable to assume that actions have internal representations just as objects and events do. A major part of the problem then becomes one of association between representations, although the representations in this instance differ in whether their tie to the outside world is through sensory qualities (i.e., attributes) or motor outputs. Miller, Galanter and Pribram (1960) provide an extended discussion of the cognitive basis of action that is quite compatible with the approach of this paper.

The Cognitive Map Revisited

In considering a possible mechanism for handling the sorts of knowledge required for survival, it seemed essential to embed the representations of objects and situations in a nonplanar networklike structure. In such a structure, any point will in general be connected to numerous other points; each of these other points is in turn connected to still other points. Thus the structure is an approximation to continuity. There will in general be several (if not many) different routes or paths between any two points. A network of this kind has a meaningful distance metric since each point is separated from any other point by a path that passes through some number of intervening points. "Direction" is also a meaningful concept here, since it is possible to move toward or away from any given point, with reference to other points in the net. Similarly, a "region" can be defined as a collection of closely associated points.

Clearly if an organism has stored in his head many possible situations and the relations between them, he has a cognitive map. It may appear at first glance that the route to this modest statement was unnecessarily arduous. Yet this attempt to establish a systematic framework leads to some conclusions that are not necessary consequences of the cognitive map concept in its more informal sense.

1. Generality: The structure underlying the spatial map of the world that people carry around in their heads is not different from the structure that underlies *all* cognitive processes.

2. Additional properties: The cognitive map (spatial or otherwise) will consist of generic representations that are motivationally coded and that are *not* related to each other in a simple hierarchical fashion.
3. Emergent functions: The same framework that is necessary for short-run, moment-by-moment decision making is also suitable for long-range planning and what might be called contemplative thought.

The first point, that cognitive processes and spatial cognitive maps do not involve essentially different structures, hardly needs elaboration. In this framework a spatial cognitive map might be viewed as a special case of cognitive maps in general. It is more likely, however, that spatial maps are not neatly separated from other sorts of cognitive structures. Thus one might expect temporal considerations, for example, to influence an individual's conception of physical space (see chap. 17). It is also likely that two points in space that have many "imageable" and thus likely-to-be-represented landmarks between them will be experienced as farther apart than two points the same physical distance apart with fewer intervening representations (see chap. 5). Whatever the detailed results turn out to be, there appear to be great advantages to the unification of the sometimes separate domains of mental maps and cognitive processes.

The force of the second point, dealing with additional properties, is that the same sort of arguments that lead to the cognitive map as a necessary structure for adaptive information processing also point to certain crucial additional properties that such a map must have. There are three major additions here: the generic quality of representations, motivational coding, and the network structure. The basic component of the cognitive map, the representation, must have a generic or schematic character. This suggests that a certain amount of the sloppiness in the maps people draw of their cities and neighborhoods may be due to generic coding (see chap. 6). Granting that this means there will be errors or omissions as far as detail is concerned, it is not necessarily the case that the map is therefore any less functional. The "goodness" of a map drawn by a subject in an experiment is perhaps best defined not in terms of the detail the investigator would like it to have but in terms of the viability of the map as a guide to the subject in whose head it resides (see chap. 15; Lynch, 1960).

The second additional property, the motivational coding of representations, does not mean that knowledge and preference are inseparable or substitutable as far as measurement is concerned (see chaps. 11 and 13). But their distinguishability should not mislead the investigator who measures only one of these facets into thinking that the other is absent or inoperative as far as the everyday behavior of the subject is concerned.

The network structure, the third additional property, must be sharply distinguished from a tree or hierarchy. There are, of course, many hierarchies embedded in it, and the process leading up to the solution of any given problem may well be a hierarchy. But the mind, like the city, is not a tree (cf. Alexander, 1965). In both cases the complexity, richness, and potential for diversity of the less neat structure argue for its adaptive advantage.

The final point concerns an emergent property of the cognitive map. This structure, which may have evolved as an efficient mechanism for moment-to-moment decision making, also provides a powerful basis for contemplation and long-range planning. The key concept here is borrowed from the artificial intelligence literature. Just as Samuel's (1963) checkers-playing machine must consider possible next configurations of the checkerboard and their successors, so any adaptive device must have some capacity for lookahead. The minimum requirement for decision making, predicting possible next events, can be viewed as the limiting case of a lookahead. But the same structure required for looking one step ahead also makes possible far more extensive search into possible futures when required by the complexity of the problem and permitted by the availability of time. Thus man is not inherently short-sighted. He has the capacity to look beyond the end of his nose under appropriate circumstances.

Within the proposed framework it is possible to identify what the appropriate circumstances might look like. Man can look ahead when he has available the required representations. It may be that future-oriented behavior is more likely when the pertinent representations have a vivid sensory quality; that is, when they are images (see chap. 2). Abstract information seems to be less useful, less motivating. As studies of the reaction to natural hazards indicate (Kates, 1962), it is hard even to think about an impending disaster when one has had no previous relevant experience.

Undoubtedly man is best at looking ahead to, anticipating, and planning for future events that he has experienced with reasonable frequency in the past. If this is all man is capable of looking ahead to, however, he is surely doomed. At present man faces the possibility of disasters that are unprecedented. The inundation of the earth with people and the destruction of the earth's resources have never been experienced before. Nor can they be experienced in order that man develop vivid images of what he must at all costs avoid. Perhaps writers and artists will be able to provide the required images. In medieval times artists were highly successful at providing images of a never-experienced and never-to-be-experienced future. That dire future was called "Hell"; perhaps an analogously vivid term will be created for our potential future.

Concluding Comments

This paper began with a consideration of the possibility that man might have a nature, and that his nature might be intimately associated with his dependence on knowledge. The subsequent discussions of efficient perception and decision were directly concerned with this issue, even though the connection was not made explicit. Once the cognitive map is seen as a pivotal concept in understanding perception and thought, a host of other issues are raised, each of which places constraints on man's nature. While space does not permit a discussion of these issues here, a few comments might be in order to provide further perspective.

The representation as the heart of the perceptual process provides the speed and capacity to identify objects given limited information. It also suggests that prejudice and the tendency to jump to conclusions are a part of man's nature. The discussion of decision and prediction indicates that man views the present in terms of possible futures and possible payoffs. This does not smack of altruism, but neither does it imply that man is inherently self-destructive. Man is, in fact, likely to take long-range considerations into account, to be provident and foresightful as long as he has the requisite experience or, possibly, adequate substitutes.

Beyond the mechanisms explicitly discussed here are the problems associated with developing, extending, correcting, and using cognitive maps. Even without delving into specific mechanisms it seems reasonable to assume that man could not be born with the maps he needs; likewise, he could not afford to wait for an emergency before he begins to develop them. Rather man must be born with a tremendous propensity to make and extend maps; that is, to explore and to learn (see chaps. 3 and 12). Man would have to be a curious, restless animal, not, as some theories have suggested, an animal that lies down and goes to sleep when his primary needs are satisfied.

Man must also have a bias toward action, toward making up his mind quickly and getting on with it. As was indicated previously, satisficing (Simon, 1957) plays a role here. But so too does "intolerance of ambiguity." The very possibilities for elaborate and extensive representation of present and future circumstances require a strong bias against too much internal rumination.

Some will complain that this does not paint a particularly attractive picture of man. This is a difficult argument to answer. Man's history is not particularly attractive either. If man is as angelic as some would assert, it is difficult to understand how he managed to generate his history. In these times of Rousseauian rebirth some are prone to point an accusing finger at man's institutions rather than man. Yet it seems quite remarkable that man manages to come up with the same sort of institution time and time again.

Man is not a "tabula rasa." He is a kind of animal that has lived by his wits over millions of years. He is a kind of animal that lives by what he knows and by what he can guess and by the plans he makes. He is a restless, searching animal. He has been selected for speed; he is quick to perceive and quick to decide. He tends toward oversimplification, toward prejudice, and toward going off half-cocked. He has been called aggressive and territorial, and probably correctly. But he is also loyal to his group, and the size of the group to which he is loyal may vary widely. He is fascinated by violence and intolerant of boredom. He is quick and efficient, at his best under difficult circumstances, eager to learn, to explore, and to act. He seeks and creates order.

It has come to the point where man himself is one of his greatest dangers. If he is to survive, I believe it will require his recognition of how much he is dependent on knowledge, and how limited his knowledge is. He must come to the realization that he, his society, and his environment are complex, difficult, and delicate systems. In dealing with these systems, we must use much more of what we know, and we must start to know more very soon.

II

Cognitive Representations

Introduction

All research areas pass through the classification stage in which phenomena are discovered, grouped, labeled, and described. This stage, though only the forerunner for the more significant stages of analysis, modeling, and theory building, is complex and confusing. The complexity and confusion come from hasty and ill-defined groupings, misleading or ambiguous labels, and the use of "imported" labels which have unwarranted connotations. The field of cognitive mapping is no stranger to these problems. There is a fundamental confusion over terminology which we must discuss in order to understand cognitive representations.

We are faced with a series of terms—cognitive map, mental map, spatial (or environmental) image—which refer to a cognitive representation of the nature and attributes of the spatial environment. Initially, we followed Boulding (1956) and Lynch (1960) in the use of the term "spatial image." However, just as we had dropped the term "environmental perception" (in its geographical, not psychological usage) in favor of *environmental cognition*, we also dropped the restrictive term spatial image for the more general concept of cognitive representation and the specific term *cognitive map*. Spatial imagery was a misleading and restrictive label, and we must clarify the sources of confusion, in addition to exploring cognitive representations themselves.

The confusion is two-fold, centering on the form of cognitive representation implied by the term spatial imagery, and on the relationship between the use of the term in a spatial context and in psychology. Although there is confusion over the form of representation implied by spatial imagery, its function is clear. Both Boulding (1956) and Lynch (1960) are explicit about the *adaptive* function of imagery in determining spatial behavior, a function illustrated as follows:

> (Vicki Folsom) once applied imagery to help her master left-hand driving before touring England and Scotland. In flight and just before falling asleep, she saw herself in a car designed for left-hand drive. She drove the roads, imagining she was coming out of a one-way street, entering into complicated turns or traffic patterns. "And," she added with a grin, "the system really works." (Wells, 1971, p. 24)

Lynch (1960, p. 8) argues that the image is used to interpret information; to guide action; to serve as a broad frame of reference within which a person can organize activity, belief, and knowledge; to serve as a basis for individual growth; and to give a sense of emotional security.

As a heuristic, the term spatial (or environmental) image has been of inestimable value in focusing and organizing research. Since Lynch's *Image of the City* has generated the most subsequent research, we return to him for a description:

> Environmental images are the result of a two-way process between the observer and his environment. The environment suggests distinctions and relations and the observer—with great adaptability and in the light of his own purposes—selects, organizes, and endows with meaning what he sees. The image so developed now limits what is seen, while the image itself is being tested against the filtered perceptual input in a constant interacting process. (1960, p. 6)

In this description, however, lies a major source of confusion. It is unfortunate that Lynch emphasizes the "seeing" aspect of imagery, leading some workers to equate imagery with the cognitive equivalent of vision, and yielding as one result confusion between environmental perception and cognition. A cognitive spatial representation (or image) depends upon more than visual input—it is an integrated, multimodal representation (see chap. 1). The confusion resulting from this misplaced emphasis is increased by Lynch's definition of imageability as:

> that quality in a physical object which gives it a high probability of evoking a strong image in any given observer. It is that shape, color, or arrangement which facilitates the making of vividly identified, powerfully structured, highly useful mental images of the environment. It might also be called *legibility*, or perhaps *visibility* in a heightened sense, where objects are not only able to be seen, but are presented sharply and intensely to the senses. (1960, p. 9)

The research tradition generated by Lynch is represented in contributions by Francescato and Mebane, and Orleans (chaps. 7 and 8). A major criticism of Lynch's original study centered on the nature and selection of the sample of respondents. Francescato and Mebane, in their comparative study of two Italian cities, obtained samples two to four times the size employed by Lynch, and divided their samples by age level, socioeconomic class, and length of residence in the city.

Orleans brings a much-needed sociological perspective to research on urban cognition in his study of Los Angeles, in which residents of five different socioeconomic areas were asked to draw maps of the *entire* city. Five different sets of responses were obtained, and Orleans relates these differences to existing sociological theory and knowledge of urban spatial behavior. He discusses the relevance and problems of using Lynch-type mapping procedures in research on differences in environmental cognition.

Having discussed one source of confusion over the term "imagery," we must turn to the second source and distinguish between imagery as used in psychology (see Bower, 1970; Holt, 1964; and Richardson, 1969), and in geography and planning. This is important since the geographic and planning use is frequently made without reference to the existing theory and findings in psychology.

The traditional psychological referent for an image was a *perceived* stimulus or complex of stimuli, and hence such phenomena as "after-images" could be traced to *physiological* processes. Holt (1964) argues that imagery is a generic term for all conscious, subjective representations of a quasi-sensory but nonperceptual character, and does not include "after-imagery" within his definition. Richardson (1969) supports Holt, and claims that no qualitative or quantitative attributes definitely differentiate images from percepts, since psychologically "normal" people sometimes cannot make the distinction. Richardson (1969) distinguishes among forms of imagery which have traditionally been studied—after-, eidetic, memory, and imagination imagery—along the dimensions of vividness, controllability, and stimulus dependency.

In contrast, the spatial or environmental image has remained a vague and ill-defined concept, having neither a clearly physiological nor a purely perceptual base. Its major components are memory and imagination imagery, and it is, by empirical definition or usage, a pictorialization of the cognitive map. But a more detailed breakdown of what Boulding (1956) refers to as the "image of man" is possible. Although we are only concerned with certain aspects of this image, Boulding has identified seven components:

> The *spatial image*, the picture of the individual's location in the space around him . . . the *temporal image*, his picture of the stream of time and his place in it . . . the *relational image*, his picture of the universe around him as a system of regularities . . . the *personal image* [placing] the individual in the midst of the universe of persons, roles and organizations . . . the *value image* [consisting of] ordering on the scale of better or worse [different] parts of the whole image . . . the *affectional image*, or *emotional image*, by which various [parts] of the image are imbued with feeling or affect. (1956, pp. 47–48; emphasis mine)

Boulding (1956) also identifies three dimensions of imagery: (1) certainty to uncertainty; (2) reality to unreality; and (3) public to private. These deal with the clarity of the image, its correspondence with "outside reality," and the degree to which it is shared by others.

One reason why our ignorance of the structure and function of cognitive representations dissipates slowly is that the present primitive state of theory and methodology allows us to get at parts of the representation but not at the total. The model for cognitive mapping research has thus far viewed the individual as a "black box" and examined the output to the exclusion of input variables. We know something about the response, but nothing about the stimulus. We have no idea how spatial information is processed (although we know something about the products of the transformation), how environments are learned (although we know something of the development of spatial cognition), and of inter- and intracultural differences in representation. Applications to environmental design and decision making have been scarce; exceptions include Appleyard (1969a) and Steinitz (1968). These questions will remain unanswered if the terminological confusion surrounding the classification stage of research into cognitive mapping is not resolved.

We shall distinguish between cognitive representations, or images, *generated* by behavior and those *generating* behavior (Stea, 1969). Generated maps or representations are those of the Lynch variety; people's perceptions and actions result in differing cognitive representations of the environment. Generating maps are those which, intentionally or unintentionally, cause people to view the environment in certain ways. The classic example of the generating cartographic map is the Mercator projection centering on the Western Hemisphere, and the consequent difficulty that many adults experience, because of childhood exposure to this projection, in cognizing Greenland as smaller than South America. Similar cartographic conventions, such as locating North at the top of the map, cause difficulty, and make polar projections practically unrecognizable to the general public.

Generated Cognitive Representations

A number of investigators have followed the lead of Lynch, producing studies of the "urban images" of cities throughout the world. Kates (1970) has provided a tabulation of these studies and their results (see table II.1).

A distinction must be made between two empirical questions concerning generated representations of urban areas; how these representations are formed and what is represented. The studies summarized by Kates (1970) produced results predominantly addressed to the latter question; only a few (e.g. Follini, 1966; Appleyard, 1970; Gittins, 1969) have explicitly addressed the former. Of the other selections in this section, the contributions of Saarinen (chap. 9) and Lee (chap. 5) are also of the "what" variety.

Saarinen's approach is methodologically close to Lynch. His technique of requesting free-hand maps or the identification of areas on an outline map of the world resembles Lynch's map-drawing and place-naming, but he adds another class of interpretation to that employed by Lynch. Saarinen investigates *locational* imagery, asking whether things are "correctly" placed, whether they are in conjunction with other things; in other words, he is questioning the *accuracy* of the location as well as the nature of attributes.

By asking subjects to draw the boundaries of their neighborhoods, Lee makes important inroads into problems of areal imagery and lays the groundwork for his own later investigations (Lee, 1968) and those of Steinitz (1968) and Stea and Wood (forthcoming). Lee's paper is one of the most imaginative and thought-provoking approaches to environmental cognition, and is included because it deserves greater recognition than its original publication source allowed.

Cox and Zannaras (chap. 10) take the United States as their cognitive area. As in Gould (see chap. 11), the resulting map is generated by the researchers from the subject's nonspatial responses to locations; however, neither study assumes or attempts to assess the subjects' total cognitive maps. Unlike Gould, Cox and Zannaras generate a map from an index of cognized similarity rather than preference. From subjects' judgments as to which three states of the United States are most similar to a given state, they arrive at locational classes for student "schemata" of the U.S.A., and apply the same technique to an analysis of sixty cities of the United States and Canada.

Generating Cognitive Representations

Maps indicate much more than cartographic truth or environmental imagery of the distant historic past. Today, they serve as powerful indicators of nationalism, self-image, attitudes, and aspirations. Variations are introduced in the existence or weight of lines used to indicate national boundaries or in the contrast between hues used to represent different nations. Most of the discussion in this book has been restricted to images of a single nation, subnational area, urban region, or city. The way in which national images and aspirations come into play cannot be understood without considering cartographic representations of larger entities: multinational groupings.

Guatemalan maps of Guatemala have depicted for more than 100 years what other nations call "British Honduras" as the Guatemalan province of Belize, usually labelled as "illegally held by Great Britain." There is also a difference between Israeli maps of Israel, and Egyptian, Syrian, or Jordanian maps of the same area, which sometimes label Israel as "occupied Palestine." The map is a powerful national symbol—an image of the image, so to speak (see Wisner, 1970). But the Soviets (*Newsweek*, February 2,

TABLE II.1 *Characteristics of city image drawn from published studies*

Principal Investigator and Year Published	*City*[1]	*Interview Sample (Number and Predominant Type)*	*Importance of Urban Elements*[2] *Landmarks*	*Nodes*	*Paths*	*Edges*	*Districts*	*Investigator's Comments*
K. Lynch (1960)	Boston (United States)	30 (Professional, managerial)	●	⊙	⊙	⊙	●	One strong edge; distinctive districts; confusing paths; understand structure.
	Jersey City (United States)	15 (Professional, managerial)			⊙			Lack of character; formlessness; low imageability.
	Los Angeles (United States)	15 (Professional, managerial)	⊙	⊙	●			Less sharp image; visually faceless; but active, ecologically ordered.
D. de Jonge (1962)	Amsterdam (Netherlands)	25 (Wives of skilled and white-collar workers)		●	●			A very strong image; strong predominance of main elements; spider web structure.
	Rotterdam (Netherlands)	22 (Wives of skilled and white-collar workers)	⊙	⊙	⊙			Over-all image weaker; buildings seen more clearly; no clear boundaries.
	The Hague (Netherlands)	25 (Wives of skilled and white-collar workers)	⊙		⊙		⊙	No wide straight path; separate elements and buildings; vague as to boundaries.
J. Gulick (1963)	Tripoli (Lebanon): entire city	35 (Students, upper middle class)		⊙	⊙		●	Stresses districts geographically distinctive or nodes; buildings not a major focus.
T. F. Saarinen (1969)	Chicago (United States)	42 Area workers	●		●		●	Tightly defined areas with internal detail.
		18 Suburban students			●	⊙		Broader areas.
		12 University students	●		●	⊙	●	Broader areas and external landmarks.

H. Klein (1967)	Karlsruhe (Federal Republic of Germany)	1118 (Residents)	●	●	⊙			Rational; striking landmarks; highly linear; imaged centre moving westward.
D. Appleyard (1969)	Ciudad Guayana (Venezuela): entire city	320 (Residents of selected settlements)		⊙			●	Little 'common' urban knowledge of city; higher for local areas; higher for lower-income population.
D. Stea and D. Wood (1970)	Mexico City: city and centre	335 (Residents of selected areas)	●	⊙	●		●	Edges almost entirely absent; strong domination by major paths; district landmarks.
	Puebla (Mexico): city and centre	144 (Residents of selected areas)	●	●				Streets extremely regular; highly legible but uninteresting.
	Guanajuato (Mexico): city and centre	114 (Residents of selected areas)	●	●	⊙			Highly irregular; unstructured; binodal.
	San Cristobal las Casas (Mexico): city and centre	176 (Students 12–18 years)	●	⊙	⊙			Legible city; clear and strong pattern of spatial activity.

[1] Central areas unless otherwise stated.
[2] The importance of elements is given as follows: ⊙ Important element; ● Very important element.

SOURCES: Kevin Lynch, *The Image of the City*, Cambridge, MIT Press, 1960; Derk De Jonge, 'Images of Urban Areas: Their Structure and Psychological Foundations,' *Journal of the American Institute of Planners*, Vol. 28, November 1962; John Gulick, 'Images of an Arab City,' *Journal of the American Institute of Planners*, Vol. 29, No. 3, 1963; Thomas F. Saarinen, *Image of the Chicago Loop* (unpublished paper); Hans-Joachim Klein, 'The Delimitation of the Town-Centre in the Image of its Citizens,' in: E. S. Brill (ed.), *Urban Core and Inner City*, Leiden, 1967, p. 286–306; Donald Appleyard, 'City Designers and the Pluralistic City,' *Planning Urban Growth and Regional Development: The Experience of the Guayana Program of Venezuela*, Cambridge, MIT Press, 1969; David Stea and Denis Wood, *Un Atlas Cognitivo: La Geografia Psicológica de Cuatro Ciudades Mexicanas* (in press).

This table is reproduced from Table 2, pages 654–655 in an article by Robert W. Kates entitled "Human Perception of the Environment". From the *International Social Science Journal*, Volume XXII, No. 4, 1970.

1970) have utilized the map in a somewhat different way. While the cartographic productions cited above were designed to induce particular impressions in the minds of foreigners and of their own nationals, the Russian maps endeavor to mislead potential enemies by purposely showing cities and other important military targets at some remove from their actual locations. The process involved abandoning the previous map projection for one which is mathematically inconsistent and then adding further deformation by the random mislocation of map features with respect to the new grid.

5

Psychology and Living Space

TERENCE R. LEE

It is difficult to think of a professional relationship which calls for a more humble frame of mind and for a more tentative approach than that of the psychologist addressing himself to the architect. There is, so far, very little direct assistance or advice that he can tender, and I am sure that during the next hour or so I shall frequently feel like a salesman, showing a few samples, but with very little productive capacity to back them up.

However, the relationship has to begin somewhere, and it is gratifying that overtures have been coming with a polite sense of urgency from yourselves in the past year or two. I hope to be able to quicken these advances by describing some of the social-psychological work that has been done up to now, by conveying to you, explicitly and implicitly, the way in which the psychologist tackles applied social problems, and by making some suggestions about the future of our proposed liaison.

Some of what I shall say will overlap with what might be said by a sociologist. The difference between a social psychologist and a sociologist sometimes appears to go no deeper than the name of their first degree. However, as one moves away from the overlapping zone, there are some real differences in method of approach and in level of analysis. To take the second first: Properly speaking, the sociologist is concerned with institutions, with the family, the neighborhood, the profession, etc., as social institutions. The social psychologist is concerned with the individual as his primary unit of analysis and with the effect on the individual of being in a family, a neighborhood or a profession. But, of course, institutions are

From *Transactions of the Bartlett Society*, Vol. 2, 1963–1964. Reprinted by permission of the author and publisher.

reducible to individuals and vice versa. The best *level* of analysis depends on the object of the analysis.

In method of approach, the psychologist is more rigorous and quantitative in answering questions, but this means that the questions are often rather trivial ones that look as if they could as well have been answered in the same way by anyone with half a grain of common sense. The sociologist is more inclined to tackle large, urgent and more obviously relevant questions, but in a way that makes one less sure that they have the right answer.

What about the architect? Presumably, he has normally regarded himself as a person who carries out alterations in the physical environment in accordance with his client's brief, and who tries to do so to the aesthetic satisfaction of whatever reference group he is anchored to. In recent years, because of the increasing scale of his operations and because, perhaps, of a new awareness, he is also beginning to see himself as a controller and manipulator of human behavior. This, in fact, is the mirror I propose to hold up to you. I hope it does not sound too insidious, an offence to the dignity of the individual or an erosion of human freedom. I do not wish to imply that this power and authority has been acquired from an unsuspecting and gullible public—indeed it is freely delegated and what some architects do not seem to realize is that they have it. They do not just *facilitate* human behavior, they *change* it.

The psychologist is concerned with discovering the lawfulness and regularity that underlies human behavior, with constructing theories about human behavior from which future behavior is at least statistically predictable. He realizes, none better, that behavior is fantastically complex, but he declines to deduce from this that it is magical. Changes in the physical and the social environment are *lawfully* connected with changes in human behavior—and it is high time we found out more about the laws.

Methodology

In order to evaluate the results of psychological investigation, it is necessary to know something about the methods. To say that psychologists are attempting to understand human nature is to say very little, because this has been the preoccupation of almost everyone since the dawn of mankind. What is different is that we are trying to use the same method as those of our colleagues who study other natural phenomena with such conspicuous success—the method known as *science.* This is generally taken to imply two broad techniques—(a) the experiment and (b) the use of systematic, objective, detached and controlled observation.

There are two reasons why we have started later and made less advance than other disciplines. The first is that our material is more intractable. By this I mean that every human event is preceded by a more complex pattern of preceding events than in any other field, and the variety of possible consequent events is out of all proportion. Secondly, and perhaps even more disabling, we are continually restrained, and properly so I am sure, by ethical considerations. Let me give you an example.

No one would deny that, with the enormous investment in rehousing that is continually in process throughout the world, it would be most useful if we could have some idea what effects housing has upon human behavior, and then to go on to find out what aspects are particularly potent in producing those changes which we consider desirable. How does new housing affect physical health, mental health, the better use of leisure, juvenile delinquency, marital discord, or what have you? We do not know the answers to any of these questions. But however insistent the questions, it is unlikely that any housing or planning authority, anywhere, would permit a simple experiment to be carried out. This would entail allocating a large number of families to either an experimental group or to a control group, using some random method of selection, and then rehousing one group and leaving the others in slums. Such a procedure would violate the socially approved order of priority for new housing and would therefore not be considered. I do not wish to imply that it should be—only that these ethical considerations make our task very much more difficult than dividing a substance into two different test tubes and observing the effects of different treatments upon them. However, this does not mean that the basic method of experiment should be any different, only that it is more difficult to use it.

The main alternative technique is that of systematic observation, and most of the examples to be described later are of this kind. I am certain that planning and architecture could be enormously assisted if we could increase their range and number. But this method has two major difficulties.

The first is that it very easily slips into a kind of obsessional fact-gathering, in which the original object of the exercise becomes submerged. I refer, of course, to the endless numbers of "planning surveys," many of which have proudly emerged in five or six colors and a prodigious number of words, only to end rather ignominiously in filing cabinets and on library shelves. The reason for this is firstly because we have not gone that much further to extract *generalisations* about the effect of different environments upon behavior. Surveys *describe* existing environments or they *describe* existing behavior, but only rarely do they grapple with the question of how the one affects the other. Any planner worth his salt is presumably trying to implement some higher value and not simply to reproduce existing behavior patterns somewhere else, and so the surveys don't help him beyond the technological level. Social surveys are useful in the same way as physical surveys. They cannot substitute for theories.

But supposing we take warning from this sorry post-war experience and attempt to order our observations so that they relate to specific hypotheses, and can begin to establish laws? Here is an example.

In a study carried out in Devonshire (Lee, 1961), I postulated that if the educational planners persisted in closing and decapitating (removing the secondary age children) from rural schools, the country planners would have to contend with a continuous rural depopulation. This hypothesis was tested by a demographic analysis, going back forty years, of 263 parishes. It showed that the population growth was depressed in direct proportion to the number of years since a parish had lost its senior children to the secondary modern schools in the market towns. Parishes where the primary school had been closed showed an even greater drain. Those which had never possessed a school were worst affected of all.

But here is the major shortcoming of the method of systematic observation. No matter how systematic, how objective, detached, unbiased and well quantified, it does not tell us whether we are dealing with causes or simply with concomitance. A careful analysis of an urban area would show that the number of churches per square mile is correlated with the incidence of vice and delinquency, but one would hesitate to say the one *causes* the other. There is almost certainly a correlation between the number of traffic policemen and the number of traffic accidents in any given square mile, but policemen don't cause accidents. Very often, events are correlated because they are both caused by some third factor and really have nothing to do with each other beyond that.

To return to the rural depopulation example, there are plenty of ways of checking and testing such observations to increase one's assurance that they are causally connected—but it is here that one needs the professional researcher and statistician—and such people are pathetically thin on the ground at present.

There is one further point worth making about method before passing on to consider results. This is that great care is needed to detach oneself from existing notions of what living space *ought* or *may be expected* to be like, if there is to be much progress towards understanding what it *is* like. I would go so far as to say that this is one of the major confusions in evaluating the neighborhood unit concept in planning.

McKenzie (1921), an American sociologist, studied social visiting among 1,000 families in 1921. He found that 76.5 percent of them took part in social visiting, and that 66 percent of these did more visiting in their immediate locality than in other parts of the city. His conclusion is a rather surprising one.

> This is doubtless due to the mobility and anonymity of modern city life, where personal acquaintance and neighborhood association have largely become a thing of the past.

This is an extreme example—but there are plenty of milder ones. It is as if they are surprised when their observations show that people no longer rely on horse-drawn vehicles for transport and the village pump for social intercourse, and they proceed from this to argue that modern city dwellers do not have any territorial associations at all, and it is therefore best to build cities as one large undivided lump. This leads me to my main illustration—a study carried out in Cambridge with the object of deciding whether the town planning concept of the neighborhood unit was an appropriate one for modern urban living (Lee, 1954).

Urban Neighborhood

When I first set out to find some neighborhoods, it was with statements like the one quoted from McKenzie in my mind. There are plenty of equivalent modern declamations, in which the neighborhood unit idea is represented as "outworn parochialism," as "village green planning," and so on. A most explicit commentator was the then Minister of Town and Country Planning, Lord Silkin (1948). He said:

> The assumption is that by dividing up your population into groups of 10,000 to 20,000 and surrounding them by open spaces, main roads, and railways, you will get nice little communities living happily and socially together. On what evidence is that based?

The obvious way to seek evidence was to ask people, but I was not sure whether to ask them about an area of ground or about an aggregate of neighbors. Gradually it became clear that, so far as the people who actually live in cities are concerned, they cannot separate the two. Their neighborhood is the place where their neighbors live and their neighbors are the people who live in the place. For them, physical and social space are inextricably linked—*in the mind*—and it is only for purposes of analysis that it might be helpful to separate them.

Another somewhat unexpected finding was that, except for the few upper-middle-class professional people who turn up in a random sample, neighborhood was a really salient experience; not a vague, shifting one, but something that could be quickly and easily acknowledged and fairly readily described to a total stranger making an unheralded call at the front door. Gaining confidence from this, I took a 6-inch-to-the-mile Ordnance Survey map on my visits, and found that people could readily respond to a request to "please draw a line round the part which you consider acts as your neighborhood or district." These neighborhood maps showed a good deal of variation, and in order to confirm a growing conviction that although neighborhoods are salient, they are also highly individualistic, I sampled a terrace row of corporation houses, eight of them within about 100 yards. When their outlines were superimposed they showed almost no coincidence. This was in spite of the fact that I had chosen an area labelled as a "neighborhood" on most people's lips and on the front of buses, and which in fact comprised an old village that had become assimilated into the edge of the city.

Later, a systematic sample was taken from nineteen of the thirty-five polling districts of the city, selected to give a representation of the whole of the residential area; 219 housewives were given an extensive interview about their behavior in the city, and of these, about 75 percent drew maps of their neighborhoods. They were each given a unique photo-copied blank, about foolscap size, made so that their own house was in the centre and marked with a large cross.

When the boundaries they drew were superimposed on to one large map of the city, they appeared as in Figure 5.1.

If it is assumed that the uniqueness of neighborhoods is not accidental, but traceable in every case to a pattern of antecedents, we have first to measure the maps and then to relate these measurements to the possible causes, about which information was gathered in the interview. The causes could be (a) differences in physical environment and (b) differences in the people themselves.

The most important finding came from a simple measurement of the area of neighborhoods. This showed considerable variation, but the differences were not related in any way to the part of the city in which the subject

FIGURE 5.1.

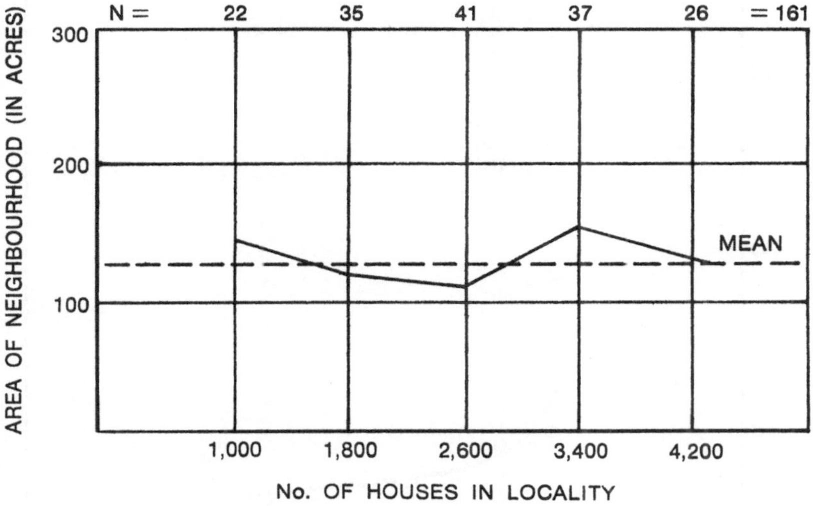

FIGURE 5.2. *The stability of mean neighborhood area with variation in housing density.*

lived. Put another way, the average *area* of neighborhoods in outer middle-class suburb and high density slum was much the same (Figure 5.2). In practical planning terms, this means that people do not delineate their neighborhoods in terms of the population they contain, as planners do, or in terms of the number of houses—but only as a space. If a planner resolved on a particular area, it could be filled at almost any population density without affecting the way in which it is perceived and reacted to by its residents.

Figure 5.3 shows a frequency distribution of areas for the total map-drawing sample.

The most clearly formulated question that planners have put to social scientists is this one of how large a neighborhood unit should be. They have frankly admitted that the size they talk about most commonly, 10,000 people, has been more or less arbitrarily determined. If we take the mode of frequency distribution shown in Figure 5.3, we arrive at an area of about seventy-five acres, which would support about 2,500 people by modern density standards. With vertical development, it could contain many more; it probably doesn't make any difference. As a space, it is much smaller than the majority of plans that have been advocated.

The criterion of "perceived wholeness" may not be the only one for determining how large a neighborhood should be, there may be economic

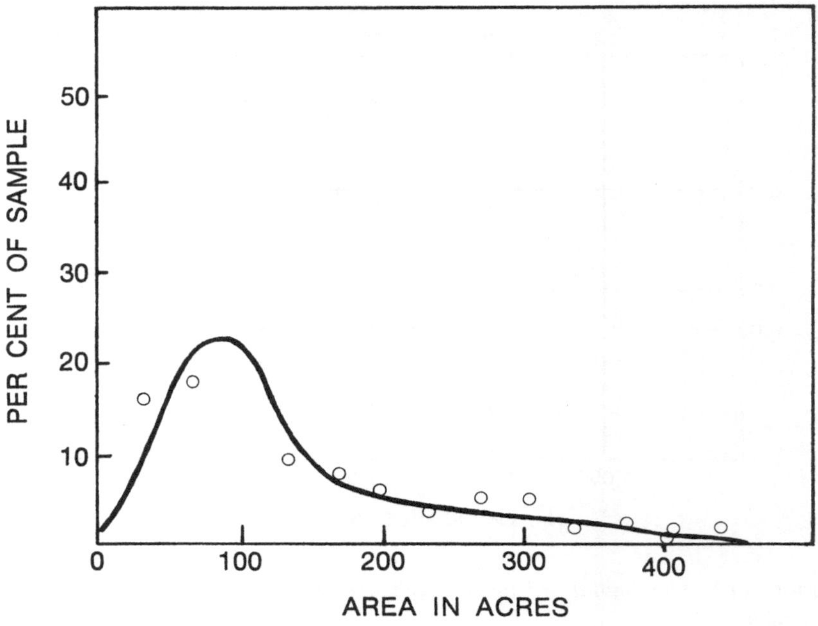

FIGURE 5.3. *Frequency distribution of the area of neighborhood maps.*

or educational or traffic considerations—but if we are to include in the values we seek the idea of "a sense of community," I submit that it is the most important criterion.

One of the measurements taken at the interview was an assessment of the number of friends a subject had, both locally and in other parts of the city. The number of local friends was closely related to the area of a subject's neighborhood but not to the density of population. It seems that even the most sparsely populated areas provide enough possible friends for everyone to reach their optimum.

Three forms of behavior were found to be associated with the more spatially extensive neighborhoods. These were the number of friends a subject had locally, the number of clubs and similar organisations to which she belonged locally, and her tendency to use local shops rather than town centre or nonlocal ones. This could be regarded as no more than a confirmation that the neighborhood map is a valid form of measurement, reflecting real differences in behavior. But there was also a more ambitious hypothesis—that the intensity of these kinds of behavior, which might be loosely grouped under the heading of "local involvement" or "community participation," are related to the *pattern* of the physical neighborhood. The way in which the differences in pattern might be dimensionalised was not only from small to large, but also from homogeneous to heterogeneous, from the small area containing mainly houses of much the same kind to the larger area containing a "balanced range" of shops, pubs, clubs, etc., a mixture of housing types and social grades—in short, to the planners' ideal conception of a neighborhood unit.

To repeat the hypothesis then, local social involvement is related to the physical pattern of the personal neighborhood.

This is not an easy hypothesis to test, because, of course, people might prefer to live in a neighborhood unit, possess the strongest of "local-involvement" attitudes, but none the less have the misfortune to live in a homogeneous desert of semi-detached's. This was overcome by expressing neighborhoods not in absolute terms but in relative ones—relative, that is, to the particular environment in which they were set. Using a Land Use Survey of the city, a complete and accurate count of the houses, shops and amenity buildings was made of two areas, for each subject. Firstly, of the area specified as the neighborhood, and secondly, of an area of half-a-mile radius surrounding the home at the centre—the locality. The first, the part of the local environment that had been "accepted," was then expressed as a ratio of the second, the environment that was "presented."

Analysis of these three ratio scores, and the raw scores of area, confirmed the hypothesis. Not only were they positively associated, individually, with local social involvement, but relations between them were critical. For example, people with a shop score that was high by comparison with their house score were usually highly involved, as judged from the various indices of their behavior. The scores were therefore combined, with weightings (derived by discriminatory analysis), to maximise this relationship.

This can be seen more clearly in diagrammatic form (Figure 5.4).

The measure was named the *Neighborhood Quotient*. It is a scale of the attitude towards local urban environments which can be applied to anyone who can outline a neighborhood.

If you will accept that the neighborhood quotient is a summary measure of the way people orient themselves within cities, the next question that needs to be asked is what is it about people that determines their neighborhood quotient. A wide range of hypotheses were tested, but the causal factors are so intricately interdependent that there is enormous scope for more precise research. The factors which did show fairly clear relationships with neighborhood were, firstly, social class. Most kinds of neighborhood attitude are to be found in each of the social classes, but there are characteristic levels. Secondly, length of residence. This is of some practical interest because it suggests that some of the gloomy forebodings about the lack of community feeling in new housing developments are just premature by existing standards. It appears that, if the sample is representative, British

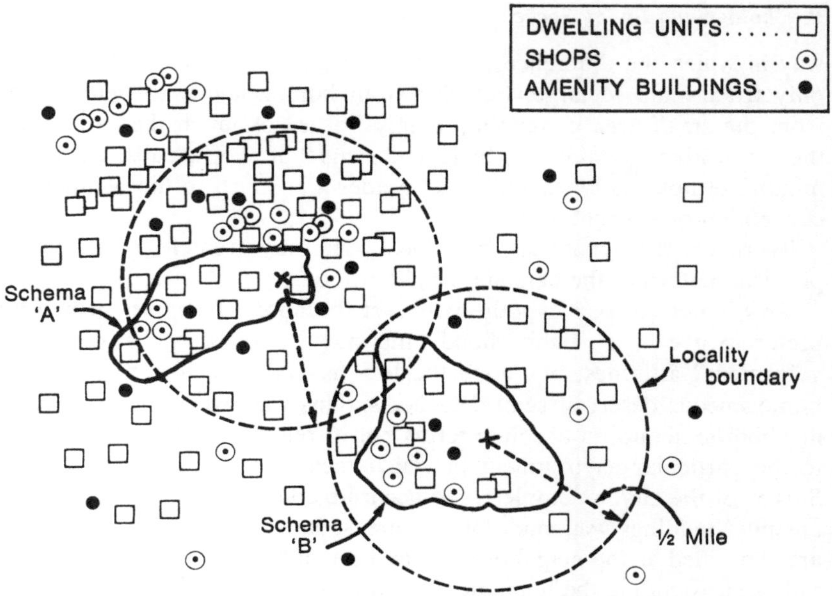

	Schema A					Schema B			
	No. in schema	No. in locality	Ratio	X D.F. constant	=	No. in schema	No. in locality	Ratio	=
Area*	.034	—	—	1.1609	·395	.052	—	—	.604
Houses	8	42	.19	−1.0000	−.190	8	20	.40	−.400
Shops	3	13	.23	.6393	.148	5	7	.71	.454
Amenity	1	9	.11	1.3870	.154	2	3	.66	.915
Total					.507				1.573

Converted to T scale, Nh.Q = 52 Converted to T scale, Nh.Q = 69

(*Area in sq. inches/100 from 6 in. Ordnance Survey)

FIGURE 5.4. *Illustrative calculations of the neighborhood quotient; an average Nh. Q subject from a high-density locality (A) and a high Nh. Q subject from a low-density locality (B).*

people do not begin to become involved in their locality for ten years or so, after which they go steadily from strength to strength. One cannot predict from this to conditions in which almost everyone arrives in a new locality at about the same time. Integration may be quicker or it may be slower, but it is time someone tested the hypothesis statistically and made what the demographers call a projection.

Thirdly, age is slightly negatively correlated with the neighborhood quotient—there is a tendency to contract one's involvement as the years advance. This finding was corrected for the fact that older people are also the people with a longer period of residence. Lastly, and perhaps more important, there is clear evidence to confirm a hypothesis that has been advanced by a number of urban sociologists in the past—that the families of men who work within the locality are more likely to have a higher neighborhood quotient, to have more friends, join more clubs and to shop locally. This applied not only to men working within the half-mile radius specified as the locality, but also, to a lesser extent, to those who worked on the same side of the city center as that in which their homes were situated.

What I wish to do next is to explain the kind of psychological theory which I think is applicable to this study, and then to go on to look at other results of my own and of other people to see how they might extend the theory and make it into a useful language for communication between architects, planners, sociologists and psychologists.

The Schema

I think the most profitable way for the psychologist to help the architect is to develop and apply a theory of the way in which the human being perceives, learns about and reacts to *space*. Almost all human behavior is determined by a combination of innate tendencies and those which result from learning. Modern psychologists are inclined to think of man as almost infinitely adaptable by learning, and certainly that is the dimension that distinguishes him most from other creatures. But we must not forget, for example, that however much he is conditioned by learning to an aesthetic appreciation of modern architecture, he can only appreciate it by looking out of the front of his head. To use a more serious example, his aesthetic appreciation is partly governed by his perception of the upright, which is a physiological matter based on his vestibular mechanism, the semi-circular canals in his ears, and on feedback from the nerves connecting with his muscles.

In my view, the concept that will help us most in studying human behavior in space is a venerable one in psychology—the schema. It was first used by Sir Henry Head (1920), a distinguished neurologist, as a model of the pattern of nerve impulses that must be built up in the brain when we carry out any complex movement with our limbs. We do not need to look to see where our arms have arrived at in performing a stroke at golf—the continuous flow of information from the nerves in the muscles tells us not only where our limbs are but how they got there. What is more, each time we repeat the stroke, we modify a kind of composite model which builds up. Sir Frederic Bartlett (1932) broadened and elaborated our ideas of the schema, particularly to explain the phenomena of memory. Most of the applications have been to emphasise the collective unity of a *temporal* sequence of actions, but it has also been applied with the same object to spatial patterns. In fact, it is very unlikely that there is much difference in the way in which the brain deals with the two arrangements, for we can only appreciate space by scanning it temporally, and we can only *act* in space successively. This must imply neural mechanisms of very similar form.

In applying the schema to neighborhood, we must say that there is a continuous input of sensory information from the physical and social objects in the urban locality, arising from our repeated transactions with neighbors, tradesmen, buildings, bicycles, parks, walls, children, shops, pubs, bus conductors and so on. These impressions are not allotted an equal-sized pigeon-hole and stored *ad infinitum* or until such time as they may be called upon. Many are rejected at once, either because they don't fit in with what is there or because they don't contribute anything new. The rest make their mark and leave their impress. They modify the schema that is building up and then they are finished in their original pristine form, living on only in so far as the schema is different because they happened. If this theory is valid, we assemble models in our heads that are constantly being modified by new and relevant experience, but which exist as entities not independent ones but with their own dynamic organisation.

If it is valid, we have a "body schema" or "body image" as the closest spatial organization and this indeed has been the subject of some interesting clinical work in recent years. After that, and partially dependent on it, we have schemata relating to familiar rooms—to the house and family, to the street, precinct, neighborhood, side of the city, city, county, country and world. Each one will depend partly on the other. All of them will depend mainly upon the unique kinds of experience that we have had as we live and play and work and go to school, and partially upon regularities that depend on our being human beings and on having had experiences that are general to our species. *Schemata are related to, but by no means coincident with, the physical reality that lies outside us.* Just how, is what the professional manipulators of environment need to know.

There is one further step. That is that all our present behavior is governed by these assemblies of selected and organized past experiences. We could not go to the tobacconist and buy a packet of cigarettes and call in at a friend's on the way home unless we had built some such model in the head. Thus the schemata are dispositional as well as cognitive.

If I may be forgiven for harking back to neurology, Sir Henry Head's most vivid illustration of the schema is the "phantom hand" which lives on in the brain of a patient after an amputation. This illustration was given somewhat piognant force by the fact that the phantom hand he was referring to had only four fingers. One had been lost about six months before the major amputation.

The best way in which I can ask you to examine your neighborhood schemata in order to appreciate its characteristics is to go back to a locality you have long ago left, or even return after a month's holiday, and you will find yourself comparing the schemata you took away with the fresh incoming bombardment of sensory impressions, and finding that they do not correspond. When you say, after a few days, that "the strangeness has worn off," I submit that what you mean is that you have assimilated the new sense impressions and modified the schema to accord with them.

A further implication for planning is that, psychologically, people do not organise the social/spatial world into "net works" or "chains" or "communalities," but into organized units which are continuous and filled, and having more or less clearly defined boundaries. They do not distinguish, normally, between social and physical space. Schemata show *similarities* in so far as they occur among the same sort of people in the same kind of environment, otherwise they are unique.

This is not to say that other ways of analyzing the spatial aspects of human behavior will not yield profit, such as, for example, regarding a neighborhood as a commonly acknowledged piece of ground or as a linked system of personal relationships. It is just that one would expect the most profit from a method of analysis that is congruent with the way in which the central nervous system seems to go about its business. It should perhaps be added that there is good evidence that schemata (those in which spatial relationships predominate at least) seem to reside in the parietal lobe of the cerebrum and that when there is injury or disease to this part of the brain the elements of a schema may remain, but the patient does not retain any idea of how the elements are related to each other—the model falls to bits (Semmes, Weinstein, Ghent, and Teuber, 1955).

If it is right to claim that the psychologist can make the best contribution to architectural and planning problems by studying sociological schemata, there are a number of major tasks that require his attention. I propose to direct your notice to these, mentioning wherever possible examples of the beginnings that have been made.

The first, in which no work is so far evident at all, is what might be called a taxonomy of schemata. We know that behavior can be classified according to the part of the environment in which it takes place, and also according to the social groupings with which the behaving subject is integrated. We have very little idea how many of such systems the average person forms, how they are related to each other, in what order they develop and in what priority of importance they are placed by him.

Another problem is how the subjective schemata develop from the dimensions of "real" space. Such studies that have been done to trace the growth of spatial representation confirm that it is not something that is immediately given—it has to grow gradually and meticulously with the child, and apparently as a result of his experience with objects, seeing them, handling them, observing their movements. As recently as 1922, studies may be found which claim to confirm the hypothesis that children have an instinctive sense of direction for the cardinal points of the compass (Hudson, 1922). This has been discredited, and various techniques developed to measure directional orientation have shown a *gradual* improvement in performance with age which suggests that learning is the process mainly responsible (Smith, 1933). Some fascinating work has been done by Piaget and Inhelder (1967) which suggests that the child's first spatial understanding is topological, that he is competent only with the properties of proximity, order, enclosure and continuity, and that it is only later that a Euclidian type of spatial relationship (that is, adding the properties of angularity, parallelism, and distance) can be appreciated. Piaget thinks that adults build a grid of co-ordinates in their heads to orient themselves in space. I think they use a set of radiating lines from a reference point in each of their schemata—but I am not sure.

When a number of primary school children in Devon were asked to estimate the distances from their homes to the major cities, and especially to London, it appeared that their spatial world was divided into various local Schemata which bore at least a detectable relationship to the physical world, but that outside this there was one total schema that might be called the "elsewhere" schema, in which physical dimensions were irrelevant. A study in the United States by Lord (1941) showed similar results. His children had a local "direct experience" schema, and beyond this they had, if old enough, a knowledge of a conventional set of mileage numbers which they had begun to associate with the names of cities. This, like a good deal of geography as it is taught at present, is rote learning and not spatial cognition.

Space and Behavior

The most conspicuously active field of research in recent years has been in showing the very powerful influence upon human behavior in various practical situations, of space and distance. It is this work that is probably of most direct interest to the planner and architect, because it shows so clearly what effect his work is capable of having. He manipulates space, and space governs behavior.

There is, for example, an inverse relation between the distance separating potential marriage partners and the number of marriages. Kennedy (1943), in a paper entitled "Premarital residential propinquity," showed that for New Haven, a city in the United States, 76 percent of the marriages were contracted between people who lived within twenty blocks of each other and 35 percent were between people living within five blocks. This is in the most mobile country in the world!

The classic study, however, is that of Festinger, Schachter and Back (1950), carried out in an ex-serviceman's housing project at the Massachusetts Institute of Technology. This was a homogeneous community, in the sense that all the residents had similar experiences from the past and ambitions for the future. They had also been allocated to their houses by random selection instead of choosing to live near their friends or income groups, so that the próject was an ideal natural laboratory for studying the effects of the spatial distribution of housing upon behavior.

There were two types of building layout. One was a cul-de-sac court containing about a dozen small pre-fab houses in a grouped arrangement. The other type was a two-storey block, converted from naval barracks, each one containing five flats on each floor. When the residents were asked to specify with whom, on the entire project, they had most sociable interaction —the majority of friendships were found to have been formed within the same block or court. But what is more, although the distance between two adjacent flats in the converted barracks was no more than 19 feet, distances of this order were found to have a highly significant effect on the pattern of relationships. The next-door neighbor was the most frequently chosen friend and thereafter the rate decreased rapidly until, in the case of the courts, there were almost no choices at all between houses separated by four others, although the distances involved were only 180 feet. In the barracks, the position of the staircases which connected the two floors influenced the traffic flow, as did also the position of the mail boxes. These two factors had appreciable effects on the social patterns that developed. Similarly, in the pre-fab courts, there were two houses at the end of the U which faced, not into the courtyard, but outwards on to the service road. Inevitably, there was less passive contact between the residents in these end houses and the others, and they became socially far more isolated, although in units of physical distance they were very close indeed.

If these potent spatial influences ended with patterns of friendship, they would be interesting and a little remarkable, but not important. After all, people want friends and they might as well make them with the usual regard for the principle of least effort. But these investigators followed the implications further, along tracks already laid by other socio-psychological research. They showed how friendship facilitates communication. How a valued group membership leads us to accept the groups' standard and opinions, partly because of limited exposure, partly because we appreciate consciously or unconsciously that the satisfactions of our membership in the group might be jeopardised if we do not "go along", and partly because when a situation is ambiguous and we lack the evidence to judge its merits for ourselves, we take strength from the "social reality" of other people's opinions. This process was studied in this instance as it affected the relatively trivial local politics involved in the launching of a tenant's organization, but there are many cases in which the most profound matters of social organization and judgement are influenced by architectural layouts.

Since this study, there have been a number of others which have shown the same unexpectedly strong and far-reaching influence of spatial factors upon human behavior in offices, in school classrooms, discussion groups and committees, children's vacation camps, farming communities and so on.

Some Further Variables

Most of these studies were made by measuring space in feet and yards, as it were, and observing the effect of variable distances upon behavior. However, if we are to learn how to make refined predictions about what kind of schema people will have in different physical environments, it will be necessary to incorporate, in our equations, various correction factors. The first one of these I wish to discuss is the means available for traversing the space—the degree of mobility. It can be illustrated with a brief reference to a study of the journeys to school made by rural infant children (Lee, 1957).

One of the most far-reaching changes in our society, in which planners are at least acquiescing, even if not actively promoting, is the closure of the smaller one-, two- and three-teacher schools in country parishes. The children are transferred by bus and taxi journeys to larger schools in neighboring villages.

It has often been alleged that this has dire consequences for the younger children, and it was this hypothesis that was to be tested. Using sets of rating scales, assessments of behavior were obtained for all the six- and seven-year-old children in fifty-seven primary schools over an extensive part of Devon, stretching from the northern areas of Dartmoor up to the coast. 883 children were involved altogether, and the assessments of their behavior were made by their own teachers who knew them well, but without any knowledge at all of the purpose for which the investigation was being carried out. The traits of behavior that were measured included concentration, response to affection, absenteeism, aggression, depression, eating difficulties, and so on. As soon as they had been made, the teachers were asked to complete detailed forms giving a full description of the journey which each child undertook in order to get to school.

It was found that almost all the kinds of behavior that were considered showed a progressive deterioration with the length of journey, and that usually a bus journey had a worse effect than a walking one. The scores for each trait were therefore combined, and Figure 5.5 shows the result of plotting the length of the journeys against the total adjustment of the child, with bus and walking journeys shown separately.

An obvious explanation for the results might be that long journeys make children tired and that tired children are that much more poorly adjusted in the classroom. But this hypothesis does not fit the data, as will be seen from a closer examination of the two curves. Although they are plotted against different scales, one of distance and one of time, they have been placed so that they roughly correspond on the time scale. That is, for example, a walk of one mile takes about 30 minutes for a child of that age. It could hardly be argued that sitting in a bus for 30 minutes is more fatiguing than walking for the same time.

What I would suggest, and this hypothesis fits the several other indications which there is no time to detail here, is that a child who lives a long way from school forms two schemata, a home schema and a school schema, with a semi-permeable barrier in between. This barrier is not, of course, a physical one. There are roads connecting his two schemata—but the link is weak (or the barrier is high) because the school bus follows a tortuous and impersonal route, much further and less direct than a flying crow, and, much more important, once the small child has been deposited, the means of crossing this barrier disappears. The bus drives away, leaving in many cases no transport at all, until nearly six hours later.

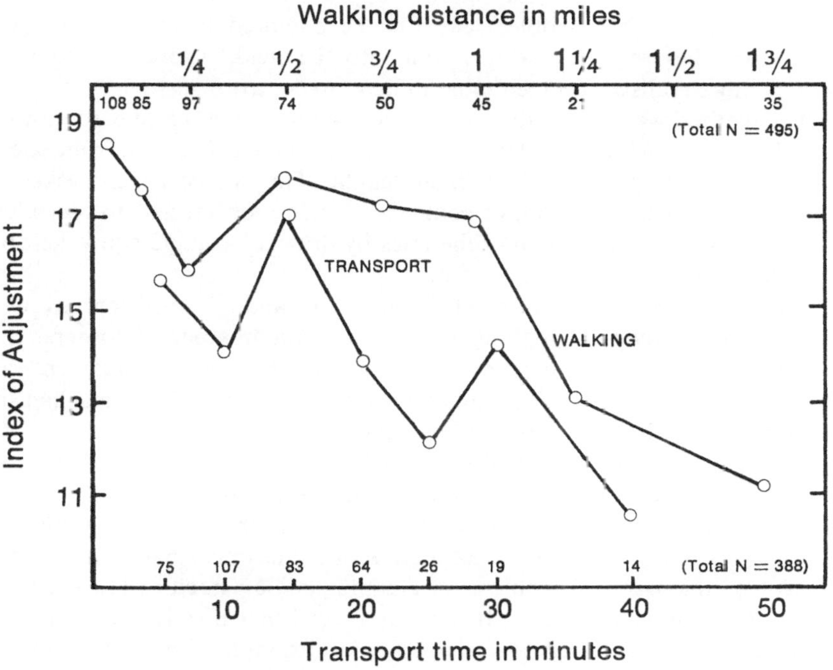

FIGURE 5.5. *The mean social-emotional adjustment of 883 rural infant children related to their school journey.*

Walking is very different. It is intimate to the environment and therefore articulates the schema—the barriers are permeable because the child knows he could cross them at any time.

I would submit that there are many similar cases in which planners should eventually be making their calculations, not in yards and miles taken from maps, but in terms of the schemata of their clients. Sometimes, as in the example I have given, it may be possible to consider deliberately modifying or re-training the schemata by means other than by altering the physical environment. In the future, we may forsee the psychologist and the architect collaborating to achieve the best fit of people to buildings.

A second variable which may have to be inserted into the equation, in attempting to express units of physical into psychological space, is the linearity of a route. This point was suggested to me by Mr. Sinclair Gauldie, the President of the Royal Incorporation of Architects in Scotland. He had noticed that business men seemed slow in making use of a car park which had been newly created near a city center, although in unit distance it compared very well with the alternatives available. The reason, he thought, was that the route, though short, was indirect.

I asked one of my students to collect some preliminary data on this, and although it is only a first and tentative approach, it gives some idea of the way in which such a problem can first be explored in the psychological laboratory, before later taking it out into the field, where the situation, though more realistic, is much more difficult to control.

The method was simply to present, on large pieces of white card, a number of lines, of one of three lengths, 6, 8, or 10 inches, but with from one to five right-angled corners along their length. The subjects were asked to examine them carefully and, of course, in random order, and to reproduce their judgments of the length of the lines by drawing a single equivalent but straight line on a paper before them.

The results of this experiment showed a clear and consistent over-estimation in the judged length as a function of the number of corners. It is very likely that this is an example of the way in which human judgments are determined by innate, built-in characteristics of the perceiving mechanism. It requires to be followed up by a field study.

In addition to the perceived means of traversing a space, and the perceived complexity of the route, there is one further extremely important factor which determines psychological space—and that is the desirability of the goal object. This can be illustrated with a study that has a most direct bearing on the practical problems of planning layouts in cities (Lee, 1963).

The problem is as follows. It is often alleged that if only social amenity buildings—community centers and pubs for example—could be provided on a sufficient scale on the new housing estates and in New Towns, the rootless anonymity that prevails in many of them could be translated into a genuine community spirit.

In opposition to this point of view are those who, pointing to the questionable success of so many such attempts in the post-war years, argue that it is wasteful to fritter away our meagre resources in scattering large numbers of small amenity buildings in Neighborhood Units. We should build something really worth while in the city centers or, at least, in centralized areas. Such a policy would make it possible to provide much better equipment; to make use, for example, of professional club leaders and, indeed, to produce a sizeable building that an architect could be proud of!

The latter case ultimately hinges on the assumption that greater travelling distances do not deter enrolment. It is a common belief that people can be roughly divided into "social joiners" and "unsociable non-joiners," and that the joiners in our modern, mobile city culture will seek out what they need regardless. The others will not be interested, whatever the planner may decide to do. Is this a correct assumption?

For the first part of the answer we may refer back to the neighborhood study. For this sample it was possible, you will remember, to make an exact count of the number of social-amenity buildings of all kinds that were to be found in a half-mile radius from the homes of a representative sample. This was the environment "presented" to them—and because of the historical development of British cities, the variation in the number of such buildings presented to people is very great indeed. How does what is "presented" compare with what is "accepted"? The actual number of memberships formed and their locations had been extracted in the course of the interview, and the relationship is shown in Figure 5.6.

There is obviously no such thing as a confirmed joiner or non-joiner. People make use of these social buildings in direct proportion as they are presented to them within a reasonable distance. If they do not have them locally, they compensate to some extent by using the ones in the town center and elsewhere, but this is only a partial compensation and does not give them a very high overall level of participation.

So far, this is merely another demonstration of the very considerable influence upon human behavior of the variable of physical space. It was suspected, however, that this influence would be very much weaker if the goal object were stronger. The large majority of the social activities involved in the study just described were what can only be named "light." I decided to do a separate study of the effect of space upon "heavy" activities. Starting from the other end, I analyzed the memberships and attendance at the part-time evening meetings of the "Social Studies Department" of a large technical college. The classes were all non-vocational, with titles like "Cathedrals of Britain," "An Approach to Literature," "The Beginners' Recorded Class," "English Period Furniture," etc. Annuli were drawn on a map of the city, half a mile in width, and the addresses of members were plotted. The enrolments in each annulus were then expressed in relation to the total population, to give a *rate* of enrolment. These rates are shown in Table 5.1.

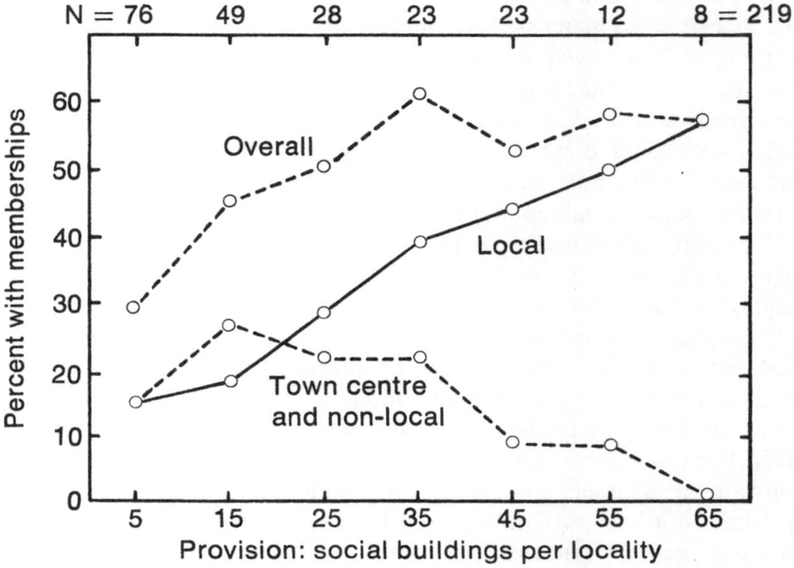

FIGURE 5.6. *The relationship between social participation and the level of provision of social buildings for alternative sitings.*

It was found that there were no consistent differences at all over a range extending from those people who, on average, had to walk 450 yards, up to those who had to travel 3½ miles.

I am primarily concerned here to demonstrate the theoretical point that psychological space depends on the desirability of what lies on the other side of it and that we must learn how to insert such components in our equations before we can predict behavior.

But I may be forgiven for inserting a comment on the practical implications of this study. It is possible that the trouble with the social-amenity-building policy of the planners is that they have been beguiled by the idealism of the community association movement into attempting to provide light and serious activities under the same roof, although their catchment areas and economics are so utterly different. The neighborhood social center should perhaps be more frankly regarded as a place for satisfying people's wish for gregariousness and good fellowship. It may then, for some, serve as a transitional community, a stepping stone between the home and the more educational/cultural institutions in the city center.

In all of the examples so far, we have been referring to one spatial schema at a time, although I mentioned earlier that they are probably interdependent, with changes or stresses in one schema leading to more or less readjustment in adjacent ones. The evidence in relation to "Brennan's Law"

TABLE 5.1. *Enrolment in educational/cultural classes as a function of distance to travel.*

	Distance in miles, less than:								
	¼	½	¾	1	1¼	1½	1¾	2+	N
Enrolment per 5,000	30	38	32	38	33	36	42	42	
Actual enrolment	22	60	57	53	68	71	44	49	424
Null hypothesis	26	57	64	53	73	71	38	42	

Chi-square = 4.22 (a coefficient of 14.07 is required for significance at the p = .01 level)

could very well be interpreted in this frame of reference. Tom Brennan (1948), a sociologist, found in a planning survey carried out in Wolverhampton that housewives tend to use shops that are not their nearest ones, but those which are in the general direction of the city center. I analysed my own data to see whether this flagrant violation of the principle of least effort occurred there, and found very strong confirmation indeed (Lee, 1962). This refers to daily food shopping, in which the range of commodities does not very much as between shopping centers near in and far out; in any case, Cambridge has not developed in a smooth gradient from the center outwards, and several of the largest sub-centers are towards the perimeter. Also, I could find no evidence that people tend to go inwards because they feel that they can break a longer journey into town. Food shoppers tend to go either to local *or* central shops, and those who go into the city center do so because they work there. In any case, none of these explanations, even if they all contributed something, seem strong enough to account for Brennan's Law. I suggest that local schemata develop a different effort-in-relation-to-distance scale on the side towards the city center because of the influence of this adjacent schema, which holds so many attractions. Subjective ease of movement becomes greater on one side than the other.

I need hardly say that this is a speculative theory, but one which has the scientific virtue of being testable. It is proposed to collect estimates from a group of subjects on the relative distances from their homes to various points within a half-mile radius. The prediction is that these estimates will be lower on the town center side for equivalent physical distances.

Conclusion

This short review of the work that has been done in the attempt to understand living space is far from complete. In particular, it has included no mention of the ergonomic studies which are becoming increasingly important in the design of the inside of buildings and it has neglected the more theoretical studies of space perception which will eventually influence our solutions of practical problems.

But even if the coverage were both exhaustive and accurate, it would not be possible to claim that there are more than a few scattered findings which can be applied by the architect here and now. There is, however, enough to suggest that three related lines of thrust would give substantial returns in the future.

The first is the integration of psychologists and sociologists into the planning team, where they could use their special skills for *ad hoc* problems, gathering information and exploring hypotheses about the people who constitute the working parts of machines for living in.

The second is that social and biological scientists should be brought in at the early formative stages of architectural training, so that they can implant the germs of organic thinking and scientific enquiry about human beings where they have most chance of taking root.

Thirdly, and in my own view most important, the basic research psychologist must aim to build a theoretical structure about the effects of space upon human behavior. I have advocated the schema as a sketch-pad outline for this, but so far the structure itself has not even reached the d.p.c.

6

Notes on Urban Perception and Knowledge

DONALD APPLEYARD

Urban perception and knowledge are special instances of general perceptual and cognitive processes that are not unified in any single theory of perception (Allport, 1955). For the environmental planner, this lack of theoretical coherence is a problem. Studies in environmental perception are fragmented, and the environment is viewed with different emphases by the various professions and disciplines concerned with it (Kates, 1966; White, 1966). The following notes result mostly from my studies of urban perception in Venezuela and Boston (Appleyard, 1969a; 1969c), together with some reading of the psychological literature. From these studies three characteristic types of urban perception appeared to be dominant: I will call them *operational, responsive* and *inferential.*

Many elements in the city are perceived because of their *operational* roles. As a person uses the city, performing various tasks, he selects particular aspects of the environment for the purpose of carrying out these tasks. The details of traffic circles, islands and intersections are often exaggerated far out of proportion in subjective maps. The noting of quite inconspicuous buildings at primary decision points appears to be part of the same orientational need. Some way of anchoring such points is essential, and the most salient element around, however small, is drawn in to perform that role. Isidor Chein (1954) has described how such an operational environment consists of goals, barriers, noxiants and other elements related to purposeful action. Miller et al. (1960) elaborated on the characteristics of operational and planful behavior. Goal seeking is certainly a dominant

This paper was originally presented at the Environmental Design Research Association (EDRA) Conference, Pittsburgh, October 1970. It is an extract from a forthcoming book entitled *The Plural Environment,* to be published under the auspices of the MIT-Harvard Joint Center for Urban Studies.

characteristic of automobile travel (Appleyard et al., 1964). Operational perception is also common in other aspects of urban perception. Children know the doorsteps of the houses where they sit, and recall the floor materials, street furniture, and pavements on which they play. This kind of perception, directed by activities and operations in the environment, is frequently neglected in environmental planning and design. Yet the environment that interferes with the performance of tasks is probably the most frustrating.

The role of operational perception may be important, too, in the currently fashionable exploration of complexity. An operationally complex environment which can be used in many ways may be much more satisfying than one which creates complex images, just as so many seemingly interesting toys turn out to be extremely dull when children try to play with them.

In other circumstances perception is much more *responsive* to the configuration of the physical environment. Bright, isolated, singular, and distinctive elements intrude on the operational search patterns of the traveler, or catch the eye of a gazing passenger. Perception under these circumstances may be more passive than active. Signs, billboards, gas stations, water, people occur within the operational view patterns of travelers on the circulation system, or are noticed in peripheral vision while the traveler is engaged in some other search. These are the "imageable" elements of the environment which Lynch (1960) has described in detail. They are not necessarily visual; they can be distinctive sounds, smells, or tactile experiences. Their imageability depends on the intensity of certain characteristics and their relative singularity or uniqueness in a particular context (Appleyard, 1969b).

Finally perception is *inferential* and probabilistic in nature. As we grow up, we develop a generalizable system of environmental categories, concepts, and relationships which form our coding system for the city—our personal urban model. When we encounter a new city, we match each new experience against our general expectations: events are "placed," never-before-seen buildings are identified as belonging to a particular class of building, functional and social patterns are inferred. Tolman (1951) identified a "placing need" as a fundamental motivation in perception. Apparently significant places are noted, while those that seem trivial are screened out.

Perception in this sense can be seen more as a cognitive decision process: fitting into categories, predicting probabilities, forming and testing hypotheses. It is well described in the writings of Jerome Bruner. The wider our urban experience and the more conventional the structure of the city, the quicker and more accurate will be our acquisition of knowledge. Those items of the environment which occur more frequently will be more accessible in our reference system and will stand a good chance of being identified. Bruner (1957b) suggests that perception depends "upon the construction of a set of organized categories in terms of which stimulus inputs may be sorted, given identity, and given more elaborated, connotative meaning." The person "builds a model of the likelihood of events, a form of probability learning."

For opposite reasons, more related to curiosity, the unfamiliar or unusual will also be noticed. This paradoxical attention to both familiar and unfamiliar features in a new environment is difficult to understand unless the multiple motivations of urban inhabitants in any situation are acknowledged.

As yet, we know rather little about these environmental probabilities, either in their objective form or in the common predictions of urban inhabitants (Brunswick, 1943). The conservative nature of this model is probably essential for our mental stability. It has some obvious limitations, however. It makes difficult the absorption of new experiences, and if the environment is ambiguous or incongruous can very often be a source of error.

> When . . . expectations are violated by the environment, the perceiver's behavior can be described as resistance to the recognition of the unexpected or incongruous. . . . Among the perceptual processes which implement this resistance are (1) the dominance of one principle of organization which prevents the appearance of incongruity and (2) a form of "partial assimilation to expectancy" which we have called compromise (Bruner and Postman, 1949).

The tensions between environmentally dominant responsive perception and man-dominant operational and inferential perception appear to be fundamental to our environmental experience. Our discourse with the environment, which is in any case a sporadic one, is continually shifting between subjective and objective, personal and environmental poles, according to our familiarity, experience, or mood, the task at hand, and the configuration of the environment.

Each kind of perception relies on certain attributes in the environment. Personal movement and visibility are attributes of operational perception; imageability, of responsive perception; and socio-functional significance, of inferential perception. In my paper "Why Buildings are Known" (Appleyard, 1969b), such attributes were scored and correlated with subjects' recall to assess their relative importance to the public. Maps of a hypothetical urban environment presented in that paper illustrate the operational, responsive, and inferential worlds of an inhabitant and the public.

While these perceptual dispositions do shift from moment to moment, it is also fairly certain that individuals emphasize one or the other as enduring dispositions. Some hold operational values high, others are more concerned with response to the environment, still others search for meaning, objectivity, and symbolism. Many of the conflicts in the design professions are rooted in these personality differences (White, 1966).

As a consequence of these varying types of perception, urban knowledge is a complex collection of variously perceived items, qualities, and events. Not surprisingly, it is a *multimodal* representation of the city. Bruner et al. (1966) assert that children pass successively through three modes of representing their environment—the *enactive*, the *iconic*, and the *symbolic*—as part of their cognitive development. From our surveys it appears that, once learned, all three modes are employed in representing the environment.

Many aspects of the city are recalled as *actions*. Certain trips on featureless roads resemble personal movements more than the roads themselves when recorded on subjective maps (Appleyard, 1970). Descriptions of journeys often concentrate on changes in alignment, stops, congestion, and other action events. Such actions and movements, when repeated endlessly, become transformed into spatial images (Mandler, 1962). The single most common attribute used by our subjects to describe buildings was their "location." Even when a person had no visual image of a building he could remember where it was. Many elements are recalled as *images*. Buildings are described by their physical size, coloring, shape, materials, style, and other imageable qualities. Lastly, elements, particularly larger districts which sometimes have no homogeneous image, are labeled by name, number, or some other *symbol* of their presence.

A particular building can thus be recalled by the activity a person engages in when he is there, by an image, or as a name, category, or graphic symbol learned through social communication, from a map or sign. Although society has developed and elaborated symbol system (verbal, numerical, and graphic) for representing and communicating environmental experience, urban experience is predominantly direct rather than symbolic. Many events are absorbed and represented in the memory as actions or images without transformation into enactive events or iconic images, while others are labeled, categorized, or interpreted for some particular purpose, through habit or experience. Our representation of the urban environment is therefore the product of two information systems; the substance of direct experience, and the indirect language of communicating that experience. We receive information directly from the environment and indirectly from several other sources: friends, strangers, the news media, maps, and books. To translate and combine these sources into a coherent network of knowledge, action sequences, associated images, and symbolic structures—many of them fragmented—must be correlated and matched. This is a formidable task that takes considerable learning ability. For those with less education it is sometimes overwhelming, and the representations of those who travel about cities solely by transit are notoriously fragmented.

Urban perception and knowledge are varyingly *schematic* in character. In an effort to organize the flux of environmental experience, we frequently, though not always, resort to methods that simplify, structure, and stabilize it. "Thinking," says Bruner, "is an act of categorization." So is perception a great deal of the time. We learn in childhood to develop generalized concepts for classes of events and objects so that new experiences can be fitted into an existing scheme of experience. In processing visual material, people use such devices as identifying regularities, grouping similar or contiguous events, or emphasizing separation, continuity, closure, parallelism, and symmetry. In situations of changing stimulus, rhythm and cycles are identified (Johanssen, 1950).

There also appear to be limits to our processing capacities which force us into strategies that reduce the number of items we have to recall (Miller, 1956). The effort has been called cognitive economizing (Attneave, 1954) though the tendency is not entirely one of skeletonizing. Nonexistent features are sometimes imported to make memory more coherent or meaningful; hence the process of remembering has been called "reconstructive" rather than "reproductive." (For a highly readable and useful analysis of memory processes, see Bartlett's [1932] classic study dealing primarily with the recall of stories. Paul's [1959] statistical analysis of the same stories substantiates many of Bartlett's earlier findings.)

The Gestalt psychologists pioneered the investigation of environmental schemata, and although Attneave (1954) and Hochberg and McAlister (1953) have convincingly suggested that perceptual organization is more a matter of information processing and cognitive economizing than of innate properties of perceptual machinery, the empirical findings of the Gestalt psychologists still stand (Wertheimer, 1923).

In cognitive representations of large cities, people have to schematize drastically if they are to gain any overall comprehension of urban structure. They extract dominant reference points, a group of districts, or a single line of movement on which to hang their recollections. These simple patterns and networks are also the common stereotypes of utopian city design.

Our representations of a city are just as likely to be incremental and *disjointed*. There exists no pure and complete processing of information; there is no neat entity as words like schema and image sometimes imply. Fragments of direct experience, more often the results of responsive perception, endure unchanged and unassimilated while other events are organized into conceptual systems. Sometimes events possess special survival value and become sharpened in recall (Allport and Postman, 1945) while others are screened out, but many linger on the edge of memory, half-recalled, to be grasped only with effort. Sometimes these representations remain in conflict. A building passed on the journey to work is given the wrong name because the name was mentioned in another context. You approach a familiar intersection from a new direction and fail to recognize it. Environmental errors like these occur all the time as we travel through the city.

Partly these errors are due to the difficulty of organizing a vast amount of discontinuous experience. Perhaps we also find it unnecessary to organize all experience. This messy characteristic of urban knowledge makes it especially difficult to investigate. At best we can only hope for glimpses of a person's store of collections, seeking to detect regularities among them, realizing that every item will not fit perfectly into a schema.

The final quality of environmental knowledge we shall discuss is its relatively inaccurate or *nonveridical* character (Bruner, 1957a). Except for known parts of the city, we depend on very little environmental information to make our inferences. Such errors are often relatively harmless orientational mistakes, but occasionally a misunderstanding can mean the loss of an opportunity, the missing of a job, and a general feeling of inefficiency. It is striking how many people express feelings of guilt when not able to respond to questions about their city. Misinterpretations of the character of other population groups were made in one of our surveys when garbage was seen on the streets, and when signs were noticed on the walls. It is easy to draw prejudicial conclusions from inadequate information. Much of our environmental knowledge is little more than rumor (Allport and Postman, 1945), and not to be trusted.

The structure of urban knowledge, therefore, is difficult to grasp and full of seemingly contradictory qualities. It can be both concrete and abstract, schematic yet disjointed, conventional yet imaginative. As a mental counterpart of the environment, it contains elements and attributes which are classed in concepts and categories and structured in spatial systems and systems of meaning; yet, given the formlessness and complexity of most cities, it is also fragmentary, partial, and inaccurate. It is evolutionary in nature and adaptive to all kinds of demands. As anyone who has given interviews will know, the aspects of knowledge revealed will depend on the question. Indeed, one of the important implications for those who try to probe the nature of urban knowledge is that interpretations should be made from a variety of responses—graphic, written, and oral—to questions which symbolically or in reality place respondents in different kinds of urban situations.

The idiosyncratic and pluralistic nature of urban experience and knowledge have not been discussed here, although they present the greatest conceptual difficulty for the urban designer. They will have to be the subject of another paper.

7

Differential Cognition of Urban Residents: Effects of Social Scale on Mapping

PETER ORLEANS

The thesis of this paper is that our knowledge of the spatial environment, the way in which we visualize and symbolize it, is a consequence of our experience in it and with it. Insofar as that experience is shaped by our physical location (and the barriers of space and time associated with that location) *and* by our social position (and the normative requirements attached to it), our cognition of the environment is likely to be a function of these factors.

Concern here is not so much with the aesthetic component of urban imagery (or symbolization, which has attracted the interest of many planners and architects) as with understanding the extent to which overlapping, and sometimes mutually exclusive, social worlds influence the way in which people see the city. Chamber of Commerce promotions and monumental architecture, not to mention folklore, undoubtedly create and sustain generalized images of specific cities (Strauss, 1961). Chicago is the hog

A revised version of this paper was originally presented at the 63rd Annual Meeting of the American Sociological Association, Boston, Mass., August 1968. I am indebted to Mr. Robert Dannenbrink of the Advance Planning Section of the Los Angeles City Planning Commission for making the details of their study available to me and to Professor Fred Massarik, of the UCLA Graduate School of Business Administration, who made available his data on the social areas of Los Angeles for 1940, 1950, and 1960. Maps 2–6 reprinted here were originally included in *Science Engineering and the City*, publication 1498, National Academy of Sciences—National Academy of Engineering, Washington, D.C., 1967. The cooperation of the Academy is gratefully acknowledged. Mr. John Bayes, Mr. Alfred Brown, Mrs. Patricia Charde, Mr. Roger Durand, Mr. Eugene Grigsby, Mr. David Kaufman, Mr. Philip Leung, Mr. Patrick Mulcahy, and Mr. Augustine Ugbolue, enrolled in Sociology 245 in the Fall of 1967, helped to design the study, carried out the field work, and executed much of the analysis. They cannot, however, be held responsible for the interpretation of the data or for the conclusions discussed in this paper.

butcher of the world; the cable car denotes in an instant cosmopolitan San Francisco; the Las Vegas strip is a testament to the garish recreation of anomic affluence: but such images bear little resemblance to these cities as they are known by their residents. The everyday experience of a diverse citizenry is not amenable to such simple symbolization. To paraphrase a common cliché, a city is many things to many people; different things to different people. It is, as Robert Park (1952, pp. 178–209) suggested long ago, a mosaic of social worlds. What are these social worlds, how consonant is the environmental cognition of their inhabitants, and if it is diverse, what accounts for discrepancies in environmental cognition from one social world to another?

The Concept of Scale and Environmental Cognition

Some of the social worlds of the city are real or rough approximations of the traditional community transplanted to the urban setting. Others—the communities of limited liability—though they may be as highly integrated, internally differentiated, and interdependent with the outside world, lack any site-specific coherence in geographic space (i.e., they are not place based—see Webber, 1963, pp. 23–56; Janowitz, 1952). They may be described as *lateralizations*, as semiorganized aggregates of individuals who participate in a way of life which is different from the various communities through which they pass (Aginsky, 1952, p. 126), or *ambiences* (Caplow, 1955, pp. 28–33). They possess a sociological coherence even though they are spatially indeterminate.

Both types of social worlds—communities, and lateralizations or ambiences—are composed of individuals who share a common fate and similar life chances by virtue of either their social similarity or their physical proximity (or a combination of the two). They can be distinguished from one another by the scope and quality of their inhabitants' (or members') involvement in the life of the city. Unfortunately, however, the notion of a spatially indeterminate community is alien to much sociological thought, receiving little attention in the literature (Schnore, 1967, pp. 82–150; Hillery, 1955).

The distinction between these two kinds of social worlds can be described in terms of the concept of *scale*. First introduced by the anthropologists Godfrey and Monica Wilson (1945), this concept refers to the range and intensity of social intercourse. Although the Wilsons' discussion of scale is limited to intergroup relations, the concept also can be applied as here, to interpersonal relations. The key terms here are *range* and *intensity*. Range refers to the number of social contacts characteristic of a population. Intensity refers to the relative dependence of a given population upon others, or the distribution of its dependences. As the Wilsons (1945, pp. 25–26) express it: "A Bushman . . . is as dependent upon his fellows as an Englishman, but the Englishman depends upon many more people than does the Bushman. The total degree of interdependence, or intensity of relations, is the same, but in the case of the Englishman it is more spread out." Holding intensity constant, scale is denoted by the range or number of social contacts. The larger the number of contacts, the greater (or broader) the scale of a population.

Given comparable ranges in two populations, an indication of relative scale is the extent, both geographic and temporal, of existing social ties. The more extensive the contacts of a population, the greater (or broader) the scale. The Wilsons maintain that this is the case even when the social relations that obtain individually are truncated and tenuous.

The Wilsons do not offer a precise definition of intensity: in fact, their use of the term is inconsistent. At various points in their discussion intensity refers to the extent of social relations—the distribution of dependences, both spatially and temporally (Wilson and Wilson, 1945, pp. 40–41). At other times intensity refers not so much to the extent of social relations as to their quality, both with respect to the frequency and continuity of contacts and the degree of commitment such contacts entail. Thus, in the Wilsons' writings there appear to be two dimensions of intensity, one referring to the extent, and a second to the degree, of dependence—what is more commonly referred to as the intensity of social relations.

The confusion arises from the Wilsons' unexplicated assumption that substantial social and/or physical distance is necessarily productive of more tenuous and less intense social ties. This is implied in their assertion that an increase in scale is denoted by "greater autonomy in the narrower relations as well as greater subordination in the wider," which suggests that the range and intensity of interdependence are mutually exclusive (Wilson and Wilson, 1945, pp. 41, 108–116). In fact, there is no reason why they should be.

For a given person, as the number of contacts increases, dependence upon any single contact is likely to decline. With an increase in the range of relations, the intensity of any particular relation is likely to be diminished. Obviously, however, there are limits. One can only interact with so many others in a given span of time. Given high physical densities of people, interactive possibilities reach infinity quickly. The practical effect of this, however, is that choice among interactive alternatives is required. The individual begins to manage his social relations, becoming more committed to some and less committed to others. Our social contacts, therefore, are rarely uniform in their intensity (Cox, 1965, pp. 38–59).

The Wilsons suggest that the number of social contacts (range) is likely to be a more suggestive indicator of scale than the intensity, or extent, of particular social ties. This would appear to be the case where environmental cognition is concerned. For example, it might be anticipated that the greater the range of contacts, the more comprehensive (though not necessarily the more detailed) will be one's imagery. The bold outline of the cityscape may be discerned and designated by range, but it is likely to be invested with content and meaning as a function of the distribution of dependencies, or the intensity of one's contacts. Thus, detailed knowledge about certain aspects of city life, or particular elements in the cityscape, is likely to be a function of intense involvement with selected others.

Planning Commission and UCLA Studies

All of this, if true, suggests that imagery of the urban environment might vary among distinctive groupings of urban residents as well as from one site (or location) in the city to another. There is some evidence to this effect. A recent pilot study by the Advance Planning Section of the Los Angeles City Planning Commission, done in consultation with Kevin Lynch, produced a series of five composite maps of Los Angeles constructed from maps drawn by distinctive samples of twenty-five respondents each, in five different locations—Northridge, Westwood, Fairfax, Boyle Heights, and Avalon (Department of City Planning, 1971). The location of these areas as well as those sampled in a follow-up study discussed here are indicated on Figure 7.1.

Respondents in the predominantly Negro sample, located in a southeastern district of the city known as Avalon (near Watts), had a rather constricted conception of the city (Fig. 7.2). Most of these respondents' maps depicted some of the major north-south thoroughfares between Avalon and the city center. Other aspects of the city were represented primarily as districts with little or no interstitial material in evidence.

An even more restricted representation of Los Angeles is the map composed from the responses of Boyle Heights residents (Fig. 7.3). This sample, selected from an almost exclusively Spanish-speaking population, conceives of the city as being composed only of Boyle Heights and its immediate vicinity, including downtown—the whole area being confined to but a few city blocks. The Los Angeles maps drawn by these respondents differed little from maps they drew of their neighborhood.

Of the five samples in the planning commission study, only the one composed of nonethnic upper class respondents from Westwood (adjacent to the UCLA campus) had a well-formed, detailed, and generalized image of the entire Los Angeles Basin (Fig. 7.4). More than any of the other composites, the Westwood map approximates what one might find on an ordinary gas station map.

The composite of the maps drawn by the middle class suburban respondents from Northridge suggests that, as in the case of the Avalon sample, their proximate environs delimit their conception of Los Angeles (Fig. 7.5). They appear to be oriented away from downtown Los Angeles. As a sign on the Ventura Freeway proclaims: "Topanga Plaza (in the Valley) is 'downtown' for over a million people." Thus, although they have

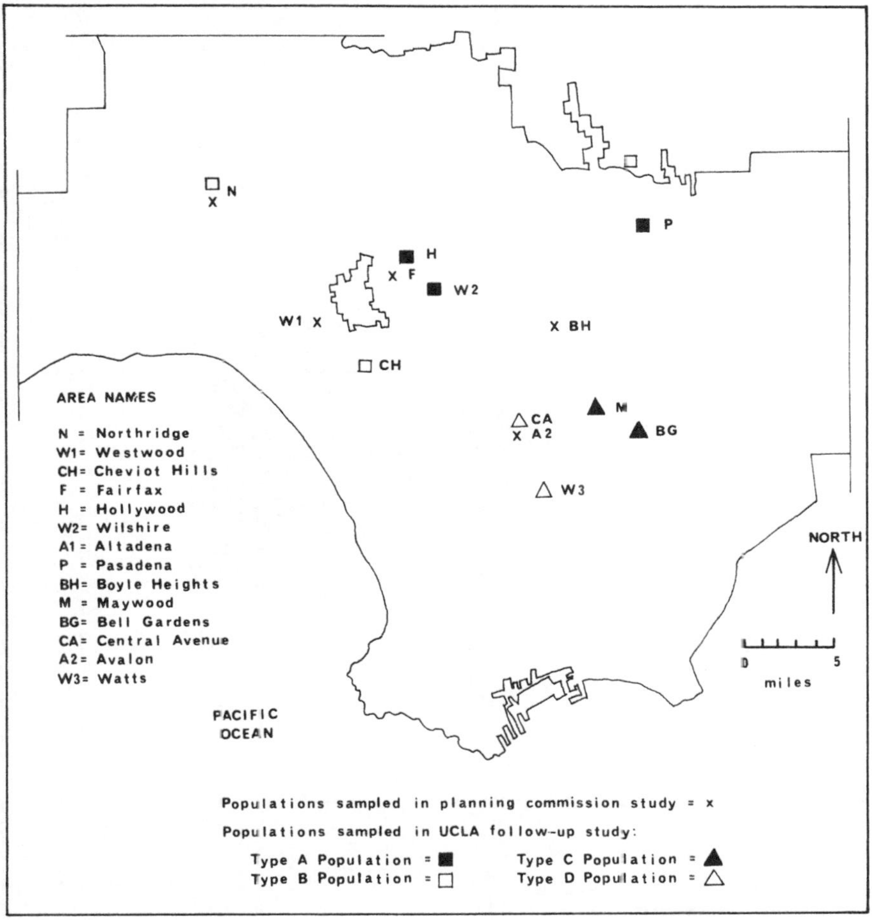

FIGURE 7.1. *Study areas in the urbanized portion of Los Angeles County.*

a reasonably detailed image of the San Fernando Valley extension of the original city, the Santa Monica Mountain chain effectively segregates Northridge residents from the rest of this sprawling metropolis.

The Fairfax sample was drawn from among residents in an older, predominantly Jewish, middle-class area. These respondents' maps are distinctive in that they record the existence of the San Fernando Valley, an area of heavy second generation Jewish settlement (Fig. 7.6). However, as was the case with each of the other samples in the study except Westwood, detailed imagery, indicated by the designation of specific landmarks, paths, and the like, is largely restricted to the adjacent area, the Wilshire corridor from Fairfax Boulevard to downtown.

These maps offer dramatic evidence of the variable nature of environ-

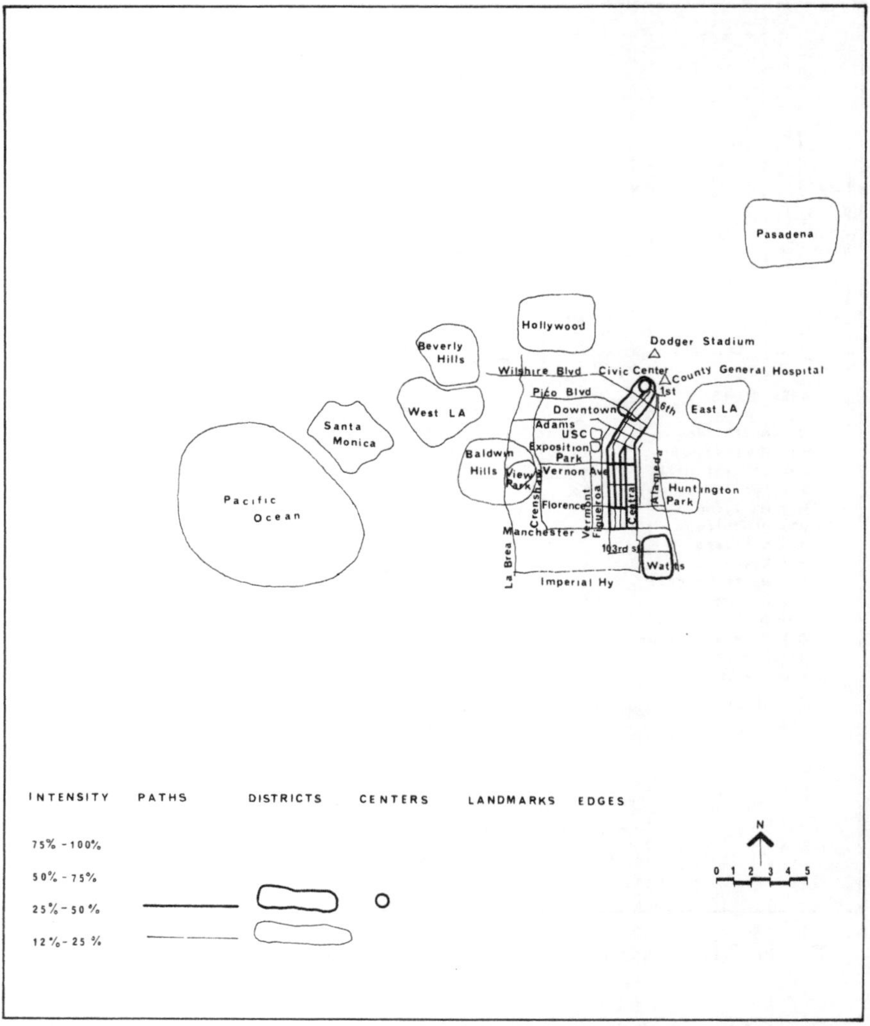

FIGURE 7.2. *Planning Commission Study, Composite City Image, Avalon.*

mental cognition. They demonstrate in a provocative and, to some, rather pathetic way the existence of discrepant social worlds which may touch, but in many instances, do not interpenetrate. The composite maps of the Planning Commission's study are fascinating to the sociologist because they indicate significant differences in environmental cognition. But the research itself offered few clues as to *why* different residents in various areas of the city have such different conceptions of it.

The sociological literature is replete with case studies which, in com-

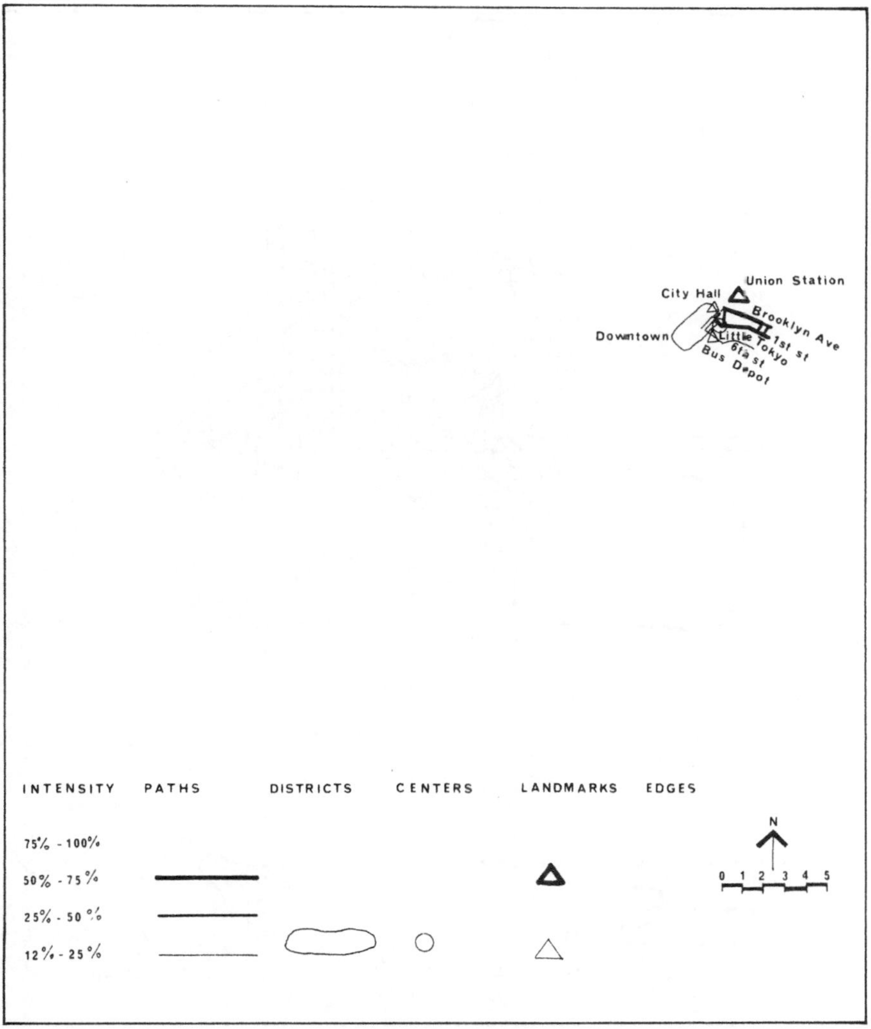

FIGURE 7.3. *Planning Commission Study, Composite City Image, Boyle Heights.*

bination, provide clues. However, because urban sociologists have stressed the disorganization of city life rather than the various dimensions of its functional integration, the evidence has never been systematically amassed to account for phenomena of this type. Although we have detailed enthnographic studies of selected urban populations (McClenahan, 1929), we do not have any generalized knowledge of the total social structure of the city or how the pieces fit together.

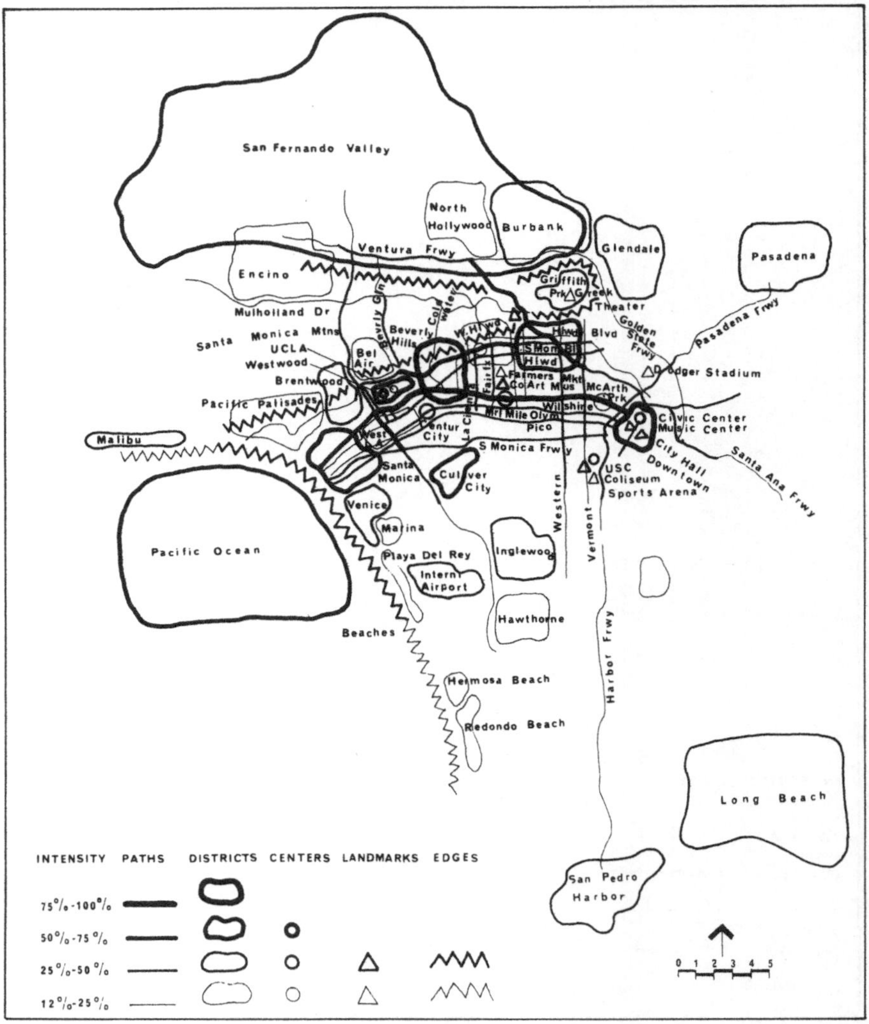

FIGURE 7.4. *Planning Commission Study, Composite City Image, Westwood.*

In a modest follow-up to the Planning Commission study done as a class exercise in a graduate seminar at UCLA, we collected information on the quality, frequency, and location of respondents' contacts to consider against their mapped impressions of the urban environment. Because of the constraints imposed by a class exercise, our sample was too small (a total of forty-eight respondents in ten locations) and our instruments too crude to allow much confidence in the data we obtained. Nevertheless, our findings, especially when considered in conjunction with the Planning Commission

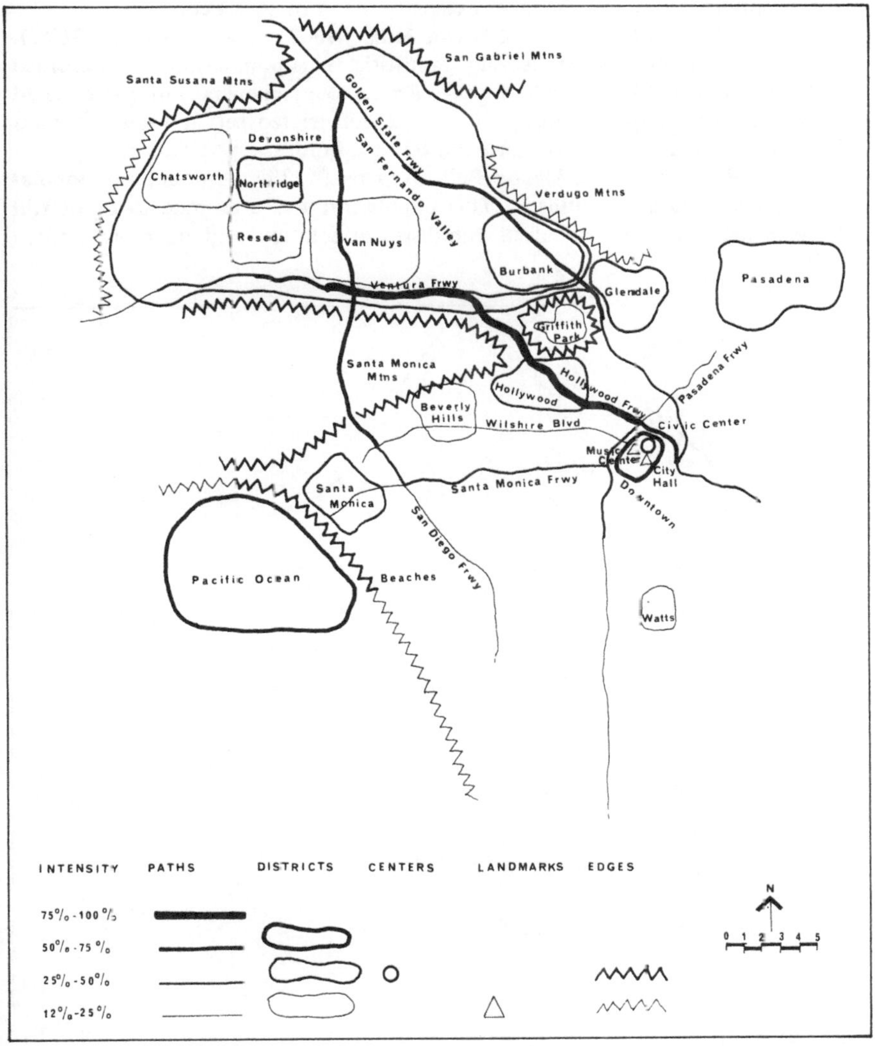

FIGURE 7.5. *Planning Commission Study, Composite City Image, Northridge.*

study, are suggestive.

Applying the technique of social area analysis to census tract data, we distinguished four population types presumed to differ in scale and widely distributed throughout the Los Angeles metropolitan area (Shevky and Bell, 1955; Bell, 1968, pp. 132–168):

Type A — Hollywood ([tract no.] 1905), Wilshire (2123–24), and Pasadena (4636). A nonethnic population, skilled and urban. This population was presumed to be of broadest scale; to have access to the greatest range of social opportunities and the most extensive array of social contacts.

Type B – Northridge (1154), Cheviot Hills (4602), and Altadena (2694). A nonethnic population, skilled and familistic. This population was presumed to be of somewhat lesser scale; to have access to a somewhat more attenuated range of social opportunities and a more constricted set of social contacts because of the more localized orientation of a familistic population.

Type C – Maywood (5333) and Bell Gardens (5339). A nonethnic population, semiskilled and familistic. This population was presumed to be of still lesser scale both because of its familistic orientation and its more limited skills.

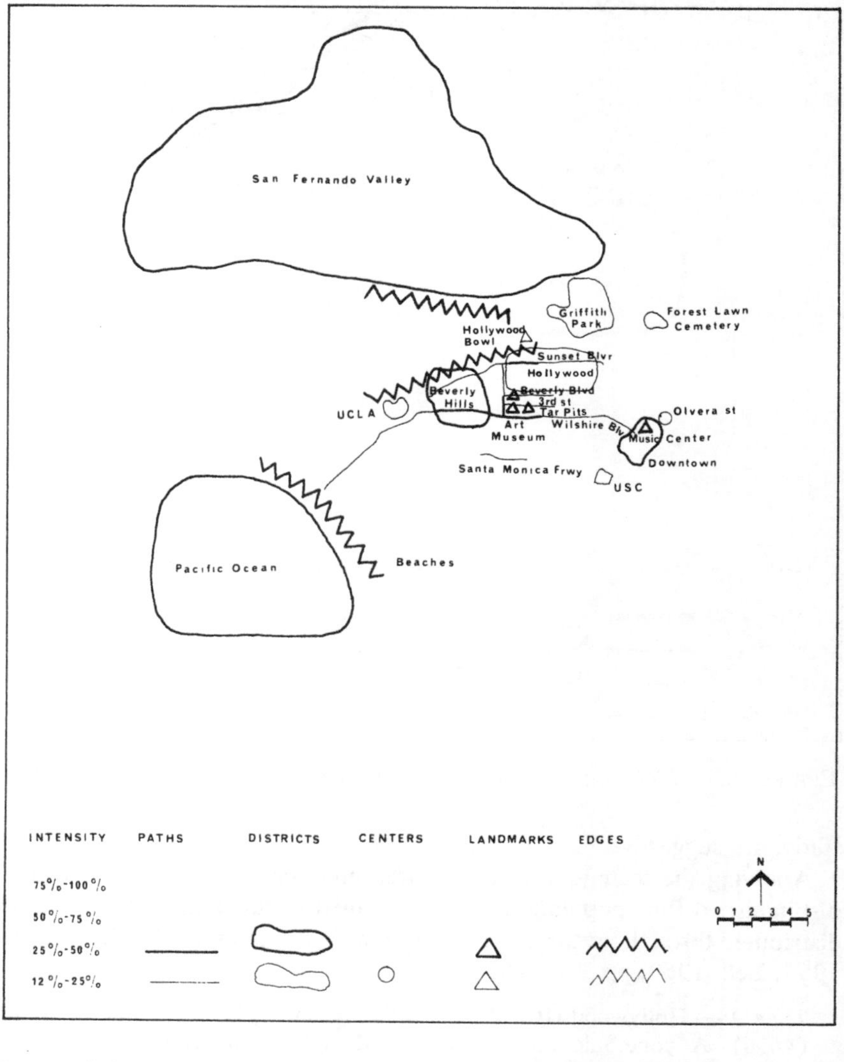

FIGURE 7.6. *Planning Commission Study, Composite City Image, Fairfax.*

Type D – Watts (2422) and Central Avenue (2265). A racially distinct population, semiskilled and familistic. This population was presumed to be narrowest in scale; to have least access to available social opportunities and a most limited set of social contacts, not only because of the lack of skills and the familistic orientation, but because of the added stigma of race.

All respondents were interviewed in their homes; the semistructured interview lasting about half an hour. The respondent was queried about the nature of his involvement, if any, in voluntary associations; he was asked a series of questions about each of his three or four closest friends; and he was asked to draw two maps, one of Los Angeles and one of his neighborhood. Background material and some additional information was also obtained.

Behavioral Manifestations of Scalar Differences

One of our concerns in the follow-up study was with the link between the scale of a population, as indicated by its aggregate characteristics, and the behavioral manifestations of scalar differences, as indicated by the friendship patterns and organizational affiliations of the respondents. We expected respondents in the broadest scale population (Type A) to be more likely to select their closest friends without respect to where they might live, to rely more on organizational and vocational contacts than on neighboring for the development of close friendships, to be more likely to belong to voluntary associations, and to belong to voluntary associations which did not meet in the immediate vicinity of their homes. As the scale of the populations narrows (from Type A to Type D), we expected these effects to moderate and to reverse themselves. In general, our expectations were borne out by the data.

By aggregating all mentions of friends (up to three) recorded for each respondent, we found that those from narrow scale populations named as close friends persons who lived in or immediately adjacent to their own neighborhood, 87 percent for the Watts-Central Avenue population (Type D) as compared with 47 percent for the Hollywood-Wilshire-Pasadena population (Type A). An exception occurred in the case of the Maywood-Bell Gardens population (Type C). Although none of these respondents could draw detailed maps of the Los Angeles metropolitan area (many declining to draw any map at all), all had travelled extensively in Los Angeles and seemed familiar with the area.

We found, as expected, that neighbors were an important source of friendship in narrow scale populations. However, our expectation that work situation and voluntary associations would provide important sources of friendship in the broadest scale populations was not borne out. Unaccountably, 50 percent of the friendships reported by the Type D population were attributable to the work situation and voluntary associations. It appears that the breadth of social contacts associated with the work situation varies from one population to another, being more restricted for less skilled people. If that is the case, it would explain both our anomalous finding and the fact that the respondents in the Type D population had a rather constricted image of the city.

Finally, contrary to our expectation, we found little variation from one population type to another in the propensity to join voluntary associations. However, as expected, 39 percent of the organizational memberships reported by population A respondents were in voluntary associations which met in the neighborhood, compared with 41 percent in the case of population B, 67 percent in population C, and 80 percent in population D. Thus, even with our crude data and limited sample, our expectations with respect to the behavioral manifestations of scalar differences appear to be borne out with minor exceptions. But, what relationship do they have to the question of environmental cognition?

Comprehensive Imagery and Site Recalcitrance

The composite maps developed by the Planning Commission from the raw data of their respondents' maps demonstrate extreme differences in environmental cognition among the populations studied. There were differences both in the bounding of the city itself and in the recognition of the existence of various amenities. But, the Planning Commission samples were differentiated both by location and type of population, leaving open the question of whether the sites or the characteristics of the populations, or both, accounted for differences in cognition.

In an attempt to separate these two elements in the follow-up study, we selected census tract populations that were comparable in composition but dispersed throughout the Los Angeles Metropolitan area. Our concern was to see whether respondents drawn from socially comparable but physically dispersed populations would generate comparable maps because of their social similarity, or whether the "recalcitrance" of physical location would subvert the emergence of any common imagery (Abu Lugod, n.d., pp. 23–26; Strauss, 1961, pp. 52–67). We also were interested in whether the comprehensiveness of the image of the city was, as we assumed, a function of scale.

We found differences in environmental cognition among our population types much as we anticipated. Respondents from the broadest scale populations, with the most extensive social contacts, had a generalized image of the entire Los Angeles metropolitan area and were able to represent this in crude map form. Respondents from narrower scale populations with more constricted social contacts had a more attenuated world view, frequently limited to an image of the city extending just a few blocks.

We did not attempt to create composite maps from the raw data. However, a careful examination of the individual maps indicates that such composites, if developed, would approximate in most respects counterpart maps from the Planning Commission study. A composite of the maps drawn by our Cheviot Hills and Altadena respondents would be more comprehensive than a composite of maps drawn by our Watts and Central Avenue respondents. The former could be similar to the Westwood composite map (Fig. 7.4) from the Planning Commission study and the latter would approximate the Avalon composite map (Fig. 7.2).

Finally, we found evidence of a phenomenon I will term *site recalcitrance*. For our Northridge respondents, much as for the Northridge respondents in the Planning Commission study, the image of the metropolitan area of Los Angeles was restricted to the San Fernando Valley where Northridge is located. They had no detailed conception of the older portion of the city located to the south of the Santa Monica Mountains. By contrast, although our Altadena respondents had a reasonably accurate conception of the extent of the metropolitan area, they tended to neglect the San Fernando Valley over the mountains to their west and instead offered a more detailed representation of the San Gabriel Valley to the southeast.

To some extent this discrepancy reflects the evolutionary development of the Los Angeles basin, with Altadena one of the early areas of settlement and Northridge a part of the most recent urban extension. However, it suggests a temporal and, especially, a spatial particularism which appears to be due in part to the impact of the immediate environment.

Future Research: Problems and Prospects

The composite maps from the Planning Commission study and the substantive information from the UCLA follow-up study suggest that significant differences in environmental cognition do characterize diverse segments of the Los Angeles population. Moreover, these data seem to indicate that differences in environmental cognition are a function of the scalar effects of both the social position and the physical location of the respondents. Both studies were severely limited in their conception and execution, and the results obtained are crude at best. Nevertheless, the findings raise several questions worthy of further consideration.

Los Angeles is distinctive among American cities, both for its geographic expanse and its rapid accumulation and circulation of population. Even in popular thought this megalith is said to lack a cultural tradition and a political coherence. Thus, it might be argued that our failure to find a coherent and comprehensive city image is epiphenomenal; that none exists because Los Angeles encourages a kaleidoscopic type of imagery (Banham, 1971). If this were the case, similar research in other cities, specifically the older, more stable, more highly articulated and traditional urban environments of the East—exemplified, perhaps, by Philadephia or Boston—might be expected to yield different results.

Yet we would contend with Walter Firey (1945) that in any urban environment, different people follow different orbits, and as a result, various sites within the city come to be invested with distinctive symbolic significance and meaning. Anselm Strauss (1961) suggests that the concept of "orbit" permits us to say something about space that earlier sociological work did not make clear or obscured: "The chief efficacy of the term 'orbit' is that it directs attention to the spatial movements of members of social worlds. Some worlds are relatively bounded in space, their members moving within very narrow orbits . . . the members of other social worlds . . . move in orbits that take in larger sections of the city as well as encompass sections of other cities." Strauss's approach differs from the one presented here insofar as he focuses on the environment which is apprehended. Our concern is more with how physical location and social position influence cognition of the environment. *Orbits* refer to the trajectories of people in the urban maze, whereas *scale* refers to the conditions which alter those trajectories or account for differences among them.

We would expect to find in any city, indeed in any differentiated aggregation of people, a mosaic of social worlds, each one productive of a distinctive environmental imagery. It cannot be emphasized too strongly, however, that we have yet to establish the generality of this phenomenon. Once it has been established, attention can be focused on differences in environmental imagery from one social world to another, as well as on the replication of various types of social worlds and their contingent environmental imageries across urban settings.

Contrary to what has just been suggested, it might be argued that a detailed unified image of Los Angeles exists in the minds of our respondents, but that our methodology was not sensitive enough to elicit the appropriate response—that the mapping data lack validity. We hold no brief for the procedures employed in our research to date. In fact it is already apparent on the basis of our interviewing experiences that a revision in method is required if more refined and reliable data are to be obtained. The notion of scale, the range and intensity of social relations, is critical to our conceptual approach, and scale is difficult to translate into researchable terms. But at this juncture, further operational elaboration of at least two kinds appears to be required.

First, a broader spectrum of more detailed data on the behavioral manifestations of scalar differences should be tapped. In this study we sought information only on acquaintanceship patterns and associational memberships, which we expected to distinguish the populations of interest (Marshall, 1968). It would be instructive to include in addition time-budget data of the type proposed by Richard Meier (1959, 1962), as well as trip information of the origin-and-destination type which has been generated in any number of metropolitan transportation surveys. It would not be inappropriate to develop gradient information (Galpin, 1915) and service area data (Foley, 1950) against which environmental cognition can be examined.

Second, the method of obtaining mapped impressions needs refinement. Because we wanted to elicit a *tabula rasa* image, our respondents were given blank sheets of paper and no cues when they were asked to draw the maps. Clearly the ability to respond to such an undefined stimulus is highly variable. It appears, as one might reasonably expect, that the socioeconomic status of the respondent is associated both with his familiarity with maps in general and his ability to draw his own map. Of course, this is at least in part what we are interested in—who can draw which maps with what detail.

Based on some of our interview experiences, it appears that a mapped imagery is not necessarily consonant with knowledge of the environment elicited in a verbal form. It might well be that a series of partially completed base maps—maps showing, for example, major streets, civil divisions, geologic features, important landmarks, and the like—would elicit a more detailed, coherent, and consistent imagery. Such maps could be administered to the respondent in a series, increasing in detail, as with a focused interview.

For some purposes it might be sufficient to ask the respondent to locate certain items on an outline map, or to piece together a cut-apart map. The procedure employed would depend upon the purposes of the research. But, this much seems clear: a blank sheet of paper as a stimulus for obtaining a mapped image of the city is more of a liability than an asset. It confounds too readily the respondents' ability to make a map with knowledge he might have of the city but cannot represent in map form.

The methodological question at issue here is not unlike the problem of determining whether a precoded question is preferable to an unstructured question. Often information can be more quickly and efficiently obtained with a precoded question, but, by the same token, the richness of the data obtained from open-ended questioning is lost, and the precoded form may bias one's results either by unwittingly providing the respondent with cues or by automatically eliminating some response alternatives.

Resolution of some of the problems involved could be accomplished, given a sample of sufficient size, through field experimentation in which different interview forms are administered to matched subsamples of purposively selected samples of each population type. Such a procedure, rarely employed in urban sociological research, has been advocated by P. H. Chombart de Lawe (1965a,b).

It is all too easy to reify the concrete manifestation of the city, to assume that its symbolic representation is equivalent for all of its residents. This is especially dangerous if the perceptions of middle-class analysts are taken as representative of the population at large. Urban life has the reputation of being essentially alienating and disorganized, even though, as William Whyte (1943) long ago made clear, upon close examination, order often emerges from chaos. It would seem even more crucial today, as we confront anew the problems of urban life and recommit ourselves to seeking solutions based upon expertise, that we continually check our own world view with that of others who share the same matrix, even though, or perhaps especially because our social worlds all too often touch but fail to interpenetrate.

8

How Citizens View Two Great Cities: Milan and Rome

DONATA FRANCESCATO AND
WILLIAM MEBANE

Introduction

In this last decade, many disciplines ranging from geography to sociology, from architecture to psychology, from natural resource management to urban and regional planning, have been involved in research aimed at understanding human behavior in relation to the physical environment. First, the designers and planners have been increasingly concerned with better understanding the human activities that the desired physical environment is supposed to accommodate (Parr, 1963; Manning, 1965; Langdon, 1966; Sommer, 1969). Second, behavioral and social scientists have been exploring human perception of the "environment," encompassing both the physical environment and what Kates (1970, p. 648) has called the "environment in the minds of men": what people think, feel, and perceive about their environment, be it sun or rain, bricks and mortar, people and things. Kates describes four research themes emerging from the growing body of literature in human perception of the environment: the use of illusion to infer the reality of visual perceptions; the quest for the image of the city; the interplay of environmental attitudes and landscapes; and the adjustment to environmental hazard as environmental behavior. Each theme presents a different environmental focus: person, city, landscape, natural resource and hazard.

Using Kates' framework, the purpose of our comprehensive study was to investigate: a) the images that citizens held of their cities, b) their attitudes toward their cities, and c) the relationships between various images and attitudes. The groundwork was laid by Lynch (1960) who interviewed a small number of residents of Boston, Jersey City and Los Angeles. From their verbal answers and sketches and the systematic field reconnaissance of trained observers, he hypothesized that a city image was composed of five element classes: paths, landmarks, edges, nodes, and districts. Lynch claimed that the subject's image gave the city identity, structure and meaning.

Lynch's work stimulated a series of essentially replicative studies (for details of these, see Kates' table reproduced in the introduction to part II.) Craik (1970a, p. 94) remarks that this body of studies, while very interesting, does not lend itself to comparative analysis because neither the selection of subjects nor the procedures of data collection and analysis have been uniform. With few exceptions, sample sizes have been small and subjects chosen from undetermined social classes. Both Craik (1970a) and Lynch (1960) recommended selecting different groups of subjects within the same city, controlling for characteristics of age, sex and socioeconomic background. Parr (1963) suggested that different age groups may have different environmental needs, since he believes that there are five stages of urbanity (city living): childhood, adolescence, adult domesticity, adult emancipation, and old age. Moreover, Sonnenfeld (1966, 1967) suggested that there are sex differences in the kind of landscapes people prefer. In the present study, we therefore selected subjects varying in sex, socioeconomic background, age, and residence status.

With regard to the data collection and analysis procedures, the methods employed in most studies to assess a city image have involved map drawing and asking people to mention "distinctive parts" of a city that they would show to a visitor. Lynch (1960) found that the similarity between an individual sketch map and the same person's interview was in some cases rather low, but that there was considerable similarity between *composite* sketch maps and *composite* verbal interviews. However, maps tended to have a higher threshold: all elements were drawn less frequently than they were mentioned (that is, items which appeared with the lowest frequencies in the interviews tended *not* to appear in sketch maps). Lynch, finding that sketch maps tended to emphasize paths and exclude parts which are difficult to draw, concluded that sketch maps were a fairly weak indicant of known connective structures.

Research Objectives and Design

In this study we decided to investigate:

1. the differences in the composite images of Milan and Rome;
2. whether there are differences in the image maps drawn by different age, sex, socioeconomic, and resident groups within each city;
3. whether mapsketching as a tool has limited applicability; that is, whether some groups such as less-educated and older people will be less able (or willing) to use a tool which may be perceived as requiring a drawing skill.

We have also been investigating other methods of assessing the comprehensive "schemata" people have of their cities and the extent to which instructional sets influence their responses. Stea and Wood (forthcoming) make an interesting distinction between "image," the people's conception of the physical form of a city (which can be gathered through graphic methods such as mapsketching), and "schema," which is the overall cognitive representation or conceptual organization of a city, including symbols, beliefs, and activities. The results of two different schema assessment procedures are briefly discussed and the entire questionnaire is presented in this paper.

Rome and Milan were chosen following Lynch's suggestion that it would be interesting to study very different metropolitan areas ranging from the very old artistic (Rome) to the old industrial (Milan).

Method

The first author and two native female assistants interviewed 118 people in Rome and Milan, approximately divided between middle-class and lower-class, native and nonnative, male and female, and 30-and-under and over-30 years of age (subjects ranged in age from fifteen to seventy-seven). (see table 8.1).

TABLE 8.1. *Sample sizes of groups in Rome and Milan*

Group	*Rome* *No. of Subjects*	*Milan* *No. of Subjects*
males	30	30
females	30	28
middle class[a]	33	31
lower class	27	27
over 30	33	33
30-and-under	27	25
natives	33	21
nonnatives	27	37

[a] Social class was determined by using occupation and education as indices. Blue-collar occupations and less than high school education were taken as indices of lower class status. High school diploma and over, and white collar occupation were taken as indices of middle class.

For each subject, age, sex, length of residence (and town of origin if nonnative), education and occupation (used as indices of social class), and location of interview were recorded. Interviews averaged about half an hour in duration and were conducted in parks, theatres, streets, shops, and railroad stations. Interviewers introduced themselves as being interested in what people think about cities and requested permission to ask a few questions. Out of 120 people who met the initial requirement of being a resident of the city for at least one year, only two refused to be interviewed. When people agreed, they were asked the following questions, whose order was randomly changed from subject to subject:

1. Please close your eyes and think about———(name of city). What do you see?
2. Pretend you are moving away from ———. What five things do you think you will remember about this city?
3. What is important to you in this city?
4. How much do you like ———? Very much; a lot; so so; don't like it; don't like it at all. Please choose one.
5. What do you like about ———?
6. What do you dislike about ———?
7. Draw a map of ——— so that a visitor could find his way.
8. List distinctive parts of the city. A distinctive part might be a street, a building, a section, or any physical feature which you feel has special characteristics which make it worthy of being pointed out to a person who wants to become more familiar with the city.
9. What would you like to change in ———?

Data Analysis

Only the sketch maps and the answers to questions 1 and 3 (which have been coded and listed in terms of frequency in tables 8.2 and 8.3) have been analyzed here. Tables 8.2 and 8.3 represent a partial component of the schemata of Rome and Milan, which will be more completely explored when answers to the entire questionnaire have been analyzed.

Each map was scored by two independent raters using the following criteria, employed in most previous studies:

1. focal point (absence or presence);
2. number of distinctive elements drawn;
3. number of occurrences of elements in each of Lynch's classes: nodes, paths, landmarks, edges, and districts;
4. existence and type of street pattern.

Results and Discussion

MAP DRAWING REFUSALS

As expected, older and lower-class people tended to refuse to draw maps even when they were willing to participate in other forms of data gathering. In Milan, 57 percent of the over-30 group versus 88 percent of the 30-and-under group, and 48 percent of the lower-class versus 9 percent of the middle-class agreed to draw maps. In Rome, the same pattern emerged with acceptance percentages of 60 percent and 93 percent for the older and younger group respectively, and 67 percent and 93 percent for the lower-

and middle-class groups. If these results are confirmed in other studies it would seem that map sketching as a tool for information gathering may be less usable with certain population groups.

COMPARISON OF THE COMPOSITE IMAGES OF MILAN AND ROME

Milan is a large, flat, industrial city with a uniform radial street pattern, whereas Rome, a cultural, administrative, and artistic center, is a hilly city with highly irregular street patterns, and is divided by the Tiber river.

The radial street pattern and its accompanying concentric circling streets were recognized by 35 percent of the Milanese respondents, although the specific concentric streets were not identified by name in many cases. A regular street pattern is absent in Rome, due to the winding River Tiber and the presence of hills.

Sixty-three percent of the Milanese drew a central focal point on their maps, with 92 percent of these respondents giving the Duomo as their focal point. Fewer Romans drew maps with a focal point (53 percent and the focal points consisted of the Tiber river, two nodes (Plaza del Popolo and Plaza Venezia), and the Vatican. About 10 percent of the maps did not have a central focus but were sketches of neighboring streets around a person's residence, exhibiting a "home orientation".

As shown on the composite image map of Milan (fig. 8.1), Milan's most

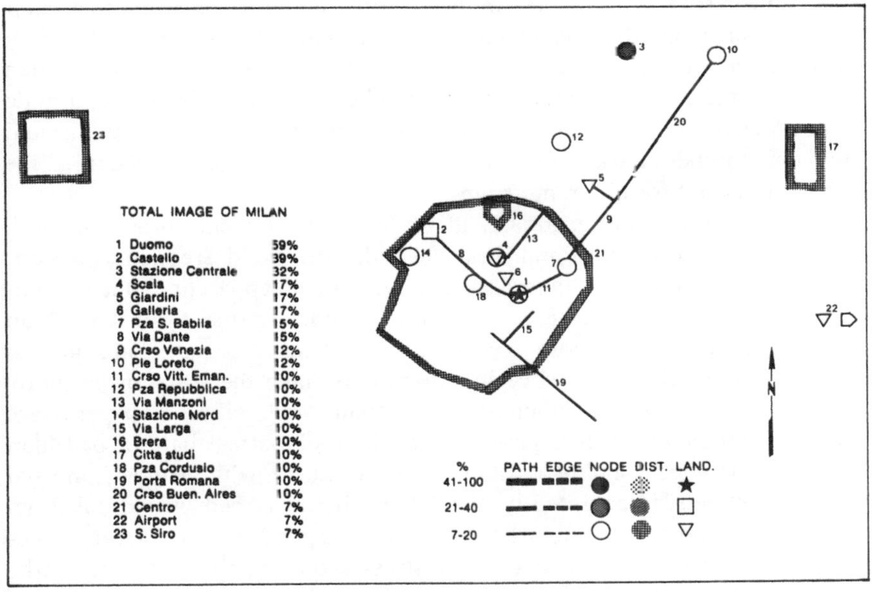

FIGURE 8.1. *Total image of Milan (N = 41).*

striking feature is the prominence of the city center. All but two of the ten most frequently mentioned elements are within the central district (element 21 in fig. 8.1), and 61 percent of the total number of elements are within this area. It appears that the uniformity of the surrounding zones, the flat landscape, and the radial street pattern all contribute to the prominence of the center. The Duomo cathedral, the most important landmark for all subgroups, is at the heart of the central district and at the hub of all radial paths. Thus, it is not surprising that it emerges as a prominent node. All of the paths in figure 8.1 lead to the Duomo, although the short streets in the immediate vicinity of the Duomo are not frequently mentioned, leaving Via Larga, Via Manzoni, and Porta Romana unconnected to the Duomo.

Only 40 percent of Rome's ten most frequently mentioned elements and 41 percent of all the elements occur in Rome's central district (fig. 8.2, no. 16). This is but one of many facets of the greater diversity exhibited by the composite image of Rome. Diversity was measured first by comparing the most frequently mentioned elements for the various subgroups of the two cities, and second by computing the ratio of number of different elements to total elements drawn. Rome's greater diversity is evidenced by a ratio of 0.41 compared to 0.31 for Milan. A greater number of different elements were drawn by the 47 Roman map drawers than the 41 Milanese (158 to 95) even if the total number drawn by the entire Roman sample (377) was only slightly higher than the total drawn by the Milanese sample (305). This indicates that each Roman was more likely to mention a greater *variety* of elements than his counterpart in Milan, even if both drew about the same *number* of elements. Additionally, the two top elements in Milan were the same two landmarks for all the subgroups, while in Rome, a node (Plaza Venezia), an edge (the Tiber), two landmarks (the Colosseum and the Villa Borghese park), and a bridge (the Ponte Garibaldi) occupied the number one spot for different groups.

While paths constitute about half of the total elements drawn in both cities, their importance diminishes when the threshold frequency is considered, and important differences emerge. As is apparent from a comparison of figures 8.1 and 8.2, paths appear more frequently in the Milan image and landmarks more frequently in Rome. No paths are among Rome's top ten elements, while in Milan, two make the list. Furthermore, all paths except one join other paths in Milan, while all Roman paths are shown as unconnected. It is possible that the radial street pattern of Milan facilitates the understanding of connective structures, while Rome's unusual pattern of streets, interrupted by the Tiber, circumscribed by the hills, and laced with important landmarks, yields less dependence on street names as aids to navigation. Lynch (1960) stresses the legibility of a city (the ease with which parts can be recognized and integrated into a coherent pattern), and believes that a vivid, integrated physical setting allows an individual ease of movement by serving as a broad frame of reference and

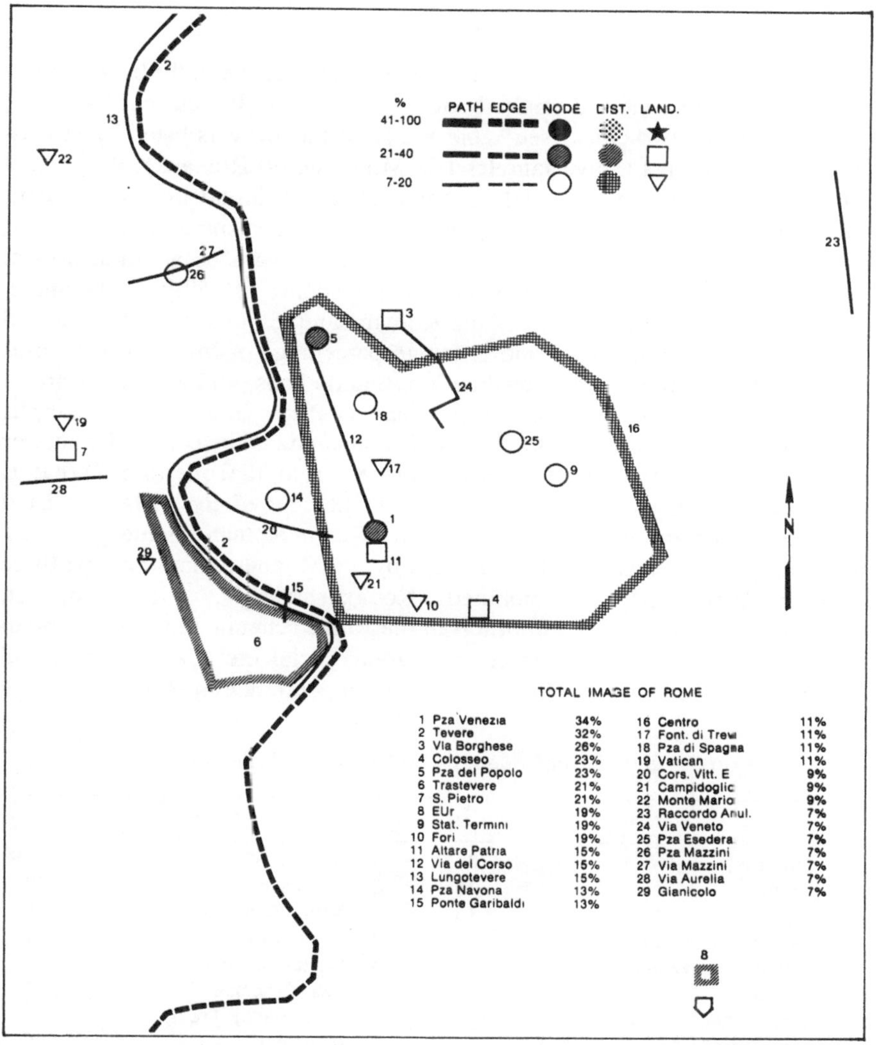

FIGURE 8.2. *Total image of Rome.*

an organizer of activity, belief, or knowledge. It appears that both Milan and Rome are highly legible cities, Milan being organized around remembered routes and Rome around a set of focal landmarks. That Rome seems also to be richer in what Lynch (1960, p. 7) has called "sensuous delight, rhythm, stimulus and choice" will become more apparent as we briefly examine the answers to two questions assessing the more general schemata of Milan and Rome.

PARTIAL SCHEMATA OF MILAN AND ROME

The composite remembered visual schemata, elicited verbally by the question "Close your eyes and think about ———. What do you see?" are very different for the two cities (see table 8.2). Of the answers listed by at least 5 percent of the respective samples (58 Milanese, 60 Romans), 91 percent of the Milanese responses involved nonlocational components such as concepts of social dynamism or such phenomena as chaos, smog, and crowding, while only 30 percent of the Roman responses were of a nonlocational nature. The Duomo was the only locational response listed by the Milanese, while the top seven responses of the Roman sample are specifically located landmarks or nodes. Furthermore, a high percentage of Milanese responses (such as traffic, fog-smog, crowding, noise and chaos, and greyness) are of an undesirable nature whereas this emphasis is absent in the Roman sample.

The importance component of the cities schemata is the composite answer (at the 5 percent cutoff level; see table 8.3) to the question "What is important to you in ———?" Notably 41 percent of the Milanese listed work as important while only 17 percent of their Roman counterparts did. In fact, work was mentioned more often by the Milanese than the next three most frequent answers combined. Not surprisingly, civilization, art, museums, and history are important in the Rome schema, but almost absent in the Milan schema. As if in compensation, social life, sports, parks, and playgrounds are important in Milan but go unmentioned in Rome.

TABLE 8.2. *Visual cognitive schemata of Milan and Rome*

	Milan			*Rome*	
1	Duomo	21%	1	S. Pietro	22%
2	traffic	16	2	Colosseum	12
3	fog, smog	16	3	Villa Borghese	10
4	dynamism, energy, activity, movement	14	4	Gianicolo	7
5	crowding	10	5	Castel S. Angelo	7
6	noise, chaos	7	6	Pza. Venezia	5
7	cars, trains	7	7	Fontana di Trevi	5
8	work opportunities	5	8	foreigners	5
9	cement, asphalt	5	9	antiquity	5
10	greyness	5	10	sunrise and sunsets	5
11	roads and houses	5			

Responses elicited by asking: "Please close your eyes and think of (Milan or Rome). What do you see?"

Percentages represent the proportion of the total respondents giving the indicated answers.

Milan N= 58
Rome N= 60

TABLE 8.3. *Importance schema*

	Milan			*Rome*	
1	work	41%	1	work	17%
2	Duomo	14	2	civilization	12
3	Castello	12	3	art, museums	12
4	social life, sports	12	4	history	10
5	parks, playgrounds	10	5	family	10
6	family	5	6	lifestyle	10
7	"nothing"	5	7	"nothing"	8
8	Brera Museum	5	8	world center, universality	8
9	good standard of living	5	9	friends	5
			10	"everything"	5
			11	St. Peters	5
			12	shops	5
			13	Pope, Christianity	5

Responses elicited by asking: "What is important to you in (Rome or Milan)?"
Percentages represent the proportion of the total respondents giving the indicated answers.
Milan N= 58
Rome N= 60

COMPARISON OF GROUP IMAGES

For group comparison, the average number of total elements drawn, the frequencies of the various types of elements, the agreement ratios, the average number of maps with focal points, the amount of specific element overlap, and four group image maps are presented.

For all groups there was great similarity in the rank order of Lynch's elements, paths appearing most frequently, followed by landmarks, nodes, districts, and edges. The only exception, the Roman lower class, reversed the ranking of districts and edges.

TABLE 8.4. *Mean number of elements drawn by Milan groups*

Group	*Number of Subjects*	*Mean Number of Elements*
nonnatives	25	5.3
natives	16	10.7
males	21	5.7
females	20	9.1
30-and-under	22	8.5
over age 30	19	6.1
middle class	28	8.7
lower class	13	4.4

TABLE 8.5. *Milan group differences in frequencies of Lynch elements*[a]

Group	*Number of Subjects*	*Agreement Ratio*[b]	*%Landmarks*	*%Nodes*	*%Districts*	*%Paths*	*%Edges*
nonnatives	25	20/64 = .31	40	10	10	40	0
natives	16	31/76 = .41	23	29	13	35	0
males	21	17/48 = .35	41	24	11	24	0
females	20	27/70 = .39	26	26	15	33	0
30-and-under	22	19/64 = .30	11	21	11	57	0
over 30	19	18/43 = .42	33	23	11	33	0
middle class	28	35/86 = .41	23	23	17	37	0
lower class	13	36/42 = .85	14	17	3	66	0
total Milan	41	23/95 = .24	26	26	18	30	0

[a] Refers to differences in frequency of Lynchian elements drawn by at least 7% of each subgroup

[b] Ratio of number of different elements drawn by at least 7% of subjects to total number of different elements drawn by each subgroup

TABLE 8.6. *Mean number of elements drawn by Rome groups*

Group	*Number of Subjects*	*Mean Number of Elements*
nonnatives	21	10.1
natives	26	6.3
males	23	10.6
females	24	5.9
30-and-under	30	9.2
over age 30	17	6.1
middle class	31	9.0
lower class	16	6.0

In both cities, male and female groups display few image differences. Although Roman males drew more elements than Roman females, this relation is reversed in Milan and there are no great differences in the frequencies of elements in Lynch's classes drawn in either city.

To the 30-and-under age group, paths are more important than landmarks, while to the over-30s, landmarks are of equal or greater importance than paths. In Milan the 30-and-under age group drew paths 57 percent of the time and landmarks only 11 percent of the time, while the over-30 group drew both with equal frequency (33 percent). In Rome the same pattern emerged with the younger group drawing 53 percent paths and 25 percent landmarks, and the older group 29 percent paths and 35 percent landmarks (see tables 8.5 and 8.7). It may be that the Italians aged 30-and-under learned the city by car and are therefore more street conscious than older Italians who knew the cities before automobiles were widely available. An additional explanation might be advanced in accordance with Parr's (1963) five stages of urbanity. A youth, presumably in the period of learning the city, will be concerned with learning to navigate in the city and therefore will give more importance to the main navigation aids: the streets and street patterns. It may also be that parks, gardens, and other landmarks become more important as one grows older.

In both cities the middle-class groups drew a greater average number of total elements than their lower-class counterparts (see tables 8.4 and 8.6). The spatial extent of the middle-class maps is greater than that of the lower-class maps (as shown clearly in figs. 8.5 and 8.6), which could suggest that the lower class is less mobile: being exposed less frequently to the outlying districts and regions, it therefore draws them less often.

An alternative explanation of these findings may be found in the greater tendency of the lower-class map drawers to have a home orientation. The home orientation produces maps covering a very small area of the city, favoring the inclusion of streets immediately adjacent to the respondent's home and excluding landmarks, nodes, and districts situated throughout the

TABLE 8.7. *Rome group differences in frequencies of Lynch elements*[a]

Group	*Number of Subjects*	*Agreement Ratio*[b]	*% Landmarks*	*% Nodes*	*% Districts*	*% Paths*	*% Edges*
nonnatives	21	41/78 = .53	34	15	10	39	2
natives	26	24/97 = .25	17	17	8	54	4
males	23	38/108 = .35	39	16	11	42	2
females	24	27/77 = .35	37	15	15	30	3
30-and-under	25	48/101 = .47	25	10	10	53	2
over 30	22	17/82 = .21	35	18	12	29	6
middle class	31	42/111 = .38	29	17	9	43	2
lower class	16	12/70 = .17	35	17	0	50	8
total Rome	47	29/158 = .18	34	23	12	28	3

[a] Refers to differences in frequency of Lynchian elements drawn by at least 7% of each subgroup

[b] Ratio of number of different elements drawn by at least 7% of subjects to total number of different elements drawn by each subgroup

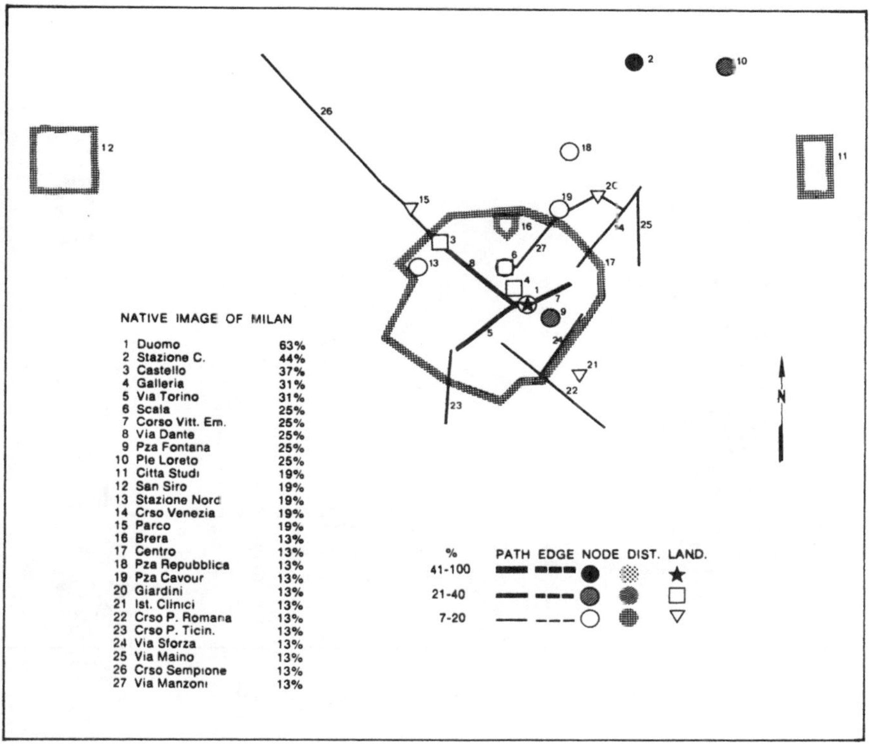

FIGURE 8.3. *Native image of Milan.*

entire city. This is an expression of Italian lower-class life style, with its emphasis on home and neighborhood; we will be able to explore this hypothesis further as we analyze verbal responses assessing what is important and liked about a city by the various subgroups.

As one would expect from the frequency differences, the amount of overlap was low. The Roman lower class was very concerned with bridges, these paths representing 25 percent of the total responses, with the Ponte Garibaldi being the most frequently mentioned element. Middle class Romans mentioned only one bridge among forty-one elements.

In Milan, the average number of elements drawn by natives was greater than their counterparts, while in Rome the natives drew fewer elements. In Milan, as shown in figures 8.3 and 8.4, nonnatives drew three elements—Idroscalo, Sun Highway, and Lake Highway—which are outside the given map area, while natives, even if drawing many more elements, concentrated them in the inner area of the city. Natives appear to have a more conservative image of the old Milan (before the recent expansion), while nonnatives

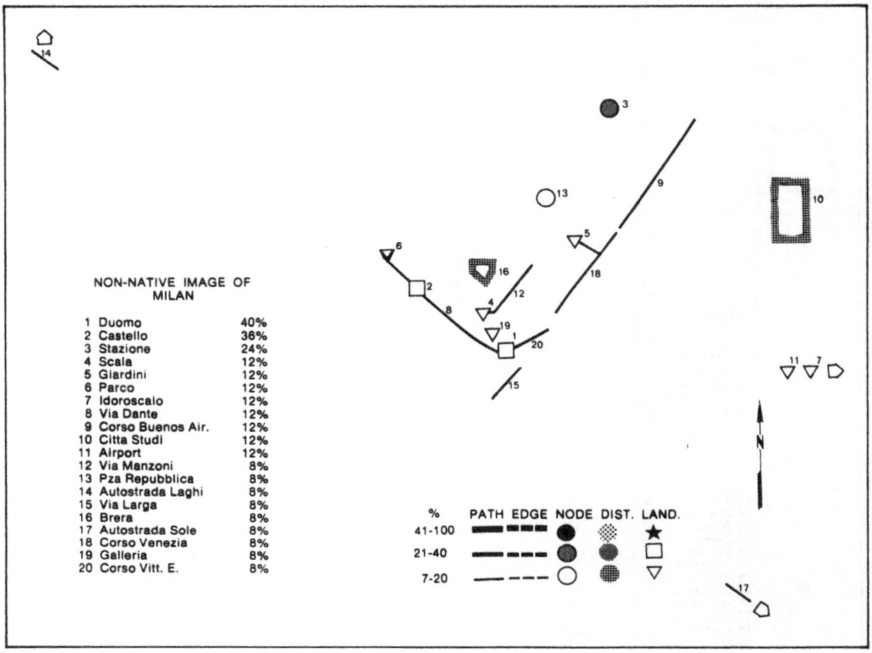

FIGURE 8.4. *Nonnative image of Milan.*

include in their image "escape routes" from Milan (Idroscalo, Sun Highway, and Lake Highway), which are means of access to recreation. This result will perhaps appear more clearly when we analyze the answers to the question which taps how much different groups like their city. We hypothesize that more natives like Milan, and therefore find more enjoyment in it. Consequently, they are less likely to include in their schemata elements associated with getting out of the city (to the lakes, the sea, etc.) In Rome the same trend emerged, with natives focusing their attention upon the core of Rome and nonnatives including many surrounding highways and other elements. Nonnatives seem to be less satisfied with the city than natives; we may be able to explore this hypothesis further as we complete the analysis of all our verbal data. Natives in both cities also showed a deeper and more individualized knowledge of their city, drawing a greater variety of elements than nonnatives.

Summary and Conclusions

Examining the results of this study yields certain tentative conclusions: residents of different cities are sensitive to and aware of the diverse physical

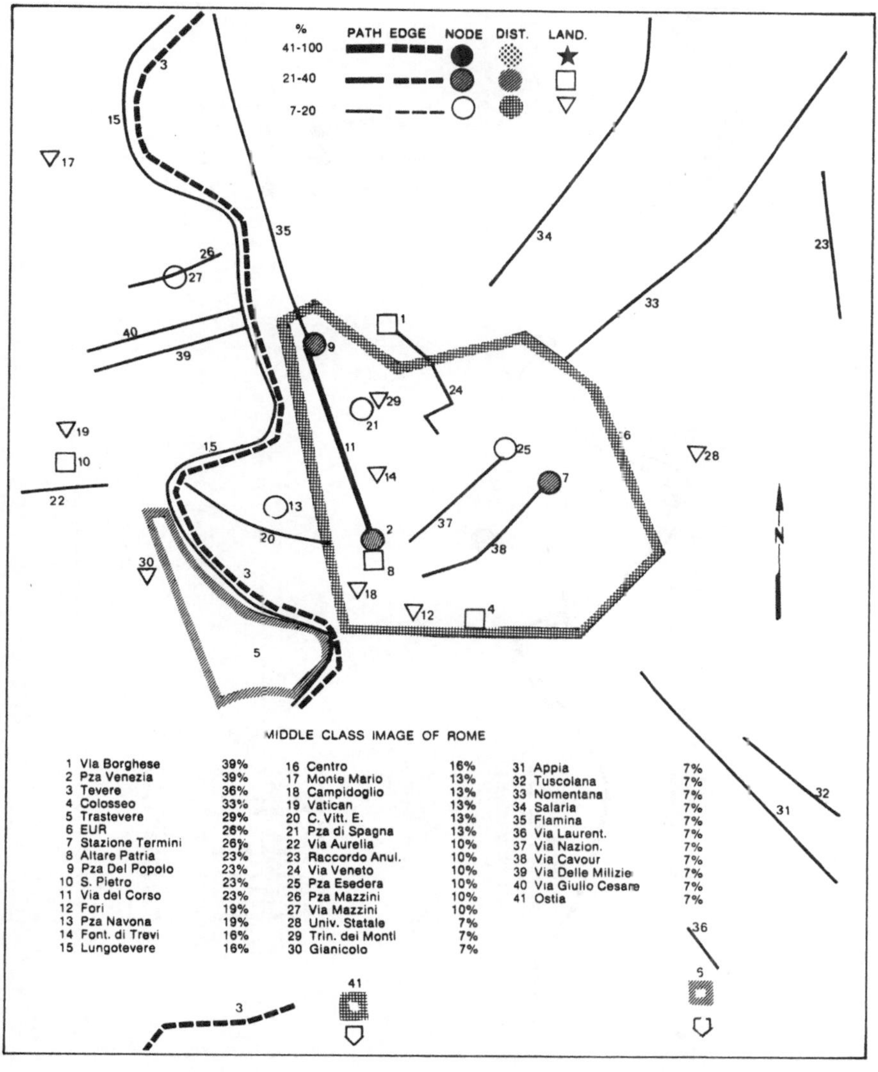

FIGURE 8.5. *Middle class image of Rome.*

environment they live in. People in Rome and Milan included in their image the dominant physical features of their respective cities: river, hills, and historical landmarks in Rome; a radial street pattern with a central focus in Milan. Within each city, there were group differences in city images between middle and lower classes, natives and nonnatives, younger and older residents. The 30-and-under age groups gave more importance to

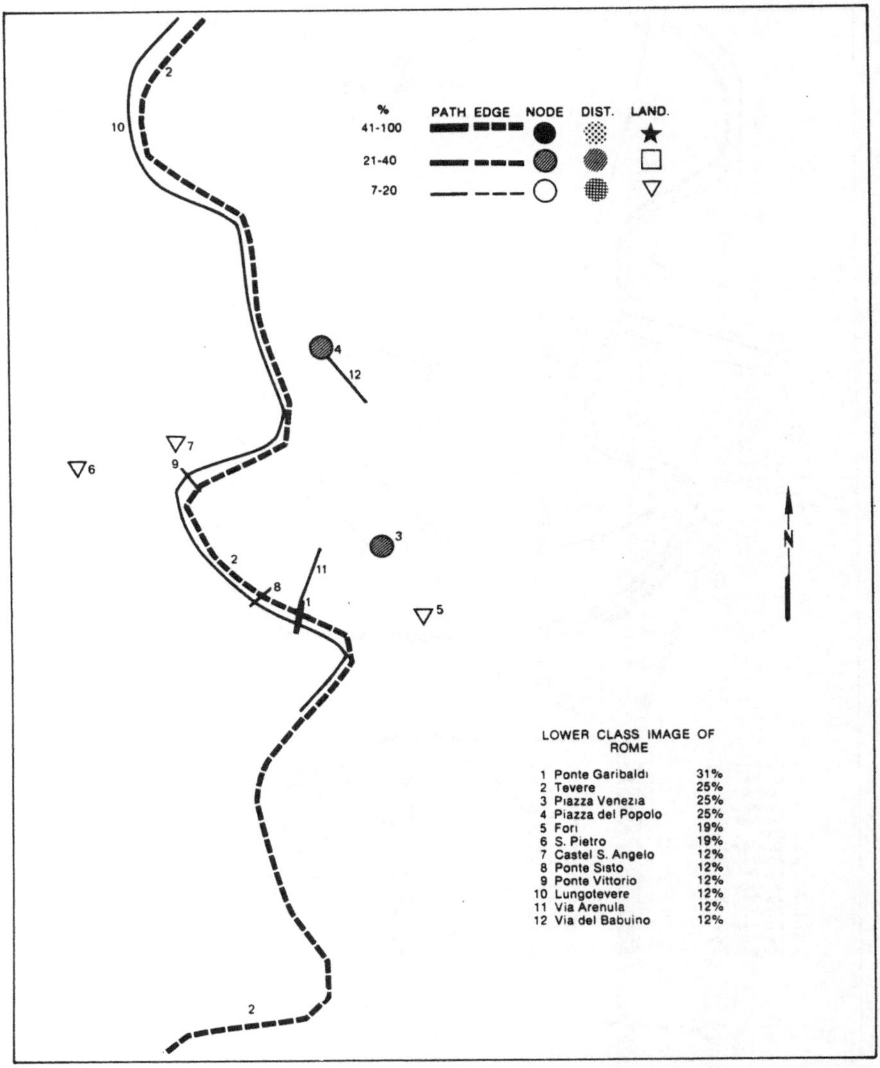

FIGURE 8.6. *Lower class image of Rome.*

streets than landmarks, while their counterparts gave the two equal importance or reversed the relationship; the middle-class groups placed less emphasis on streets and more emphasis on districts than the lower-class groups; and nonnatives stressed paths existing from the city more than natives. No remarkable sex differences appeared in this study.

Our preliminary analysis of the verbal schemata shows that the physical environment played a different role in Milan and Rome, the physical locational environment (Lynch's elements) being much more prominent and more favorably perceived in Rome than in Milan. Only the complete analysis of the verbal responses will elucidate what importance physical elements have in the total picture of Rome and Milan, especially since verbal methods are less skewed toward favoring physical locational elements than mapsketching methods. Clearly, what people can locate on a map are streets, nodes, districts, landmarks and edges; they cannot put down how they feel about a city, what they normally do in it, what they like, dislike, or deem important. All these activities, attitudes, and feelings are part of what Stea and Wood (forthcoming) have called the "opportunity surface" of a city, the totality of what a city offers (or allows) its citizens. What people select out of this totality to form their city images and schemata may be a result of the frequency of experience that inhabitants have with different aspects of their city. The more city dwellers use an element or experience an event such as heavy traffic, the more frequently that component ought to be mentioned in schemata or drawn in maps. However, this is as yet an untested hypothesis, since elements that are infrequently used but widely known or greatly enjoyed may also be included. For example, St. Peters may be frequently mentioned but infrequently visited, or a Milanese citizen may take great pride in La Scala but rarely attend its functions. What kind of role direct experience with different parts of a city plays in one's image and schema is therefore an open question. We feel that the inclusion of usage data (questions about how often somebody goes some place, etc.) along with opportunity and schema measurements will produce a better understanding of city imagery.

It is to this wider inquiry that future studies of cities should address themselves. We will then be better able to assess what relative role the physical environment (a city's image) plays in the total schema of a city and what kinds of physical environments have the strongest effects upon the lives of their inhabitants.

9

Student Views of the World

THOMAS F. SAARINEN

Introduction

This chapter is designed to explore in a preliminary fashion student views of the world as reflected in freehand sketch maps. In *The Image of the City*, Lynch (1960) indicated the utility of such sketch maps for obtaining insights into how people mentally structure the city and which elements are perceived as important. Such information is not readily obtainable by other means, which perhaps accounts for the wide application of this essentially projective technique. It has been used in studies of cities (Appleyard, 1970; Gulick, 1963) and portions of cities such as the Central Business District (Saarinen, 1968) and neighborhoods (De Jonge, 1962; Ladd, 1970). In this chapter an application of the freehand sketch map technique on the world scale is discussed. It was expected that the distinctive ethnocentric views of different areas would be likely to emerge on such maps. It seems important in a world continually upset by international conflicts to try to gain an understanding of variations in world views (Klineberg, 1965; Kelman, 1965; Stagner, 1967). This would seem to be an essential starting point in education for international understanding.

Four groups of high school students in the United States, Canada, Finland, and Sierra Leone were given a blank sheet of paper and asked to sketch a map of the world, labeling all places they considered to be interesting or important. They were allotted about half an hour for the task. In addition, students in several introductory geography classes at the University of Arizona were asked to fill in the names of places on an outline map of the world with only the *political* boundaries shown. Analysis of

The author wishes to acknowledge the help of Hannele Rikkinen, Curt Albertson, Marty Radbill, and Carl Saarinen who provided the samples.

these data should indicate which regions or countries are most included or omitted and whether these combine in characteristic patterns for each group.

The analysis begins with a broad discussion of the frequency with which certain types of items and major world regions are included on the four sets of maps. Tables of frequency for types of items mentioned are included, and both aggregate maps and examples of individual maps examined. Later, the analysis turns to a search for explanatory factors. A number of hypotheses are examined including the roles of proximity, of size, of shape and location, of current events, and of cultural factors.

Data Collection Procedures

Five main sets of data were collected. Four represent samples of high school students from different countries (see table 9.1).

TABLE 9.1. *Numbers in sample by age, sex and country*

Age and Sex	*Finland (Helsinki)*	*United States (Tucson, Az.)*	*Canada (Calgary)*	*Sierra Leone (Makeni)*
16-year-old male	—	—	—	12
17-year-old male	25	31	30	9
17-year-old female	28	31	21	—
18-year-old male	16	11	16	9
18-year-old female	11	7	8	—
Totals	80	80	75	30

An attempt was made to match sample size in terms of age, sex, and current education level. However, there was no control as to quality of the school. In the cases of Canada and the United States, or more accurately, Calgary, and Tucson, the students were enrolled in public schools which were reasonably representative for their city. The Helsinki school was private but student performance is about average for that city. The sample school in Sierra Leone was considered to be better than average as it was a private boys' school. The Sierra Leone sample poses certain problems. It is smaller than the others and consists entirely of boys, many of whom are a year younger than any from the other areas.

The representativeness of the sample could be questioned on several counts. First, the places sampled do not cover a wide spectrum of world locations whether judged spatially, culturally, or by socioeconomic levels. Three are from white, wealthy, industrialized nations and the sample from the only black, poor, nonindustrialized nation is deficient in the respects noted above. In addition, the samples might not be representative of the nations in which they were obtained. This is especially true in the cases of Canada and the United States where size alone suggests great regional variation. For these reasons the data will be treated as samples of their particular locations, probably reflecting broad national biases, but to a degree not clearly known. This is not of the utmost importance since the purpose of the paper is a broad exploration of the kinds of insights regarding world views derived from freehand map sketches. Thus, discussion of common features will take precedence over the unique qualities of the separate samples.

Some people have criticized the use of sketch maps because of individual differences in drawing skills. My experience in university classes indicates that for such a group the sketching of a map poses no great problem. Drumheller (1968), in a study concerned with maturation and map skills, suggests that students may be ready to start conjuring up a map by junior high school age. But to test for possible distortions in freehand sketches a control group was used. A map of the world containing only the outlines of political boundaries was issued to 180 students in introductory geography courses at the University of Arizona. They were asked to fill in the names of political units, the task requiring only place recognition, not motor skills. If the pattern of places included differed greatly from the pattern of places included by the Tucson High School students in their freehand sketches, a case could be made for the importance of motor skills. A drawback was the difference in age and education between the two groups.

Group rather than individual differences will be the main concern of the following discussion. However, it should be noted that a great range of individual differences was present in the sketch maps collected. Some maps were so covered with accurately located places that space seemed to be the major limiting factor. Others were so crude that only a few places were named and even major continents were omitted or grossly distorted in size, shape, or location. In spite of these individual differences in general there were great similarities among sketch maps from the same place.

General Description of Tabulated Data and Maps

Apparent in a cursory examination of student sketch maps of the world is the predominance of the nation or country as the most frequent feature class (see table 9.2). On the world scale this seems to be the most convenient size of building block, corresponding to states or provinces on the scale of a nation or blocks within cities.

Over half of the names placed on the maps were those of countries. Minor political units, such as states or provinces within the larger countries, are often as large as many nations (e.g. the Canadian provinces and certain of

TABLE 9.2. *Frequency of types of features included on world sketch maps*

Features Noted	*Sierra Leone*		*U.S.A.*		*Canada*		*Finland*		*Total*	
	N[a]	%[b]	N	%	N	%	N	%	N	%
nations	85	28.5	1162	58.9	1926	60.7	1727	47.4	4900	53.9
cities	5	1.7	109	5.5	127	4.0	1191	32.7	1432	15.8
minor political units[c]	2	.7	187	9.5	475	15.0	177	4.9	841	9.3
continents	133	44.6	260	13.2	173	5.5	131	3.6	697	7.7
oceans and seas[d]	43	14.4	99	5.0	233	7.3	123	4.5	498	5.5
others[e]	33	10.1	153	7.9	239	7.5	298	8.0	717	7.8

[a] N = times noted
[b] % of total features noted
[c] states or provinces within a nation
[d] also includes capes, gulfs, straits and bays
[e] made up of such varied features as rivers, mountains, deserts, lakes, types of regions, islands (including Greenland), and geographical coordinates.

the United States). These minor political units make up close to 10 percent of the total. Continents are almost too large to be labeled *except* where the students appear ignorant of smaller-sized features. The best sketch maps generally did not include the names of continents and oceans while the poorest ones contained little else. The thirty students from Sierra Leone who, in general, included the fewest features on their maps, had a higher total number of continents noted than the eighty Finnish students who produced the most detailed maps of all groups. For the Africans, continents and oceans make up half of the total features labeled. While the cruder maps labeled only oceans, the better ones often omitted them but tended to note the names of smaller seas, gulfs, and straits. The only features of relatively small size included frequently were cities, indicating that the students perceive the importance of cities. This emphasizes another factor common to almost all maps, the predominance of *human* over *physical* features. Haddon noted the same tendency when he asked students to write, in short words or phrases, the thoughts and images that were conjured up by the name of a particular country (Haddon, 1960). Rivers, mountains, deserts, and lakes are included only occasionally in the results of the current study.

For each of the sample areas a composite political map was constructed based upon the percentage of the sample which included some mention of each nation. The resulting map patterns (figs. 9.1–9.5) are presented here to illustrate characteristics common to all, and special features of each sample group. Figures 9.6 through 9.10 are included to illustrate striking examples of productions by individual subjects.

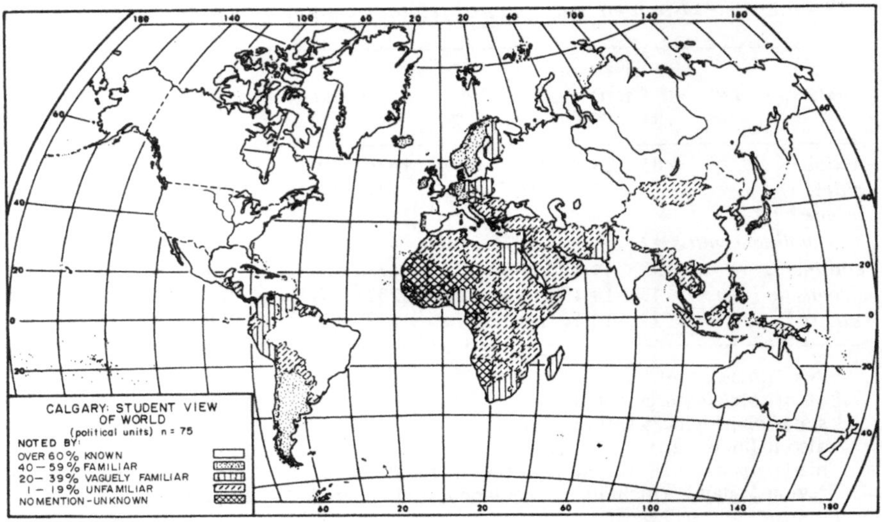

FIGURE 9.1. *Calgary: Student view of the world.*

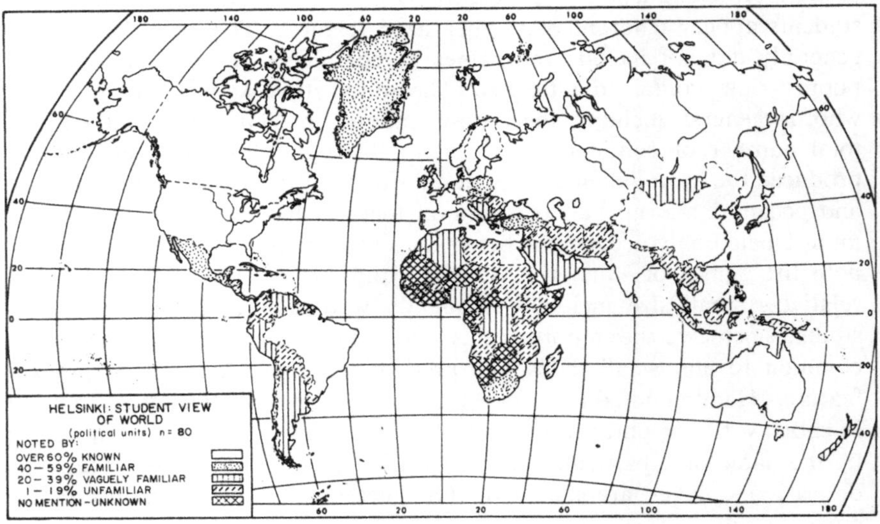

FIGURE 9.2. *Helsinki: Student view of the world.*

Factors Which Explain Results

This section briefly links some of the results obtained and indicates factors which help to explain them. Although each is discussed separately, they obviously interact, and *ceteris paribus* is an underlying assumption. The factors are stated positively but further research, as suggested in the final section, is necessary to establish their validity and relative importance.

From the *factor of proximity*, one would expect that the home area and areas immediately adjacent to it are the most familiar and hence the most

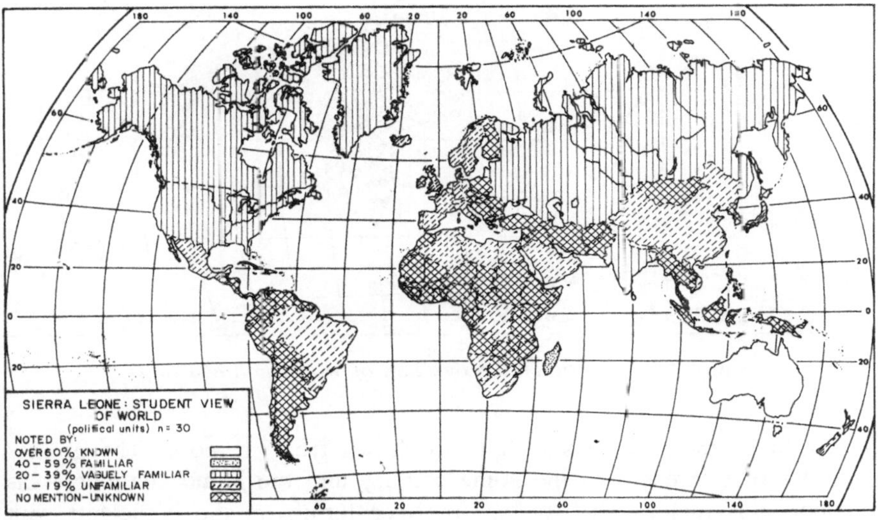

FIGURE 9.3. *Sierra Leone: Student view of the world.*

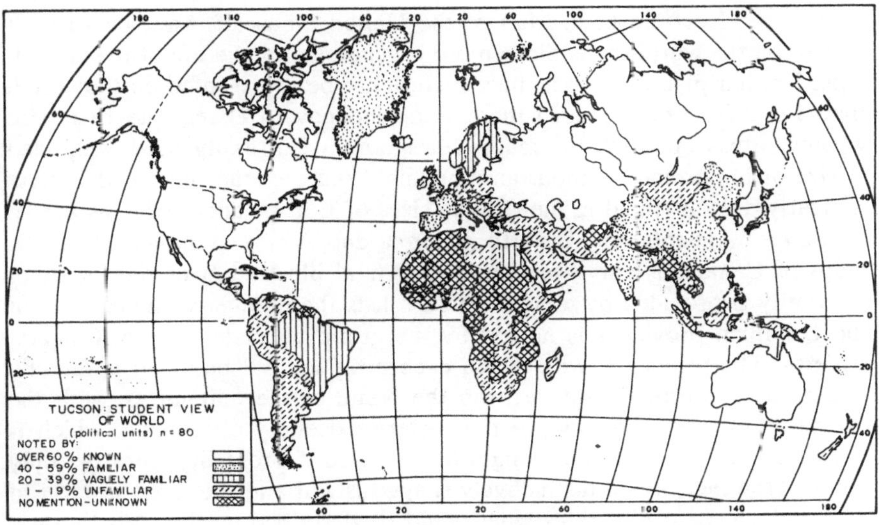

FIGURE 9.4. *Tucson: Student view of the world.*

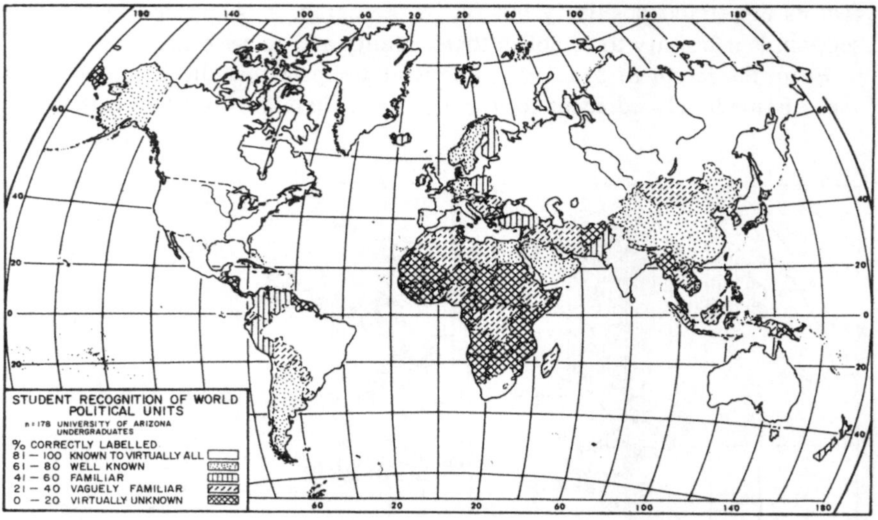

FIGURE 9.5. *Student recognition of world political units.*

likely to be included in sketch maps drawn from memory. The greater number of references to the home country and the home continent bias support this factor. The student maps centering on the homeland with progressively fewer features and greater distortion with distance provide a further example of the working of the proximity factor (see figs. 9.6, 9.7, 9.8, and 9.10). The same set of factors which produced the distortions seen in the New Yorker's Map of the United States of America (1936) appear on the world scale. Since more is generally known about the closest, most familiar places, there is more material to be represented on the sketch map for these areas. Hence there is a tendency to exaggerate their size whereas areas of which the mapper is ignorant can easily be downgraded in size or importance without any qualms. Whatever the reason, the effect is clearly demonstrated in the relative sizes of Europe and Africa as drawn by Finnish students, or of North America and Africa as indicated on the maps of Canadians and Americans. One final illustration of the factor of proximity is provided by table 9.3 which lists the number of times each of the Canadian provinces is named by the group of students from Calgary, Alberta. The frequency with which each province is mentioned is directly related to its ordinal distance from the home of the mapmaker, with the exception of Newfoundland, at the eastern extremity of Canada, which is named more frequently than might be expected if proximity were the only factor. However, it is a distinctively shaped island and the most easterly in a spatial series. It illustrates what could be described as the "Vladivostok effect" or the "Perth principle," that is, the tendency of sketch mappers to

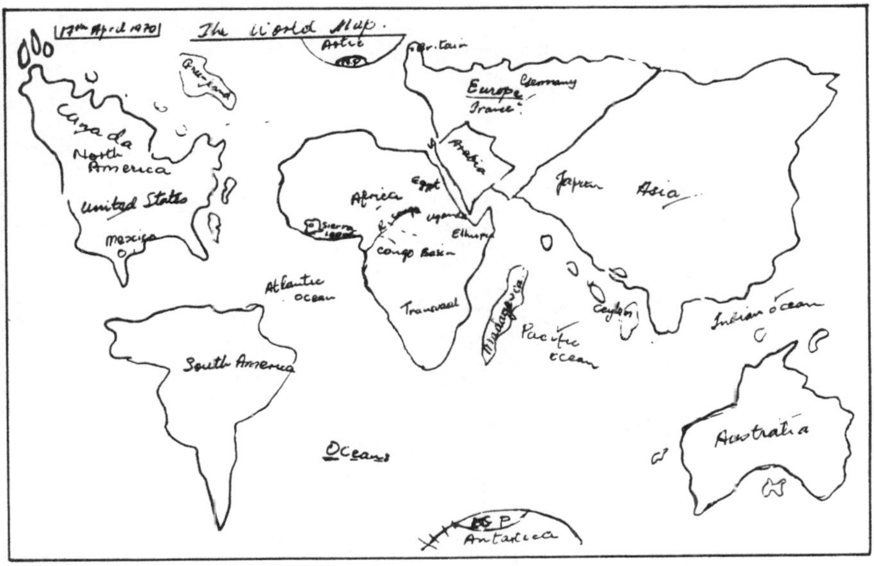

FIGURE 9.6. *An African version. Note the centrality of Africa and the vague notion of the shapes of other continents; i.e., North and South America are separate islands.*

include such remote, isolated places in order to mark the edge or end of a country or continent especially when there is no other good reason for inclusion. Other frequently included examples of this effect are Tierra del Fuego and the Cape of Good Hope. In connection with table 9.3, note the effect of the United States—Canadian boundary. Montana, the closest state, was mentioned only ten times by the Canadian students, less than one-third the frequency of mention of British Columbia or Saskatchewan (see Mackay, 1958, for discussion of measurement of border effects).

The *factor of shape* is important in influencing the rate of inclusion. Chile, with its shoestring form, is more often included than other South

TABLE 9.3. *Frequency of mention of Canadian provinces by Calgary students*

Province[a]	*No. of times named*	*Province*[a]	*No. of times named*
British Columbia	32	Quebec	23
Alberta	33	New Brunswick	4
Saskatchewan	30	Prince Edward Island	10
Manitoba	26	Nova Scotia	9
Ontario	25	Newfoundland	27

[a] the provinces are listed in order from west to east

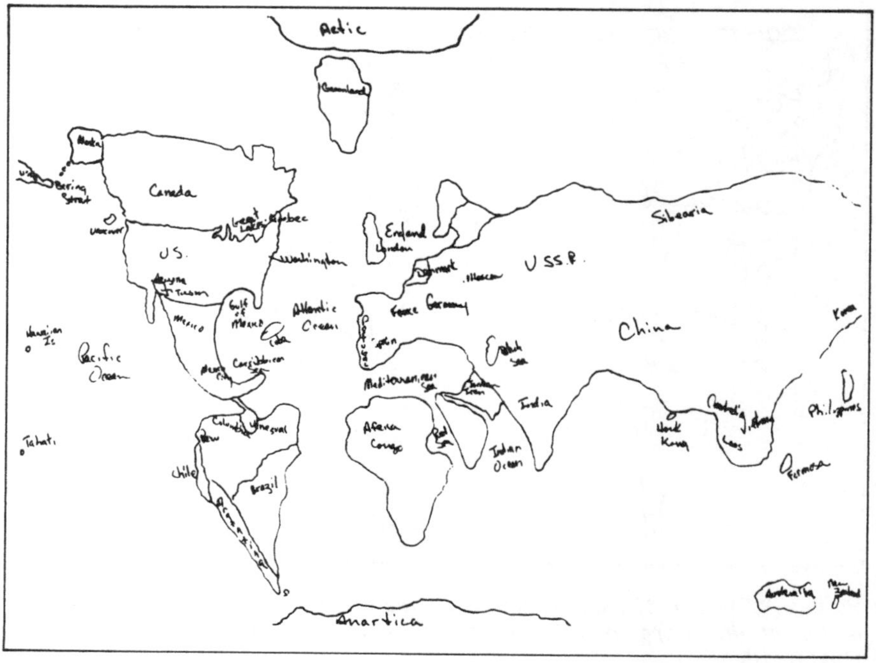

FIGURE 9.7. *An American version. Notice the size of North America and Europe compared to Africa and Australia; also the exaggerated size of Mexico, the nearest neighbor for Tucson students.*

American countries greater in size or population (see figs. 9.2 and 9.5). Italy's striking boot shape no doubt helps students remember it for in one case toes are included (fig. 9.9). Three groups of countries of different sizes were compared in terms of the total number of times noted in the student sketch maps. In each group, the size range was selected to include an island since this was considered the most distinctive of forms and therefore most likely to be remembered and included on world sketch maps (see table 9.4). In each group the island ranks highest or close to the top in frequency of mention. Greenland is second only to Mexico in Group I, Madagascar follows three European nations in Group II, and Iceland and Cuba stand out in their size range. Landlocked countries appear to provide the least memorable image: thus the Sudan is rarely included even though of great size, and Afghanistan, Botswana, and the Central African Republic are seldom cited. Where many countries with varying outlines are adjacent no clear image emerges (e.g. Guatemala, Honduras, French Guiana, Liberia, Cameroons, Dahomey, the Central European nations, and the Balkans). Even large islands may lose their individual identity if surrounded by others; in Indonesia, Borneo, New Guinea, and Sumatra are rarely singled out.

TABLE 9.4. *Distinctive shape and size and frequency of mention*

Group I[a]	*No. of Mentions*	*Group II*[b]	*No. of Mentions*	*Group III*[c]	*No. of Mentions*
Mexico	156	Spain	147	Iceland	89
Greenland	147	France	138	Cuba	85
Saudi Arabia	52	Sweden	103	Portugal	79
Argentina	51	Malagasy Republic	57	Hungary	19
The Congo	34	Morocco	26	Bulgaria	16
Algeria	33	Kenya	11	Jordan	16
Sudan	6	Afghanistan	6	Honduras	5
		Botswana	1	French Guiana	5
		Somali Republic	1	Guatemala	4
		Central African Rep.	0	Liberia	2
		Cameroons	0	Dahomey	0
total	479		490		320
average	68.4		44.5		29.1

For each group the areas selected were *all* those within the appropriate size range in the table on Principle Countries and Regions of the World in Espenshade, E. B. Jr., *Goode's World Atlas, Thirteenth Edition* (Chicago: Rand McNally and Co. 1970, p. 189). Exceptions were provinces or states with nations.

[a] Includes those from 750,000 to 1,000,000 square miles in area

[b] Includes those from 170,000 to 250,000 square miles in area

[c] Includes those from 35,000 to 45,000 square miles in area

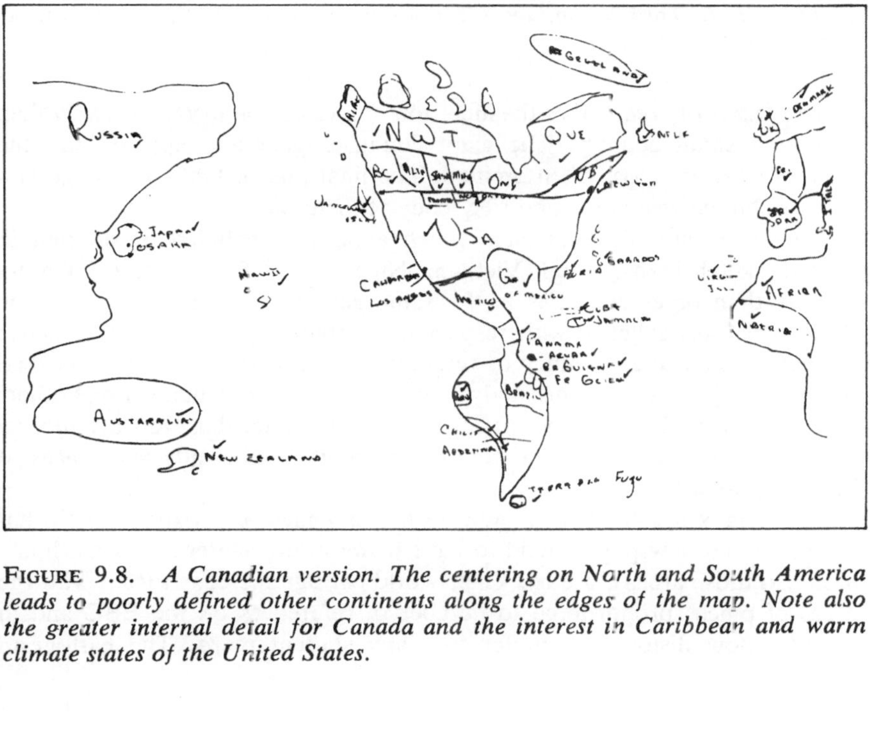

FIGURE 9.8. *A Canadian version. The centering on North and South America leads to poorly defined other continents along the edges of the map. Note also the greater internal detail for Canada and the interest in Caribbean and warm climate states of the United States.*

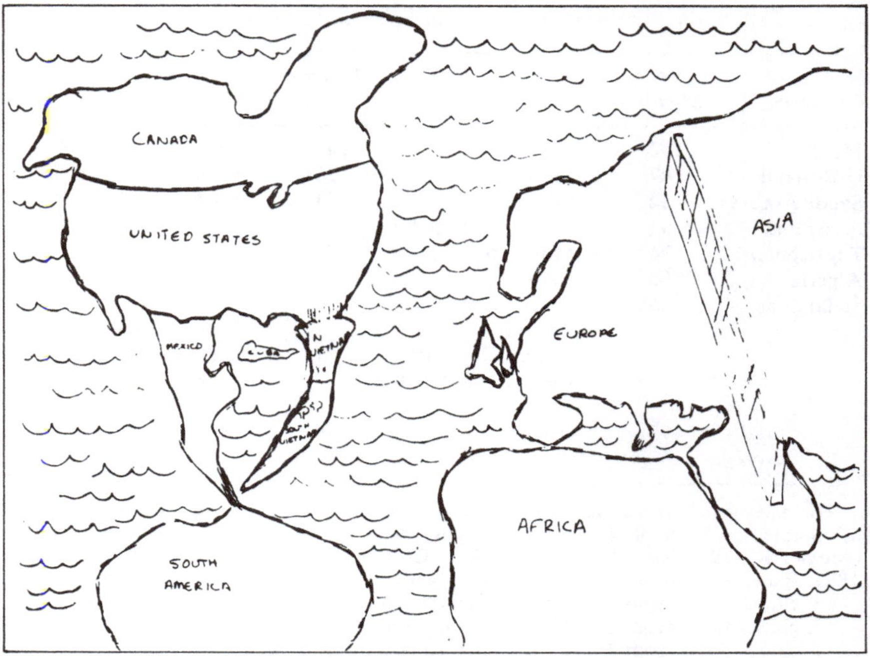

FIGURE 9.9. *An American version. The grim humor of this map signed in shaky script. Richard Nixon, age 17, a boy, brings one world problem close to home.*

The *factor of area* makes the inclusion of a large country more likely than that of a small country. The giant countries generally appear, and the smaller ones are most often omitted. The final line of table 9.4 suggests a relationship between area and frequency of mention.

Current events also influence which areas are included or omitted in student sketch maps. Israel, Vietnam, Nigeria, and Egypt stand out more sharply than other countries in the same general areas and of about the same size. This salience with respect to surrounding countries is a better indication than total number of references since certain world areas tend to be overrepresented in this study (Europe and North America) and others underrepresented (Asia and Africa). Current events thus add a temporal effect so that in different time periods one might expect different places to be highlighted.

Cole and Whysall (1968) indicate the influence of *proximity* such that the radio and newspapers tend to have many more references to the home area and those areas close by. Their world cartograms constructed with the areas of places proportional to number of references on the BBC and in Pravda show distortions similar to many student maps. The cartogram

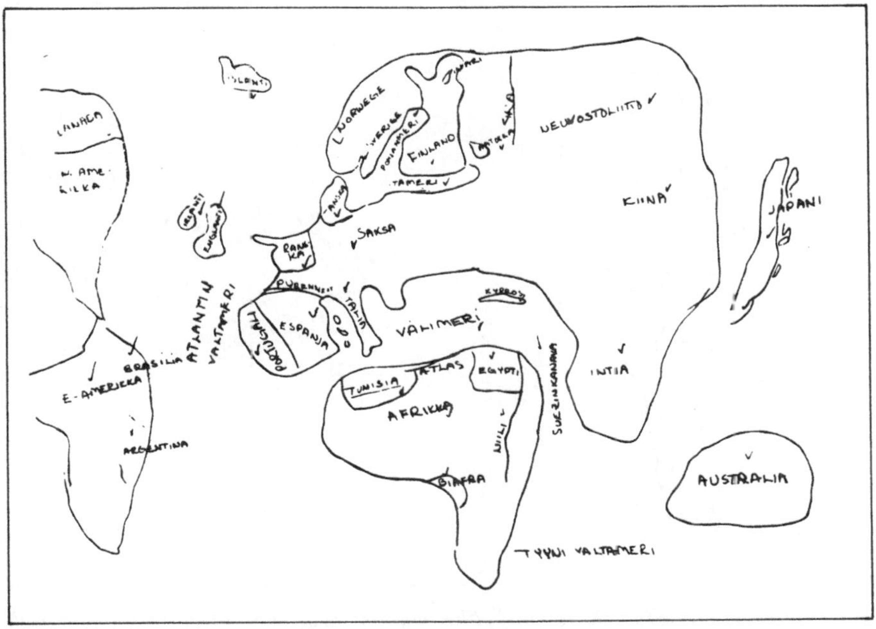

FIGURE 9.10. *A Finnish version. A beautiful example of home country exaggerated. Notice that Finland is larger than Canada and Europe is as large as Asia.*

based on BBC news has Great Britain as the largest nation, and the one based on Pravda news items has the USSR largest.

Cultural factors are important in determining which countries will be included by each group. Most of the countries sampled are wealthy, urban, industrial nations and the maps tend to include like areas and exclude areas of a markedly different character unless distinguished for some special reason (size, shape, proximity). Since the basic building block of the sketch maps is the nation, political factors are most easily examined. On the sketch maps of Canadian students, 32 percent of all features noted were within the British Commonwealth. This compared to approximately 15 percent of mentions by each of the other groups (U.S.A. 15.9 percent, Finland 14.1 percent, Sierra Leone 15.5 percent). One reason why the percentage is so high for Canadian students is the large number of references to the home country. But if the references to Canada are removed the percentage of map features devoted to Commonwealth countries is still close to 20 percent higher than for any other group. Clearly, the Commonwealth association has helped Canadian students to gain an increased awareness of these former British colonies. While well over half of the features noted by students in Sierra Leone, another Commonwealth member, were continents

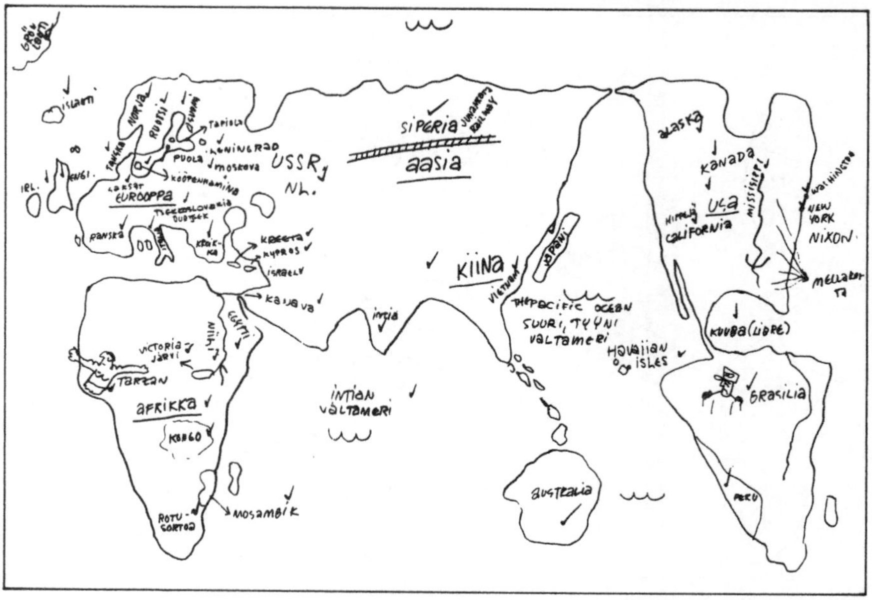

FIGURE 9.11. *A Finnish version. This student uses the old cartographic trick of filling in the empty spaces with drawings. Note the unusual placement of the New World on the right of the map.*

or oceans, the proportion of Commonwealth to total national references indicates the same strong bias toward other Commonwealth countries. An analogous effect for students from the United States was an increased awareness of several world areas where heavy American military commitments have been made.

Other factors may have important effects on student sketch maps of the world. The role of the particular projection to which a subject is most often exposed has been cited. Greenland, for example, is exaggerated in size on the Mercator projection. North American students, familiar with classroom maps centering on the Western Hemisphere, often center their sketch maps accordingly, splitting Asia in two (see fig. 9.8). The maps of Finnish students suggest a nostalgia for the balmy Mediterranean while Canadian students tended to name West Indian Islands more often than any others, and among the American states they cited most frequently were Hawaii (29), Florida (19), and California (17). Do wistful residents of northern climates tend to look more frequently at map areas featuring balmy climes?

Conclusions and Suggestions for Further Research

There are a number of features common to the world sketch maps of all groups' samples and some unique qualities dependent on the place of origin of the maps. Some of the similarities and differences have been discussed in terms of such explanatory factors as proximity, shape, size, current events, and cultural factors. It may be that common features are overemphasized here because the largest portion of the sample were from essentially similar groups within the wealthy, urban, industrial, Western World. It would be interesting to test this by careful selection of samples along such dimensions as developed-undeveloped or communist-capitalist areas (Robinson and Hefner, 1968). Would people from communist countries have markedly different mental maps of the world? If so, in what ways would they differ from those discussed here? Ginsburg (1968) outlines the unique qualities of the Chinese perception of a world order which owes its origin to the special circumstances of China's geography and history. How many other distinctive views of the world might be expected? An analysis of voting patterns in the United Nations by Russett (1966) revealed six different coalition groups which he has labelled the Western Community, Brazzaville Africans, Afro-Asians, Communist Bloc, Conservative Arabs, and Iberians. This is a much more complex and subtle view of world political groupings than is generally found. It seems likely that each such group would have a different mental image of the world. Boulding (1959) suggests that to avert dangers in our unstable international system more sophisticated world images are required. To what degree do people in various parts of the world share a common image of the Earth? In what ways do they differ? Research employing the simple technique of analysis of sketch maps from various areas might contribute toward answering these questions. Such research could help clarify deficiencies in current geographic education and thus suggest possibilities for development of more sophisticated views of the world.

10

Designative Perceptions of Macro-Spaces: Concepts, a Methodology, and Applications

KEVIN R. COX AND GEORGIA ZANNARAS

Introduction

A brief perusal of some recent psychological literature suggests that psychologists are increasingly interested in the influence which perceptions of the surrounding environment have upon man's behavior. Geographers have recently evinced a similar interest but largely because such perceptions of space are ultimately reflected in human behavior which has locational implications. This paper is written with the latter viewpoint in mind, but it draws on a concept developed largely in psychology—the concept of the *spatial schema* (Lee, 1969; Stea, 1969; chap. 5). This concept has also gone under the names of *mental map* and *cognitive map*.

Specifically we have three purposes in writing this paper. First, to specify in more concrete terms the notion of spatial schema; second, to present a methodology capable of capturing at least some of the attributes of the schemata which people have; and third, to apply this methodology to two specific content areas: the spatial schemata which people develop from the states of the United States and the spatial schemata which they use when thinking of the larger cities of North America. These applications are presented less for their own substantive interest than for exemplification purposes and the identification of some critical research questions. Each of these tasks is taken in turn.

The Concept of the Spatial Schema

As a result of our sensory and verbal contact with the surrounding environment, we receive a great variety of data which form the basis of our perceptions of that environment. Such perceptions refer to particular places and in employing the concept of *place* it is important to recognize that we are referring to locational entities at a variety of geographic scales from the individual lot or neighborhood to the nation as a whole.

Some of these perceptions concern nonlocational aspects of places such as their business concerns or their racial composition: others concern their relative location to other places. In order to use that information in navigating, searching, and generally manipulating our environment, we clearly need some way of ordering it, eliminating its redundancies, occasionally extracting (via a more-or-less reliable inference procedure) more knowledge than at first seemed possible. The problem of ordering is considerably aided by learning the similarities and differences between places. Such similarities form the basis of a *classification* of locations in both locational and nonlocational terms: we form a set of *locational pigeon holes* in which we can place incoming information. A simple example is provided by the inner-city–suburban dichotomy which most of us use in, say, searching for a residential site in a new city.

Terence Lee (1969, p. 12) has defined the concept in similar terms as:

> an inner representation of space that is originally built from the disposition of desired objects, but which later becomes an abstraction, a series of meta-concepts which organize the objects, facilitate navigation and allow us to coordinate our behavior with that of others. They have been described as spatial schemata. They are obviously constructed from past sensations, myriads of "wherenesses" which have become organized into a structure in the form of an image, that is, something that is available for examination when required. New perceptions from the external environment are given meaning by their connection to it.

All movements in space call for some type of schema. We have schemata for our home, our neighborhood, the town in which we live, and the nation of which we are citizens. It is possible to conceive of a range of schemata from those applicable to micro-spaces to the macro-spaces represented by nations.

While the resultant classification or *spatial schema* is based on actual sensory and verbal contact with the environment, it clearly has an eventual existence of its own with a power to affect the information which will be selected for ordering. Information which cannot be fitted into the schema tends to be discarded until, presumably, the utility of the schema appears to be prejudiced by its inaccuracy or lack of comprehensiveness. The organized structure of this spatial schema, therefore, has a tendency for self-perpetuation; earlier perceptions are considerably more important than later perceptions since they provide the basis for a core of meaning in the schema to which later perceptions must be assimilated.

In discussing the characteristics of spatial schemata, it is useful to discuss at the outset certain basic dimensions along which they can be classified. In particular we would like to distinguish two classificatory dimensions in our typology of schemata. First, we can classify in terms of the type of perception which forms the basis for the schema: there are schemata based on purely *designative* perceptions of places—such as location, social composition, climate, topography—which are devoid of all evaluative content. It is schemata such as these with which we will be primarily concerned in this paper. There are also schemata based on *appraisive* perceptions. Such schemata contain a very strong evaluative component and have been already quite extensively tackled in the geographical literature (see chap. 11). Operationally this literature has focused upon mental maps of residential desirability which provide some insight into the values which people put on the places in their different locational pigeon holes. The schemata revealed in this way are presumably based on appraisive perceptions of the environment around us.

The second dimension along which we can arrange schemata is locational specificity. On the one hand, we develop schemata applicable to highly *specific* content areas: we have schemata of the United States, for instance, or the city in which we happen to live and, indeed, of the world as a whole. On the other hand, we also develop a more *general* schemata applicable to a wide variety of specific content areas. Perhaps one of the best examples which we have of this is the general schema which most of us have of the city and which we apply in moving around in a city which is unknown to us. We tend to have a schema which classifies locations within the city into *concentric* zones: higher order retail functions in the downtown areas, the Negro ghetto not too far out from there, a zone of older housing structures inhabited largely by working class whites, proceeding out eventually to the ring of white, middle class, and generally low-density suburbs. The reality of this general schema is illuminated very readily by our use of it when searching for a new residence in a new town. In like manner we have a general schema which classifies urban places: we tend to expect larger cities to be further apart and to have a greater variety of retailing, service and entertainment functions. Again anyone who doubts the validity of this should consider his own behavior when planning overnight stops for an automobile trip across the United States.

All schemata, however, whether general or specific, appraisive or designative, tend to have certain characteristics in common. First, they tend, like all classifications, to have a marked *hierarchical structure*. Thus while we may think of the United States as divisible into a certain number of regions, we also tend to think of each region as divisible into states; and our schemata may also envisage the further subdivision of at least some of those states into smaller regions: Upstate New York versus Downstate New York, for instance. An interesting example of such hierarchical structure concerns the degree to which the fineness of the hierarchization for a specific schema is dependent upon the location of the observer: it seems highly likely that we tend to subdivide areas closer to us in space in much greater detail than areas further away. We tend to project this aspect of schemata on to other people when communicating with them. This is apparent in cocktail party conversation which revolves around the question "Where are you from?" If the anchor location is Columbus, Ohio, responses to this question will range from "Springfield, Ohio" through "Southern Ohio" to "Kentucky," "Pennsylvania," etc., to "England." The Englishman is most unlikely to respond with the answer "The Midlands of England" since it is unlikely to be meaningful in terms of the schemata prevalent at the cocktail party. Still less is he likely to respond with the answer "Leamington Spa." In brief the fineness of the hierarchization of schemata deteriorates with distance from the observer and this is intuitive in our handling of communication with others.

Within this hierarchical structure of the schema, each class can be seen to have certain properties based on perceptions of the constituent places. Given that each class consists of *places*—towns, states, areas, counties—each class has certain *internal locational characteristics* relating the constituent places of that class one to another: it may be, for example, that the member places in a class are all perceived as geographically close to each other. Or on the other hand the class may be seen as locationally discontinuous, as with a schema which groups towns by size. The internal arrangement of places within a class may also produce certain characteristic shapes such as the doughnut image which most of us have of that class of places which we call "the suburbs."

We also tend to think of the classes in our schemata as having certain *external locational characteristics* relative to each other. If the states of the U.S.A. form the classes of our schema, we may perceive Missouri as being *contiguous* with Mississippi. In the "New Yorker's Map of the United States" (Tobler, 1963), there were a variety of locational errors of this type which exemplify the perceptions of relative location which we apply to the classes of our schemata.

Finally, we may tend to perceive the locations in our classes as having certain *nonlocational attributes* in common which serve to tie them together in bonds of similarity. Hominy grits, poverty, and a highly stratified social system may serve to link places in the Southern states into a class or region in our schema. We also use a distinct set of social, economic, and housing-density criteria to differentiate suburbs from inner cities.

Clearly these characteristics of the classes in our schemata need not be accurate spatial images of the real world. We have already referred to the distortions of the relative locations of states inherent in such cognitive maps as the "New Yorker's Map of the United States." Identifying the constituent places in the class, however, allows us to relate the spatial image to the real world. Such deviations of course are of considerable value in understanding behavior which departs from that predicted on certain assumptions about behavior within a real world context as opposed to behavior within a perceived context.

In summary, the schemata which we have contain classes of places organized in some hierarchical manner; the classes in our spatial image have certain internal locational characteristics, external locational characteristics, and some nonlocational attributes. These characteristics of the spatial image derived from certain perceptions of places may be more or less accurate in terms of the real world characteristics of the places constituting the class.

While these schemata are probably unique for individuals, it seems likely that there is considerable similarity among individuals within identifiable groups sharing certain sensory and/or verbal experience of the environment. Groups based on life cycle, social class and ethnic status seem particularly relevant in this respect. Orleans (chap. 7) has drawn attention to the contrasting schemata which blacks and whites have of the Los Angeles area. The schema of the less travelled black is spatially much more constrained and incomplete than that of the white.

It seems highly likely that the criteria employed in developing schemata change with alterations in travel behavior and in vicarious verbal experiences of places. An interesting contrast, for example, concerns the contemporary tendency for people in the more developed societies to view their nation as divisible into regions which have a locational status with reference to the nation as a whole: the Midwest, the West, the South, the Midlands, the Northwest. Such classes contrast markedly with those characteristic of an era of more limited locomotion and spatial experience: New England, Genesee, the Home Counties, Provence. As criteria change, so we would expect schemata to change.

The final comment concerns some possibly speculative relationships between the essentially discontinuous image of space implied by the schemata and perceptions of a more continuous space. It seems likely, for instance, that the fragmentation of spaces into classes strongly defined by proximity may exercise an effect on our perceptions of the distances to various locations: if two places are in the same class of our schema, then they may be perceived as geographically closer together than two places which are in different classes respectively. Thus, if one perceives Ohio and Indiana as being part of the Midwest and Pennsylvania as part of the East, one might underestimate the distance from Columbus, Ohio to Indianapolis relative to one's estimation of the distance from Columbus to Pittsburgh. Likewise the average Britisher may perceive a Welsh county such as Flint to be closer to another Welsh county such as Cardigan, than to any English county which is objectively closer to Flint than Cardigan.

Again, the locational and hierarchical characteristics of schemata may afford some insight into the usual tendency for people to overestimate shorter distances relative to longer distances. Given the hierarchical structure of schemata and a tendency for the hierarchy to be more finely developed in the immediate vicinity of the observer, it seems plausible that the psychological principle of closure would tend to bring more distant places perceptually closer to the observer; this would be a result of the perceptual contraction and resultant distance distortion as apparent in the "New Yorker's Map of the United States."

Given this conceptual overview, the remainder of this paper adopts a more restricted focus: briefly that of a methodology for identifying the schemata which people have of specific macro-spaces. We shall also confine ourselves to those schemata which are based on designative rather than appraisive perceptions since that latter problem has already received considerable methodological attention. We now proceed to a consideration of methodology and then to some applications.

A Methodology for Identifying Specific Schemata for Designative Perceptions

The methodology developed here is derived from the work of a group of political scientists at the University of Michigan who have been carrying out extensive questionnaire surveys into the structure of international attitudes in the public at large (Hefner et al., 1967; Robinson and Hefner, 1967; 1968). One aspect of the investigation has been the identification of perceptions of cross-national similarities:

> Questions about the perceived similarity of countries were asked concerning 17 countries, chosen so as to represent all major regions of the world. Each country in turn served as the anchor country and the respondent was asked to indicate which three of the remaining 16 countries were most similar to the anchor. . . . It was hypothesized that this rather unstructured task would give us meaningful insights into the perceptual and cognitive maps of interrelationships among countries typical of people with various educational and social backgrounds. Further, we felt that multidimensional analysis of this data would indicate meaningful attitude structures, subject to dimensional interpretations. (Hefner et al., 1967, p. 141).

The consequent analysis did indeed reveal the existence of dimensions of perceived variation, in particular a Developed-Underdeveloped dimension and a Communist–non-Communist dimension.

Given this basis for a methodology, assume that a study group is given the names of subareas of some larger area such as a nation—the departments of France, for example. They are asked to take each subarea in turn as an "anchor subarea" and identify the three other subareas which they regard as most similar to the anchor subarea. The resultant choice of the study group can then be placed in a matrix in which Cij gives the number of times that subarea j has been chosen as similar to anchor subarea i. Division of each cell entry by $3M$, where M is the size of the study group, yields Pij, the probability that subarea j has been chosen as similar to anchor subarea one.

Manipulation of the data tableau in some of the ways suggested by Hefner and his associates might plausibly provide evidence of the classes used in the schemata or cognitive maps of which, presumably, the choices are a function. Factor analysis of the *Pij* matrix, for instance, might permit the identification of *dimensions of choice* such that subareas loading high and in the same direction on a particular factor would frequently be chosen together as the most similar subareas to particular anchor subareas. In this way it would be possible to identify sets of classes of subareas which are regarded as similar either by a simple mapping of the loading for each factor or by a multivariate grouping applied to the loadings for a set of factors.

Clearly the methodology has limitations from the viewpoint of providing a comprehensive view of the properties of schemata characteristic of groups of respondents. Most importantly, it assists in establishing only the classes which are employed in schemata. The results of such an analysis would tell us nothing conclusive about the locational characteristics of schemata or about the attributes employed in deriving these schemata; we could only infer the criteria used by examining the real world characteristics of the places constituting the classes.

Nevertheless we believe that the identification of such classes is important in the overall process of identifying the properties of schemata. They afford a reasonably objective means of defining the classes which people use and they provide a control in further investigation over variance in the classes characteristic across a set of individuals. The significance of this can be readily grasped by considering the problems of asking a group to draw cartographically their spatial images of the United States; almost certainly there would be some differences in the sets of classes identified. The investigator could, of course, specify beforehand the classes to be used, but without the procedure which we have outlined above, such classes might bear little relationship to those which constitute the spatial image of the average respondent.

What we envisage in investigations of spatial schemata, therefore, is a two-stage research process. The initial stage would establish classes in the manner which we have outlined above. The second stage would ask for cartographic depictions and/or verbal descriptions of these classes from the respondents involved. It is with the first stage that the remainder of this paper is concerned.

Application of the Methodology

Partly by way of example and partly to point out problems and possibilities in the application of this research methodology, we have applied it to two specific content areas for which people develop schemata: 1) the states of the United States; 2) the cities of North America. The limited question which we have asked is, What are the classes which constitute the schemata which students have of these specific geographical aggregations? Perceptions of similarity allow us to establish such classes.

THE STATES OF THE U.S.A.

The methodology outlined above has been applied to the perceptions of state similarity among the states of the U.S.A. presented by a group of 40 undergraduates at The Ohio State University with little or no geographical background. A list of states was presented to each student who was instructed to: "Take each state in turn and select the three states most similar to it. Similarities should be evaluated in as many ways as you think states differ." Students were given twenty minutes to perform this task.

The resultant choices were then taken off the schedules and inserted in a *Cij* matrix. The cell entries were converted into proportions of row totals. The high proportion of zero entries in this matrix suggested that for most anchor states there was considerable group consensus as to the most similar states.

In order to identify the classes of the cognitive maps held by the students, the *Pij* matrix was factor analyzed. Prior to this, however, the diagonal cells of the matrix were inserted with values equivalent to .33: technically this was done in order to maximize the correlations between the variables representing choices for different states; conceptually the approach can be defended on the grounds that the state *most* similar to a state is the state itself.

The variables that are factor analyzed, therefore, are state similarity choices with *one variable for each of the forty-eight states*. The *observations* are the *forty-eight anchor states*. Loadings are on state choices while scores are computable for anchor states. The loadings are presented in table 10.1. In interpreting the factor loadings, states which have similar loadings on the same factor can be regarded as being perceived as similar and belonging to the same set; states which have very different loadings on the same factor can be interpreted as being perceived as correspondingly dissimilar and in different sets of the schemata involved. In this treatment only loadings in excess of $\pm$ .4 have been employed in identifying classes of states: these classes have all been mapped on to figure 10.1 in order to conserve space.

The first factor appears to be bipolar in form, differentiating areas on an East-West basis. In one sense the Mississippi River proves an almost perfect boundary line with almost entirely negative loadings to the west and positive loadings to the east. Closer inspection of the loadings, however, reveals that the map has two epicenters: a cluster of relatively high positive loadings in what is commonly regarded as the East, stretching from Maryland in the south to Vermont and New Hampshire in the north; and a cluster of relatively high negative loadings in the Mountain States and

TABLE 10.1. *Factor loadings on the states of the U.S.*

	Factors					
Location	*I*	*II*	*III*	*IV*	*V*	*VI*
Alabama		−.66			−.40	
Arizona	−.45					+.40
Akansas		−.48		−.40		
California						
Colorado	−.61					
Connecticut	+.55	+.52				
Delaware	+.50					
Florida					−.42	
Georgia		−.70				
Idaho	−.53					−.53
Illinois			−.82			
Indiana			−.70			
Iowa	−.40			−.53		
Kansas	−.47			−.65		
Kentucky		−.48			+.54	
Louisiana		−.60			−.44	
Maine		+.47				
Maryland	+.48					
Massachusetts	+.50					
Michigan			−.81			
Minnesota			−.57			
Mississippi		−.63				
Missouri				−.51		
Montana	−.59					−.49
Nebraska	−.47			−.60		
Nevada	−.60			+.40		
New Hampshire	+.47	+.52				
New Jersey	+.48					
New Mexico	−.42					+.49
New York	+.44					
North Carolina		−.52				
North Dakota						
Ohio			−.74	+.41		
Oklahoma				−.41		+.43
Oregon						−.64
Pennsylvania						
Rhode Island	+.54	+.50				
South Carolina		−.51				
South Dakota	−.42					
Tennessee		−.57			+.49	
Texas						+.44
Utah	−.60					
Vermont	+.44	+.50				
Virginia		−.40		+.40	+.60	
Washington						−.66
West Virginia				+.41	+.54	
Wisconsin			−.61			
Wyoming	−.61					

For purposes of clarity, only loadings greater than +0.4 or less than −0.4 are shown.

adjacent Plains States. In the light of the roles which the so-called "East" and "West" play in popular symbology, the importance of this basic dimension in the conceptual spaces of the students is perhaps not surprising. In figure 10.1 the locational identities of these classes are referred to as the *Northeast* and the *Mountain* or *West* respectively.

The second factor describes a second dimension of the cognitive maps superimposed upon the East-West factor and differentiates between two of the classic regions of the U.S.—the *South* and *New England* (see fig. 10.1). The regional images, moreover, reveal considerable subtleties of judgment which accord well with more objective criteria of regional definition. There seems, for example, to be a clear perception of a Southern Borderland with such states as Kentucky, Virginia, Arkansas and North Carolina receiving

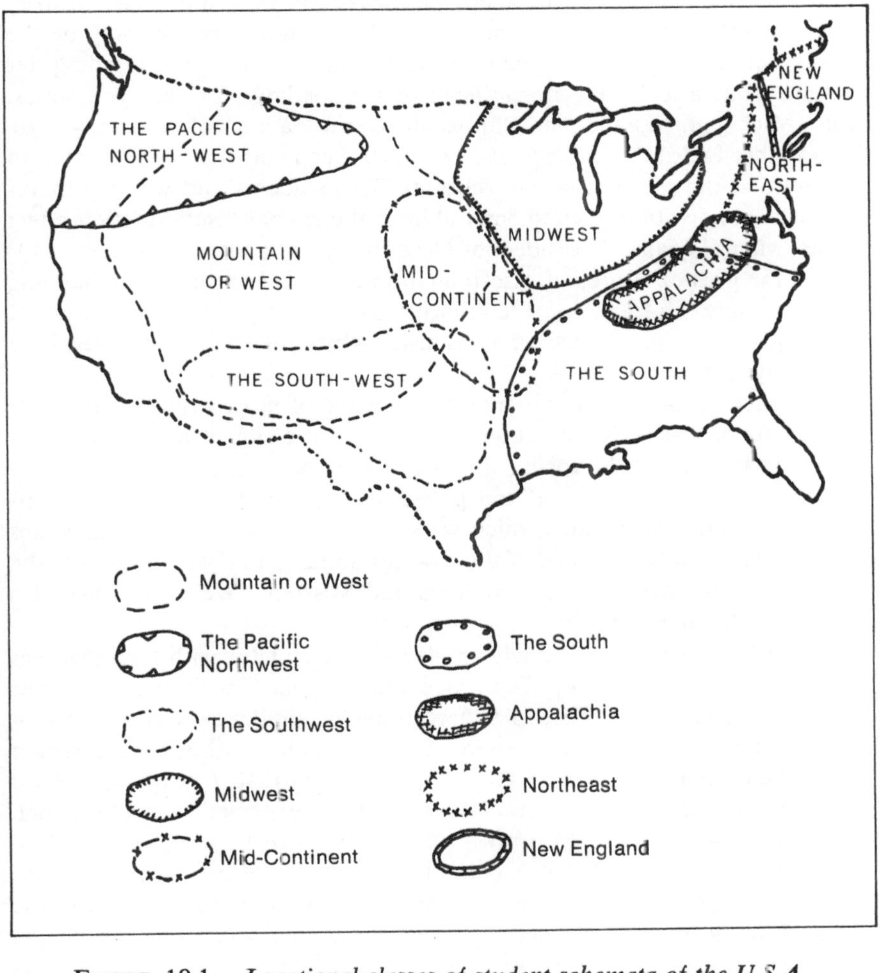

FIGURE 10.1. *Locational classes of student schemata of the U.S.A.*

rather more modest loadings than the Deep South of Georgia, Alabama, Louisiana and Mississippi. Interestingly enough the atypicality of Florida, Texas, and Oklahoma was also recognized, but the firmly Deep Southern character of South Carolina appears to have been less clear to the students.

The third factor, in contrast with the first two factors, is not bipolar. Loadings on a few contiguous states in the Great Lakes area are very high and negative, but are not counterbalanced by high, positive loadings elsewhere in the U.S. What is being depicted by this factor is a perceived area of homogeneity which we might call the *Midwest.* As with the South, the core area consists of states with loadings in excess of —.5 (Wisconsin, Minnesota, Illinois, Indiana, Michigan, and Ohio). This core area is surrounded by another tier of states all with negative, but low, loadings; the residue of the nation is characterized by relatively low loadings.

Of particular interest in this factor, however, is the high degree of clarity with which the regional core is perceived. This is indicated not only by the large differences in loadings between the core area and the surrounding tier of states, but also by the magnitudes of the loadings for the core states. Ohio, Michigan, Indiana, and Illinois all record loadings of at least —. 70. On no other factor do loadings exceed —.70. Such clarity may be related to the fact that the perceptions are those of Ohio students and we may therefore be dealing with a phenomenon akin to the spatial discrimination effect in the "Mental Maps of Residential Desirability" literature (see chap. 11). Just as the greater degree of discrimination accorded to states surrounding the home state is a function of familiarity, so the high loadings on factor III may suggest a high degree of consensus which likewise originates in a familiarity factor.

The three remaining factors may be disposed of more briefly. Factor IV appears to be extracting a rather more surprising dimension of the conceptual spaces. It would appear that a considerable degree of homogeneity was accorded by the students to a polyglot group of states in the Midcontinent, including some states which we would regard as Great Plains states —particularly Nebraska and Kansas—and some states which are on the periphery of the Midwest such as Iowa and Missouri. We have called this class the *"Midcontinent."*

Factor V appears to be largely an *Appalachian* factor with high loadings for West Virginia, Kentucky, Tennessee and Virginia. This is also of interest in that it suggests that there is a perception of similarity accorded to an area of the nation not usually regarded in geography textbooks as a region in the classic sense. Factor VI is also surprising in that it draws a contrast between two regions which one would not have expected to present such clear images—the *Pacific Northwest* and the *Southwest.*

These notes suggest that it is possible to conceive of a system of classes or regions corresponding to the schemata of the students by placing the areas of homogeneity revealed by the factors into some sort of simple hierarchical classification: the West, which can be broken down into the Pacific Northwest, the Mountains, the Southwest and the Midcontinent; and the East, comprising the South, the Northeast, Appalachia, the Midwest and New England.

THE CITIES OF NORTH AMERICA

The procedure used in examining state similarity choices was also applied to a sample of sixty cities of varying size located in North America; the selected cities are shown in table 10.2. In this instance, the variables factor analyzed are city similarity choices with one variable for each of the sixty cities, and the observations are the sixty anchor cities. The factor loadings on the city choices are presented in table 10.2.

TABLE 10.2. *Factor loadings on North American cities*

	Factors					*Factors*			
Location	*I*	*II*	*III*	*IV*	*Location*	*I*	*II*	*III*	*IV*
Albany					Macon	+.51		−.44	
Albuquerque		+.55			Manchester				
Atlanta			−.44		Memphis	+.56		−.58	
Baltimore					Minneapolis				
Binghamton					Milwaukee				
Boston	−.41				Mobile	+.47		−.56	
Buffalo		−.47			Montreal	−.41			−.65
Cedar Rapids					New York	−.58			
Charlotte					Omaha			+.43	
Chicago	−.50	−.45			Ottawa				−.73
Cleveland		−.57			Philadelphia	−.43			
Dallas		+.50		+.40	Pittsburgh		−.55		
Denver		+.59			Providence				
Detroit		−.56			Salt Lake City		+.43		
Des Moines			+.48		San Diego				
Duluth			+.41		San Francisco	−.53			
El Paso		+.59			Seattle				
Erie					Shreveport				
Evansville			+.46		South Bend	+.41		+.48	
Fort Wayne	+.43		+.50		Spokane				
Grand Rapids					St. Louis				
Harrisburg					Tampa				
Houston		+.55			Terre Haute			+.44	
Huntington					Toledo		−.41		
Indianapolis					Toronto	−.43			−.78
Kansas City					Tucson		+.57		
Lexington	+.50				Vancouver				−.72
Little Rock	+.45		−.41		Wheeling	+.47			
Louisville	+.50		−.40		Wichita				
Los Angeles	−.54				Winnipeg				−.78

For purposes of clarity, only loadings greater than +0.4 or less than −0.4 are shown.

The factors on the city data were more difficult to interpret than those obtained from the state choices. Each factor represented a greater complexity than is usually encountered, and this complexity raised doubts regarding any simple classification of city choices. Students seemed to use a combination of attributes in selecting similar cities, and, consequently, only the first four factors could be interpreted with any degree of confidence. As with the American states material reported above, only the loadings greater than +.4 or less than —.4 were used.

The first factor can be described as a Cosmopolitanism-Provincialism dimension of differentiation. As can be seen from table 10.2, the factor is bipolar in form. It summarizes a complex cluster of characteristics. Geographically the negative loadings outline a peripheral distribution of the largest cities in the sample (see fig. 10.2); an examination of these cities

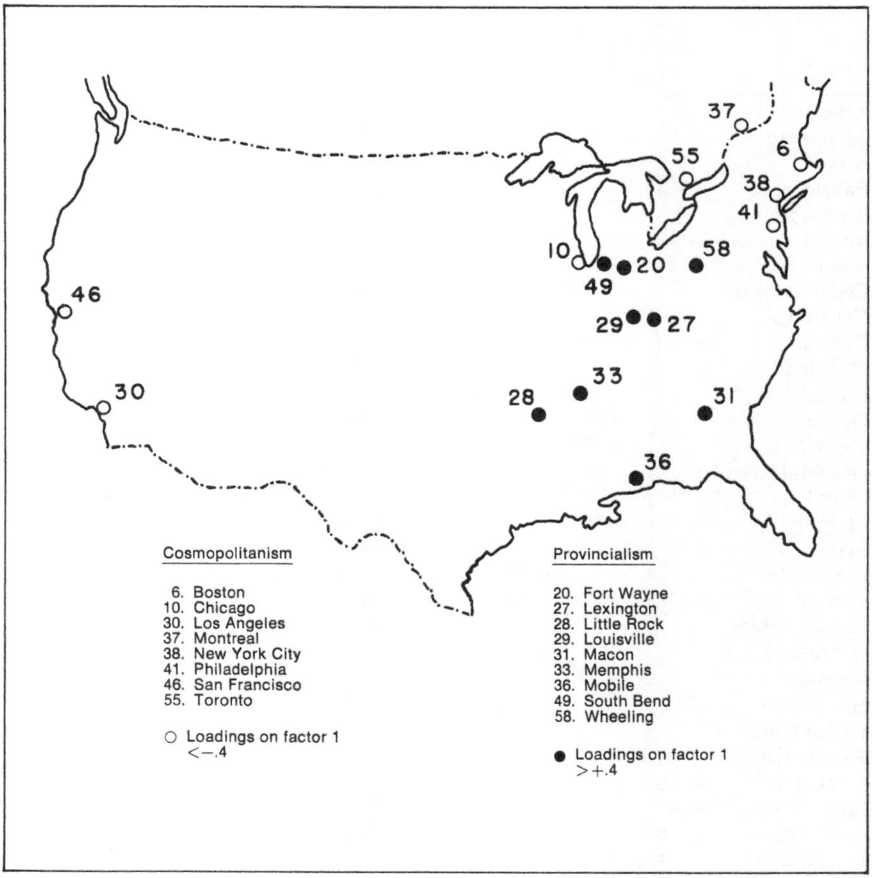

FIGURE 10.2. *Cosmopolitanism-provincialism: the first set of locational classes emerging from the factor analysis of city similarities.*

suggests that they are the more cosmopolitan of North American urban centers and that they perform service functions for wide surrounding areas. The positive loadings relate cities which are small interior regional centers. These cities (Fort Wayne, Lexington, Little Rock, Macon, Memphis, Mobile, South Bend, Wheeling) have a considerably more provincial image than the cities picked up by the cosmopolitan end of the dimension. The cosmopolitanism of the larger cities is illuminated and emphasized by the way in which the international boundary is ignored. Clearly, however, the cities at the opposite poles of the dimension can be differentiated in more ways than their relative degrees of cosmopolitanism and provincialism.

The second factor is also bipolar in form and can be interpreted as a factor which differentiates between two urban systems: the Great Lakes Urban System and the Southwest Urban System (see fig. 10.3). This factor is also complex. The first distinguishing characteristic is locational, the positive loadings clustering in the Southwest and the negative loadings

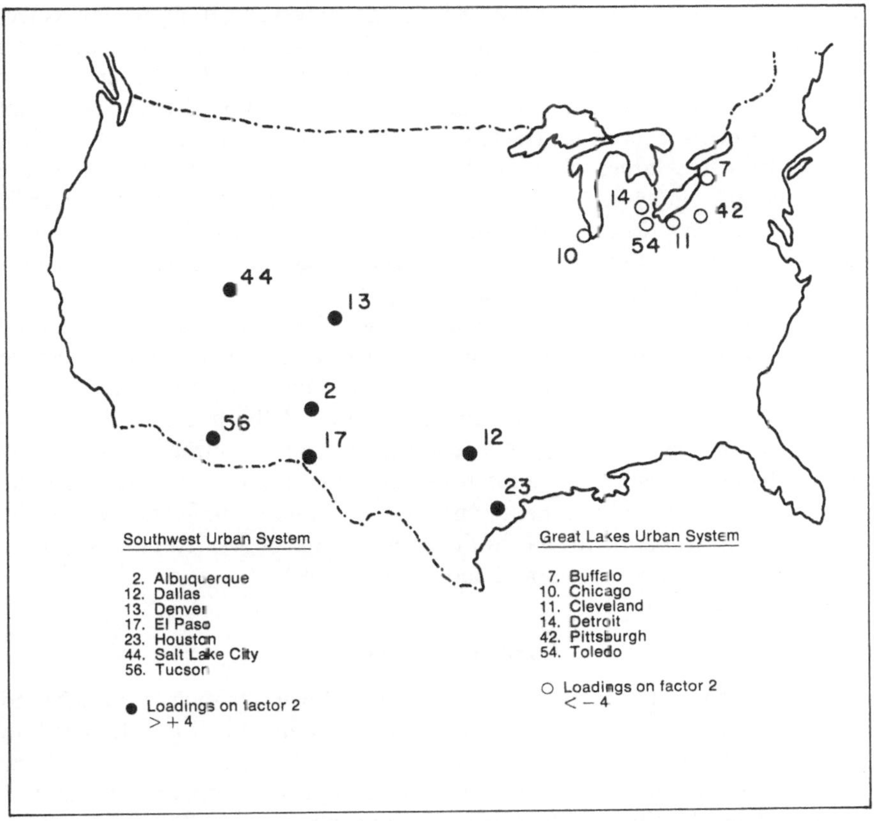

FIGURE 10.3. *The Great Lakes Urban System-Southwest System: the second set of locational classes emerging from the factor analysis of city similarities.*

clustering in the Great Lakes area. Distinctive sets of communication linkages underlie this proximity criterion, the Great Lakes cities in particular being tied together commercially by a network of water transportation. Studies of airline transportation have underlined the degree to which southwestern cities are integrated into a network focusing on Dallas-Fort Worth (Taaffe, 1962).

A further interesting feature of this factor is the possibility that it contrasts the environmental quality of the two areas. The cities in the Southwest are more open in physical structure, less polluted, and less industrialized than those around the Great Lakes. The factor seems to differentiate in fact between a set of cities characterized by nineteenth-century forms of industrialism and a set of cities which are more contemporary in their industrial composition.

The third factor is again bipolar, but it is perhaps the simplest of the factors. It is certainly a factor of location: the negative loadings summarize cities in the traditional South, while the positive loadings relate cities in the Midwest. In examining the positive loadings, however, an interesting aspect emerges. Only the smaller, more provincial Midwestern cities appear to have larger loadings on this factor: the big cities of the region such as Minneapolis, Kansas City, Milwaukee, Chicago, and St. Louis are absent and seem to have been abstracted from the surrounding region. It is as if people perceive larger cities as having an existence independent of the surrounding region, while smaller cities are closely associated with their regional context. Consistent with the idea is the derivation of the more cosmopolitan cities from a variety of regional contexts, spreading in particular over the international boundary.

The fourth and final factor which could be interpreted is a Canadian factor. It is more unipolar than the other factors. There is only one positive loading greater than .4 (Dallas), although a number of other cities in the Southwest have loadings in this direction. Four of the five Canadian cities in the study have high negative loadings, though Montreal is noticeably absent from the list. The factor testifies to the strength of political units as the bases of spatial schemata, and this recalls the significance which the English-Scottish boundary has in the mental maps of residential desirability of British school-leavers (Gould and White, 1968).

Concluding Comments

In this paper we have progressively narrowed down our focus from a broad conceptual treatment of spatial schemata to a methodology designed to capture a limited but important aspect of those schemata: the classifications of locations upon which they are based. Specific applications of this methodology do allow us to extract certain dimensions of variation along which students appear to differentiate places. While it might be difficult to define the exact boundaries of locational classes, we are at least given a good idea of their conceptual relationships: we know, for example, that Iowa is much more likely to be grouped with Missouri than with Ohio.

The methodology raises certain questions, and two in particular concern us here: first, To what extent do such measures constitute a valid measure of the dimensions of differentiation employed by students? Second, What are some of the important avenues of research which need to be developed in the future?

With respect to the first question, it is clearly important that there be a close relationship between the geographical units which form the building blocks of spatial schemata and the geographical units employed in the methodology. In the case of the students we assumed that the states of the U.S.A. provided meaningful units. Counties would certainly not have provided such units, but the use of states in mass communication and informal interaction in the U.S.A. suggests the individual observations upon which the classes of schemata are based. Clearly this is an important consideration in any research probing the nature of schemata.

With respect to the problem of validity, there is the question of *forcing* classification where it doesn't exist in the schemata. It is possible, for instance, that such forcing was present in the methodology as we applied it. We did specifically call for three similar units when, in fact, three may not have existed in the individual schema, or there may have been more than three. We also employed a sample in the case of the North American cities such that the student might have faced a situation in which the cities he thought most similar to a specific anchor city did not appear on the list. In retrospect, it is probably advisable to employ universes rather than samples, and to adopt a looser arrangement as to the number of places selected as similar.

Major research questions for the future include the problem of the criteria which people use in classifying places. The two applications presented in this paper suggest that relative location is very important, but why should this be so? Does location itself provide a cue to the individual, or is it a perception of spatial autocorrelation in the attributes used as criteria? The Great Lakes cities could have been grouped together in one class because they are located near to each other; alternatively, the raison d'être of the class might reside in similarities of industrial structure and urban environment among those cities. Alternatively, it might be a mixture of the two mechanisms that determines the group, the perception of locational proximity providing a cue for assumptions of spatial autocorrelation.

Whatever the criteria employed, we clearly have a set of puzzling problems related not only to the classifications which do emerge, but also to those that do not. We fully expected urban size to be a major criterion in the North American cities case, for instance, but apparently it was not. In investigating this problem, we are in a sense professionally handicapped:

as geographers, we are likely to project on to the students' perceptions geographical differences which we think are important, but which from the layman's viewpoint are insignificant, esoteric, or both. What is probably required here is an informal interview tool, possibly using the semantic differential technique to identify the major dimensions underlying the variations in viewpoint. This is at least one future and possibly fruitful line of investigation into the characteristics of spatial schemata.

III

Spatial Preference

Introduction

This section is an anomaly within this book, since it contains only one contribution. But Gould's "On Mental Maps" has generated as much thought, research, and controversy as any other paper in modern geography (see the author's "Postscriptum 1970"), and is a classic in its own right. Most of the work that followed was cast in the same conceptual and methodological mold, and "On Mental Maps," long unavailable to the reading public, is still the exemplar of this approach to spatial preference.

Why "preference"? Gould himself uses the term in his very brief description of methodology. Clearly, because the subjects rank ordered entities such as the states of the U.S.A., countries of Europe, or administrative districts in Africa, they are also *discriminating* among places on the basis of residential desirability. The U.S. students, asked to imagine themselves married and settling down with a family given a complete freedom of location, may have been vicariously *choosing* a future home. But it is precisely these terms—preferring, choosing, discriminating—which have been attacked by some psychologists (e.g. Skinner, 1950) as referring not to behavior but to processes in a different dimension (i.e. to unnecessary "mentalistic" intervening constructs.) The attack was a formidable one, but the terms remain in common use not only among "mentalistic" social scientists of the type represented in this book, but also among those with a behaviorist bias (e.g. Restle, 1961), and even in papers on operant psychology (e.g. Stea and Bower, 1963).

All of the terms refer to *learned* processes—one learns to discriminate, to prefer, and to choose. How such learning occurs in a geographic context is an unanswered question. Gould calls for research on "the role of the information that people receive," and, in his paper on Swedish school children (see part IV), presents an analysis of developmental changes in the amount of information about a particular country. Sonnenfeld (1966, 1967) linked preferences for environmental characteristics to early, prolonged experience with a "native" environment, while Golledge (1967) has linked the habitual use of spatial objects such as shopping centers with spatial learning processes. Obviously, these represent tentative attempts to unravel a complex series of learned processes which are generally held to be determinants of our spatial behavior. However, the predictive value of knowledge about preferences is uncertain: the belief that preferences are a major factor in determining spatial behavior is not proven.

It is hard to resist the urge to contrast Gould's approach with that of Lynch (1960), another early "mental mapper" who also generated a major school of research (see the introduction to part II). Superficially it appears that Gould and Lynch adopted opposite approaches: in Lynch's work, the graphic representations were originally produced by the subjects, and in Gould's by the investigator. But in another and more significant sense, the composites compiled by Lynch from subject-produced cognitive maps (which indicate the frequency of mention of features in each element-class) and the preference surfaces statistically derived by Gould from subjects' rankings of places are both investigator-produced. The term "cognitive map" has too often been identified with a *graphic* output from the cognizing individual. Thus, the sketches produced by Lynch's subjects seem to be more "cognitive-map-like" than the purely verbal rankings of Gould's subjects, but they need be no more veridical as cognitive *representations*. The two response modes have differing forms but identical functions—that of representing information about the nature of the spatial environment.

On the side of the "dependent variable"—the investigator's composite product—both studies are concerned with spatial preference, although to differing degrees. The *preferences* of Gould's subjects are explicit, although we can only make inferences as to why a particular place is preferred. Lynch's subjects *choose* (albeit unconsciously) to remember a place and to emit its name or to draw it upon request: *why* it is remembered is an unanswered question, although suggestions occur in the papers by Hershberger (1970), Steinitz (1968), and Appleyard (1969b). Lynch attributes the selective memory of urban places to visual potency, but later investigators (Stea and Wood, forthcoming) have stressed that preference in terms of historical evaluation or utility is also significant.

Gould's contribution is both conceptual and methodological and, more significantly, inspirational. It has been an invaluable heuristic, in line with Gould's original intentions:

> At the start of this limited inquiry into the mental images that men have of geographic space I offer no theories or even explicit hypotheses. On the contrary, you will find only unstructured, intuitive hunches. . . . These speculations . . . will have served their purpose, and more, if they stimulate others to replace them with better notions. As usual, what we need are more penetrating questions.

This statement implies a more "limited" inquiry than the reader receives, and yet it sums up the current state of the art. We find that we are faced with a series of unanswered and frequently unasked questions: What is the role of information in preference formation? What is the predictive value of knowledge of spatial preferences? Are there cross-cultural and within-cultural differences? Gould's paper still offers our best introduction to these questions.

11

On Mental Maps

PETER R. GOULD

Can geography be mixed up with psychology?
LUIGI BARZINI, *The Italians*

At the start of this limited enquiry into the mental images that men have of geographic space I offer you no theories or even explicit hypotheses. On the contrary, you will find only unstructured, intuitive hunches, and interpretations pushed, in many cases, beyond the limits that the data allow. Of these speculations, many will undoubtedly be wrong. They will have served their purpose and more if they stimulate others to replace them with better notions. As usual, what we really need are more penetrating questions, but before these can be asked we must record what we do and do not know. The boundary of ignorance is not very far away, but it seems only sensible to stake it out before we try to cross it. We know so very little about the spatial images, the mental maps, that are in the minds of men. We know even less about how they are formed, the degree to which they are unique or general, and the way they impinge upon, and are reflected in, the decisions that men make. As human geographers reach out across traditional disciplinary boundaries to the other social and behavioral sciences, it is increasingly apparent that the truly satisfying explanation they seek is going to come from emphasizing the *human* as much as the *geography*. We may, perhaps, define our subject as essentially efforts to understand the spatial aspects of Man's behavior. If we grant that spatial behavior is our concern, then the mental images that men hold of the space around them may provide a key to some of the structures, patterns, and processes of Man's work on the face of the earth. The emphasis upon the conditional tense is

For help in distributing questionnaires, computing or making materials available to me, I would like to thank: Messrs. Benneh (Accra), Bigelow (Penn State), Boateng (Accra), Fuller (Hawaii), Garner (Arhus), Haggett (Bristol), Julliard (Strasbourg), Knighton (METRA, Paris), Mabogunje (Ibadan), Manshard (Giessen), Marich (Penn State), Ollson (Michigan), Pecora (Rome), Porter (Minnesota), Pred (California), White (Northwestern), and Wilson (Alabama). *Michigan Inter-University Community of Mathematical Geographers,* Discussion Paper #9, 1966.

quite deliberate, and is only partly a result of intellectual cowardice and a general propensity to broadcast *caveats* in lieu of signing academic insurance policies. The other reason is that we really do not know whether mental maps are relevant to our problems. But the suspicion that they are is strong, and at the very least it seems worthwhile making some tentative probes along these lines.

Images and Decisions

The human landscape, in reality or abstracted and modeled as a map, is nothing more and nothing less than the spatial expression of the decisions of men. Even when we examine apparently simple geographic patterns and processes that are a reflection of these decisions, we quickly become aware of the extreme complexity that underlies them, the myriad of variables that compete for attention, and the way in which these form interlocking and convoluted structures that are numbing in their difficulty. Many of the decisions that men make seem to be related, at least in part, to the way in which they perceive the space around them and to the differential evaluations they place upon various portions of it. For the moment this is a virtually unsupported assertion, but it seems plausible that the manner in which men view their spatial matrix impinges upon and affects their judgments to some degree. For example, men *decide* to migrate, not on a regular surface of equal opportunity or desirability, but in a world often perceived in an extreme, differential manner (Wolpert, 1965). Men *decide* to grow crops and raise animals for their sustenance, not in an arbitrary way, but in part according to their particular views of the space around them. Men *decide* where to locate their industries and business activities, and with more and more "footloose" industries coming onto the scene we are finding that traditional location factors are declining in importance. Törnqvist, for example, has indicated the virtual irrelevance of transport costs for some industries in Sweden (1962, 1963), while in the United States, Harris's classic paper outlined the vast area in the American Manufacturing Belt that lies around the point of minimum transport cost with only slightly higher access costs to the market (Harris, 1954). Even traditional Weberian analysis discloses a basic characteristic of many spatial extremum problems, namely a large area of only slightly higher aggregate cost around the minimum point. What Rufus Isaacs has termed the "principle of flat laxity" seems to be operating extensively in geographic space (Isaacs, 1965), and it takes on new meaning as we become critically aware of what *satisficing* behavior in a spatial context implies. Thus, in view of the decline in importance of the more traditional location factors, might not the decision to locate be increasingly related to the image an area has in the minds of a few key people (Wilson, 1966)? More and more the quaternary industries, the research and development companies, look to the scenic and recreational facilities, cultural assets, and intellectual resources of an area.

Snow and mountains are not essential for the economic health of certain well-known electronic companies, but New Hampshire and Colorado are undoubtedly grateful for their physiographic and climatological inheritance. Similarly, large universities in pleasant surroundings are the locational loadstones for many research and development consultants. It is not difficult to think of numerous examples where the maps that are carried in men's heads might be relevant in quite crucial ways.

What Do We Know About Mental Maps?

Man's view of geographic space is extremely varied, and the views of individual men are always in part unique. Entering into the particular outlook of a particular man are a host of experiences, prejudices, and desires, some widely shared, others quite specific to the individual. In the United States, the Northerner is reluctant to be assigned by his company to the South, for he holds to a mental picture that is part of his northern cultural inheritance —an inheritance absorbed in childhood, and reinforced by his daily sources of information. The townsman, comfortable and safe amidst the roar of traffic and bustle of urban life, is reluctant to live in the green peace of the country, which he associates with the stillness of bucolic decay. The New England family, suddenly presented with greater economic opportunity amongst the tall trees of Oregon, decides to stay with the known view and the familiar friends, for "Oregon is such a long way from civilization!" Thus, the political, social, cultural and economic values held by a man blend into an overall image about the space around him, an image whose components may be particular to him or held in common by many.

It hardly seems necessary to add that we know very little about these spatial images in the minds of men, for the literature is extremely sparse. We have only a few examples where they are discussed at all, usually as interesting, but definitely peripheral, points in larger investigations. Tobler explicitly raised the question of the mental images that people have of their environment (Tobler, 1961), but his basic concern was the mental transformations of distance that people make. Lowenthal and Prince have discussed the attitudes of a people towards the visual landscape (1965), while the former, in a synthesis that has yet to be equaled, has examined the relevance of the psychological literature in this area (Lowenthal, 1961). A number of other geographers (Burton and Kates, 1964) have focused upon the perception of environmental hazard and the spatial implications that such images have for locational decisions. In political geography, only Herman seems to have moved beyond fuzzy speculation to true investigation of the changing values and attitudes of a people towards the national space (Herman, 1959). Occasionally maps such as the "New Yorker's View of the United States" appear, but their humorous context obscures the fact that such cartograms of mental images can be extremely illuminating if properly used. Getis has shown how shape distortions can focus the eye upon a particular portion of the map (Bunge, 1962), and Mackay constructed a map in the late fifties (unfortunately never published) showing Canada through French-Canadian eyes. In a related field, Lynch has systematically investigated the differential images of the urban landscape. In his most imaginative series of maps we have our only notions of these mental pictures (Lynch, 1960).

The literature in other fields is equally sparse. The psychologists, in their concern with "perception," have barely touched upon the investigation of mental images of geographic space, for many of their efforts have concentrated upon the physics and physiology of the senses, often within highly controlled laboratory conditions. For example, the space with which Sandström was concerned was far removed in scale from geographical space (Sandström, 1951), and his insights on disorientation and loss of criteria for making locational judgments under formal experimental situations are hard to carry over in any meaningful way to regional scales. Even Hull's work, though a source of extremely stimulating analogy, does not deal with the larger space of the earth's surface (Hull, 1952). Nor did the swing of some psychologists to the Gestalt outlook produce a shift in the *focus* of their concern, and the discussions of space by this particular school do not really deal with the larger area of the earth's surface which is the geographer's realm. Though much of the pioneering work of the child psychologist Piaget is directly concerned with the way in which children learn about space, the world around them, and geometrical and topological concepts (Piaget, 1963c; Piaget and Inhelder, 1967; Piaget, Inhelder, and Szeminska, 1960), it does not deal with the essentially *geographic* images that children hold or the way they learn about them. Only Trowbridge's paper on imaginary maps, written half a century ago, specifically raises the question of the spatial images people carry around in their heads (Trowbridge, 1913). Unfortunately, this line of investigation was never followed up, and his paper represents the solitary gold nugget at the bottom of the psychological pan. The rest is a residue of vaguely structured insight that hardly rewards the effort of panning it out from the ground material.

In other areas the prospect is equally bleak. The mythical space of Cassirer, though treated as a mental construct, seems less than useful to the geographer (Cassirer, 1955), and the world outlooks that have been discussed by other philosophers may be splendid flowers of many hues, but they are difficult to transplant into the hard earth with which the geographer deals.

The Question of Uniqueness and Generality of Viewpoint

Because the *total* experiences of individual men are unique, it might seem at first glance that they perceive the world around them in quite distinct, totally individualistic ways. But if this were really so, it would be impossible to say anything of general, and, therefore, scientific worth about their spatial perception. Though this statement may seem tautological, it does raise, by jarring our commonsense experience, the notion that the views of men are not, in fact, totally disparate. We may disagree with some about the desirability or undesirability, the beauty or ugliness, of a particular place, but we can usually find others whose "views from the bridge" closely parallel our own. Perhaps this is the key: a portion of our viewpoint is quite particular to ourselves, while another part is shared, or held in common, with many of our fellows.

Given some information about the preferences of a group of people for various portions of an area, we require a way of separating out the general or shared portion of their perception from that which is quite specific to them individually. Putting it another way, we would like to partition the total variation in space preferences for a given sample of people into those portions that indicate general or common viewpoints, and those that represent unique portions that may be assigned to individuals. It is for this reason that the problem has been approached through principal components analysis, to which there are numerous elementary introductions (Cattell, 1965a, 1965b; Gould, 1967a; Harman, 1960; Horst, 1965; Kloven, 1968; Rummel, 1967).

In all the examples that follow, people were asked to provide rank order listings of their preferences for various areas. The question was posed in the context of residential desirability with "all other things being equal." For example, in the United States, students were asked to imagine themselves married and settling down with a family with complete freedom of location according to their own particular views as to what was desirable. The question was similarly posed in Europe and Africa, with some modifications to adapt it to local circumstances. Thus the basic data consisted of a matrix whose rows represented areas (states in the U.S.A., countries in Europe, and administrative districts in Africa), while the columns represented people. Each column contained the rank order values that a particular person had assigned to places in the rows, so that in a crude sense each person become a variable upon which the residential desirability of a place was measured.

Clearly, if two people held very similar views their rank order lists would match quite closely. Thus, the whole basic data matrix may be summarized by a smaller matrix of rank correlations (fig. 11.1). It is upon such matrices that the principal components analyses are performed to break out the underlying structures of space preferences in terms of a smaller number of dimensions or components. By definition, such dimensions are unrelated, and they may be thought of as independent scales upon which the areas have particular values or component scores. Thus, to summarize, the maps are constructed from scores on an orthogonal principal components struc-

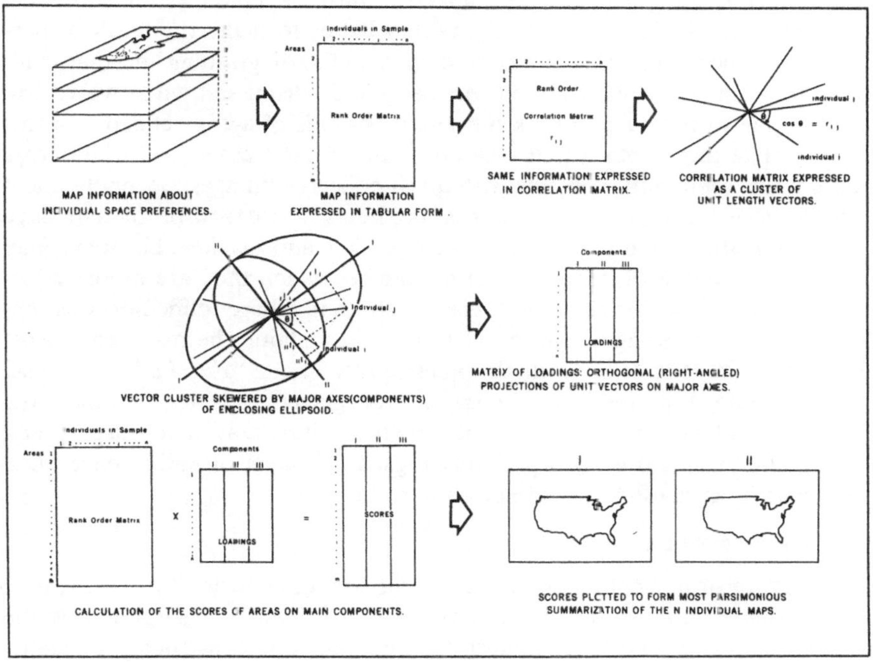

FIGURE 11.1. *Steps in the construction of the component maps.*

ture, which is merely used descriptively as a summary device, to see if we can break apart and simplify the structure underlying the views and values men hold and place upon geographic space.

The Perception of Residential Desirability in the United States

Apart from turning their attention temporarily on particular places during times of physical or human crisis, perhaps most people in our highly mobile society perceive and think about the geographic space called the United States in terms of residential desirability. At the state universities of California (Berkeley), Minnesota, Pennsylvania, and Alabama, students in beginning geography courses were asked to provide rank order lists of the forty-eight contiguous states in terms of their own, quite personal preferences. Sample sizes varied from about twenty-five to fifty. While the lower bound constitutes quite a small sample, repeated samples of this size have indicated an extremely high degree of consistency in the preferences and in the appearance of the final maps.

Two problems must be faced right from the beginning. First, the state units were quite gross, and any analysis must be made at an extremely general, macro-level. While it might seem better to make the mesh of perception a finer one, either by using county units, or gridding the map with 100-mile squares, such a notion can be quickly dismissed when we realize that people would have to rank literally thousands of areas defined in terms quite unlike those with which they are familiar. At least states, while gross areal units, are familiar objects with quasi-collective images. Secondly, most people faced with the problem of ranking items in order of preference have some immediate, and usually quite strong, likes and dislikes, but there may be a large number of items "in the middle" to which they are more or less indifferent. Thus, instructions were given that where difficulty was experienced in assigning a rank order to a state, it should be matched against the others in succession with the question in mind, "If I *had* to choose, which would I prefer?" It should be recognized that the middle area rankings will be less valid, although there is little reason to suspect any systematic bias, and we can probably regard the effect of indifference quite legitimately as random noise injected into the data.

THE VIEW FROM CALIFORNIA

The surface of perception derived from the first dimension (fig. 11.2), may be considered a general, overall view of the residential desirability of the United States as seen from California. The map was constructed by trans-

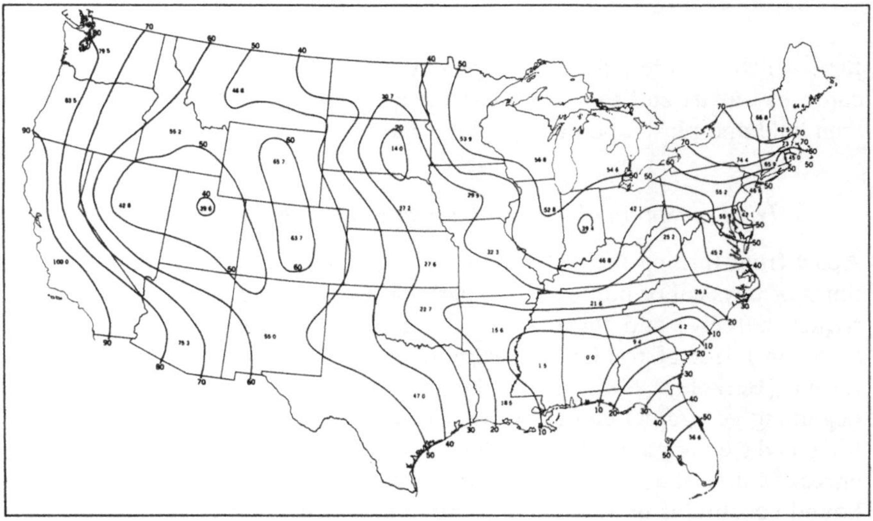

FIGURE 11.2. *The view from California: the first dimension.*

forming factor scores to percentage values relative to the highest score to make later comparisons easier. The values are plotted at the approximate center of population of each state, and isolines added to heighten the visual effect and to provide the concept of gradient.

From California a ridge of high desirability extends along the entire west coast, with the highest peak of the whole perception surface in California itself. However, the gradient to the east is very steep and there is a clear "perceptual" as well as "Great" Basin with a low point in Utah. Overall, as the view moves eastwards, there is a steady decline in desirability to the Great Plains, with the exception of a local peak in Colorado, and a quite definite "sinkhole" in South Dakota. However, upon reaching the ninety-fifth meridian, the general east-west trend of the perception surface changes radically as the overall orientation shifts by ninety degrees. A very clear discrimination is made in a north-south direction between the Midwest and the Northeast, where the surface begins to rise again, and the South, which forms the lowest perceptual trough on the entire surface. Alabama, Mississippi, Georgia, and South Carolina, with their images of civil and social unrest, are the least desirable places in the country for California's students. Only Louisiana is perceived as slightly less undesirable. Florida escapes the general Southern trend, and while Californians are not prepared to place this state as high as some other groups, possibly because of an old rivalry for the title of America's premier place in the sun, the gradient is steep from the low point in Alabama. To the north, the surface trend is upwards all the way to New England. Noticeable, however, is the way West Virginia distorts the even march of the isopercepts, possibly because of the recent emphasis upon the problems of poverty in Appalachia and the high social awareness usually ascribed to California's students. Though much of Kentucky may be in a similar economic plight, her image and her value on this first, general dimension of residential perception is bolstered to almost twice that of her well-publicized eastern neighbor. Perhaps white-fenced bluegrass pastures and sour-mash bourbon with pretensions to spiritous greatness, familiar themes in many advertisements, have conveyed an image brighter than the purely economic facts about the state would warrant! From Pennsylvania northeast to New England the rise is very rapid, reflecting an image that appears to include more than the bright lights of Megalopolis. The view of the northeastern "cultural hearth" seems to carry visions of the mountains and lakes of Vermont and New Hampshire and the quiet, rock-strewn coasts of Maine.

THE VIEW FROM MINNESOTA

From Minnesota the view of the United States is almost the same as that from California (fig. 11.3). True, the highest peak on the perception surface has shifted to the point from which the perception took place, but the high

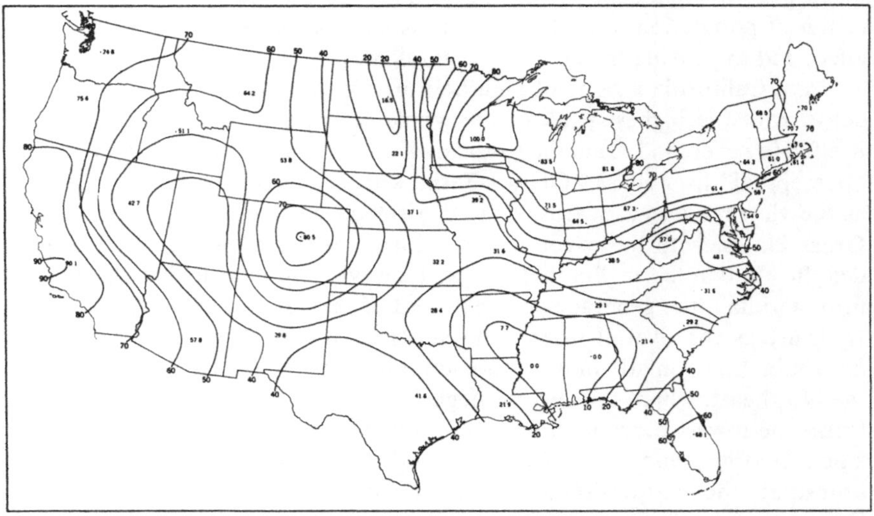

FIGURE 11.3. *The view from Minnesota: the first dimension.*

west coast ridge is still very much in evidence, together with the steep gradient to the Utah perceptual basin, the rise to the Colorado high and the fall to the Dakota sinkhole. Indeed, the Minnesotan seems all too aware of his western neighbor, and is most reluctant to trade his "land of sky-blue waters" for the dry, flat dreariness a few miles to the west. The steep decline in the surface to the flat country of Iowa reinforces an impression of spatial chauvinism, although to the east Wisconsin appears quite acceptable for residential purposes. Once again, the general west to east trend shifts ninety degrees around the hundredth meridian, and a low trough, centered in Mississippi and Alabama, blankets the South. Florida is again an exception. Northwards the isopercepts rise, twisted by the West Virginia (Appalachia) distortion, and the Midwestern and New England states bask in residential acceptability with almost uniform values in the sixties.

THE VIEW FROM PENNSYLVANIA

Perhaps more cosmopolitan than the Minnesotans, or set on edge by their rural location at the point of maximum inaccessibility, students at the Pennsylvania State University are seduced by the Californian siren to form the only sample in which the perception point is not the most preferred (fig. 11.4). Otherwise the surface is almost identical in general form to the California and Minnesota examples. Once again, the perceptual ridges and basins of the West are repeated, and the South is the lowest, or least desired part of the country.

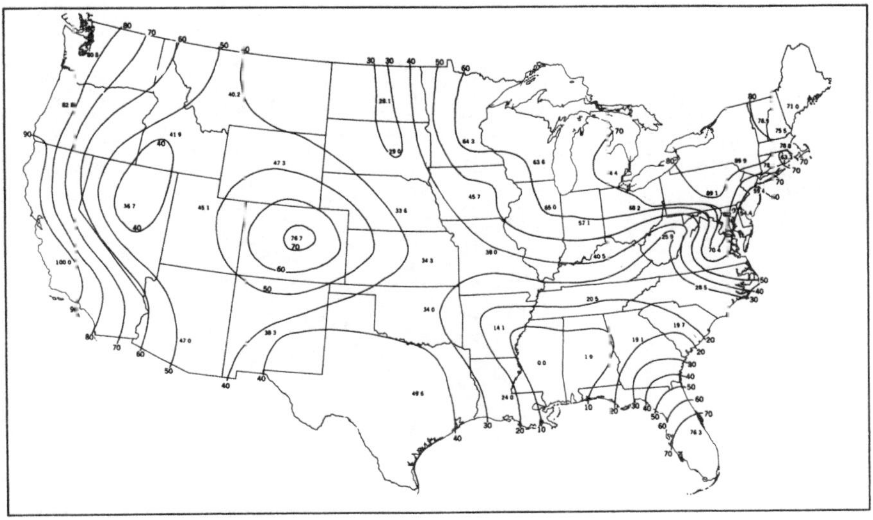

FIGURE 11.4. *The view from Pennsylvania: the first dimension.*

ALABAMA: DIFFERENT VALUES AND DIFFERENT VIEWS

Apart from local peaks of desirability, the three northern and western viewpoints are virtually identical. It is almost as though we had the same perception surface drawn for all three upon a rubber sheet and could reproduce the exact surface by moving a tennis ball beneath it to form, or reinforce, the local point. However, the view of the college student from Alabama, while sharing one or two points of similarity in the West, is generally quite different (fig. 11.5). The high peak of the surface is centered at the point of perception once again, but whereas the previous three samples tended to lump the "South" into a single low trough, Southerners appear to perceive the area with a high degree of spatial discrimination. Most noticeable is the very steep gradient down to Mississippi in the west: there seems to be little love lost between the two states that Northerners tend to place together, possibly because of the extreme violence associated with the civil rights movement in Mississippi. A fairly high degree of discrimination is also apparent between South Carolina and North Carolina. While California's politically and socially concerned students did tend to single out the former, generally the difference in desirability perceived by the northern students between these two states was not marked. Southerners, on the other hand, with a more intimate knowledge of this area, place North Carolina on a par with gentlemanly Virginia and Kentucky, while assigning the southern neighbor the next-to-lowest score in the whole South.

With the effect of the Civil War still being handed down in the minds of

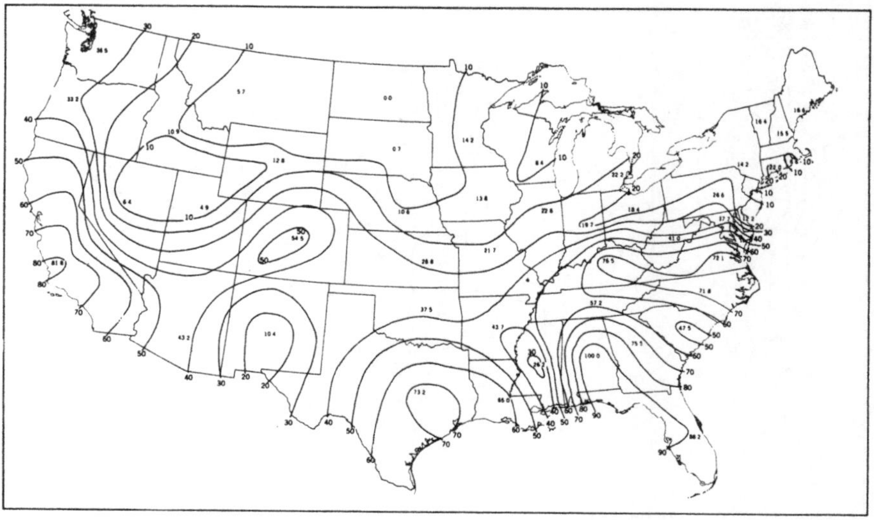

FIGURE 11.5. *The view from Alabama: the first dimension.*

men one hundred years later, the surface falls away rapidly to the northeast, with a very steep gradient along the Mason-Dixon line. North of this historic divide, there is considerable perceptual homogeneity: all Yankees are lumped together in Southern minds, just as Northern viewpoints blotted out and homogenized possible differences in the South. A disparaging view is held of the Midwest and the whole set of northern states as far as Washington and Oregon on the Pacific. Northern, remote, and blizzard-ridden, with only eye-straining horizons of waving wheat, the Dakotas are the last place in America for the college-bound Alabaman. In the West, only the Californian ridge and the Colorado peak appear, though less strongly than in the other examples. It is interesting to speculate upon the reason for the sinkhole in New Mexico, compared with the high value of Texas (still in the South), and Arizona (Goldwater country in the election year, shortly after which the sample was taken). What mental image does the word "Mexico" conjure up for the white Southerner that reduces the desirability of this state so far below its neighbors?

Questions and Speculations

The remarkable degree of similarity in the mental maps held by the California, Minnesota, and Pennsylvania samples raises the question of the degree of perceptual homogeneity within the groups themselves. The proportion of the total variation explained in each case by the first component does not vary radically (table 11.1), but the differences, if they do not simply arise from sampling fluctuations, are intriguing. Pennsylvania possesses the highest degree of perceptual homogeneity, closely followed by Minnesota, while Californian students appear to agree somewhat less about places of residence. Perhaps this is because the University of California at Berkeley draws upon a more heterogeneous population for its student body, while the state of California itself receives large numbers of migrants from

TABLE 11.1. *Variation accounted for by components I and II*

	Percentage of Variance Extracted	
State	*I*	*II*
Pennsylvania	46	16
Minnesota	41	15
California	36	15
Alabama	28	13

other states in the Union. Surprisingly, the sample with the least homogeneity of outlook is Alabama, and we might postulate a sort of spatial schizophrenia, for while the most dominant portion clings tightly to Alabama, and shares a mental map that discloses all the century-old cliches about "Yankeeism," the rest are split in their views with little agreement between their mental images. We shall meet such heterogeneity again when we examine the mental maps of French students.

Other questions also emerge from an examination of the maps based upon the first component scores. One of these is the effect of size on the ordering process. While it has been postulated that in viewing a map our minds act as a high pass filter so that small-scale features are accentuated (Holloway, 1958), it is worth noting that Rhode Island is consistently lower in overall score than its New England neighbors. Does it really have the image of a less pleasant place to live than nearby Massachusetts and Connecticut? Or is it, in fact, so small that people tend to forget about it, and in partially overlooking it assign a lower rank to it than it might otherwise receive? Another source of possible bias could result from the propensity for people to group together things that are spatially contiguous (Hochberg, 1964).

Finally, the question of difference must be raised once again. By requiring each person to insert and rank a Neutral Point in his list of preferences we may obtain at least some notion of the severity of the problem (Gulliksen, 1964a, 1964b; Tucker, 1964; Shelley and Bryan, 1964). A positive Neutral Point is injected into a rank order list when a person feels his preference changing from one of positive like to shoulder-shrugging indifference, while a negative Neutral Point is inserted at the place in a preference list where indifference changes to active dislike. In an example available for Pennsylvania students, the overall score of the Positive Neutral Point on the first component places it in the twenty-third rank. Thus nearly one-half of the states are generally perceived as positively desirable to live in, a remarkable comment upon the spatial mobility of the present college population. Indeed, the level at which a Neutral Point appears on the overall scale generated by the first dimension may be considered as a measure of the degree of parochialism of the sample group. A Neutral Point high on the first dimension, indicating that people are positively inclined to only a few states, might measure either a high degree of parochialism or a high level of discrimination, depending upon one's own attitudes towards such value-loaded words. We might hypothesize that the degree of parochialism as measured by the position of the Neutral Point would be directly related to the average age of the group, and the degree of social and cultural isolation experienced by the people in it.

The Rise and Fall of a Hypothesis

Our assumption that indifference exists in the rank orders as random noise makes the interpretation of further components somewhat hazardous. Nevertheless, the remarkable consistency (table 11.1) in the proportion of the variance explained by these second scales invites interpretation.

Remembering that the dimensions or scales that we impose in a principal components analysis are orthogonal, and, therefore, unrelated to one another, we might expect that successive maps of perception surfaces would illustrate quite independent concepts about the mental images that men have of geographic space. The view from California on the second dimension (fig. 11.6) immediately suggests that underlying the overall, general surface (fig. 11.2) there is another surface, quite independent of the first, that is strongly related to distance from the point of perception. In fact, apart from a rather awkward *contretemps* around Oregon and Washington and small distorting pockets in the Ohio, Kentucky and West Virginia areas, the correlation of the scores with raw, crow-flying distance from California is remarkably high ($r_s = .90$). Thus, there appears to be some strong evidence that a distance component is present in mental maps and that mental images of the differential desirability of geographic space cannot be meaningfully represented on a single, general scale (Blackith, 1965).

The second perception surface for Pennsylvania (fig. 11.7) bolsters the idea that a distance component is present. Once again the West Coast states prove to be exceptions, destroying the fairly regular march of the isopercepts with distance away from Pennsylvania, but even so the correlation with raw

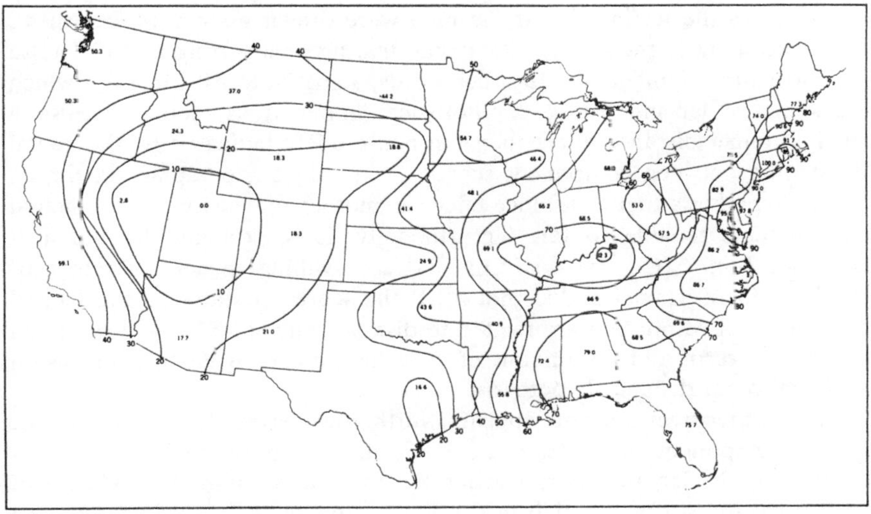

FIGURE 11.6. *The view from California: the second dimension.*

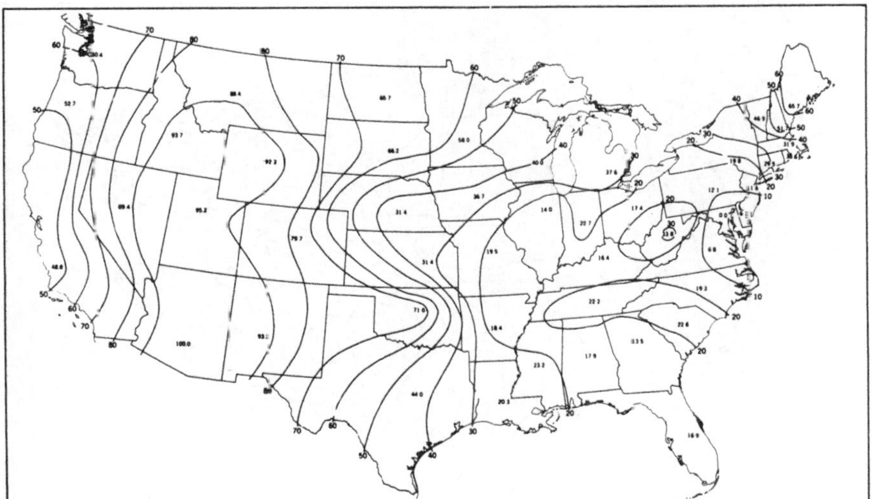

FIGURE 11.7. *The view from Pennsylvania: the second dimension.*

distance is high ($r_s = .64$). New England, too, distorts the distance effect and lowers the overall relationship, but the scores vary with such regularity that one has the feeling that if the map were drawn upon a rubber sheet, mere stretching of the space, rather than tearing or inverting it, could raise the correlation significantly. In other words, a spatial transformation, which has some rather interesting psychological implications, could disclose a much stronger distance effect than the one crudely indicated by the actual association between component scores and straight geographic proximity.

The hypothesis that a distance effect is another dimension of the mental map appears tenable so far. Unfortunately, it is exploded by the next example: Minnesota, true to form, and ever willing to demonstrate geographical inconsistencies, does not fit! If the scores of states on the second component of Minnesota are related to distance (fig. 11.8), the correlation is virtually zero, and no amount of stretching could make this dimension conform to our distance hypothesis.

But another question now appears worth considering. In all three of the second component maps (figs. 11.6–8), there appears to be some propensity for the isolines to run generally north and south. In the case of California, the surface is high in the West, dipping to low values over the Great Basin and Mountain states, and rising again slowly at the Great Plains to another high, though convoluted, plateau in the East. Similarly, the

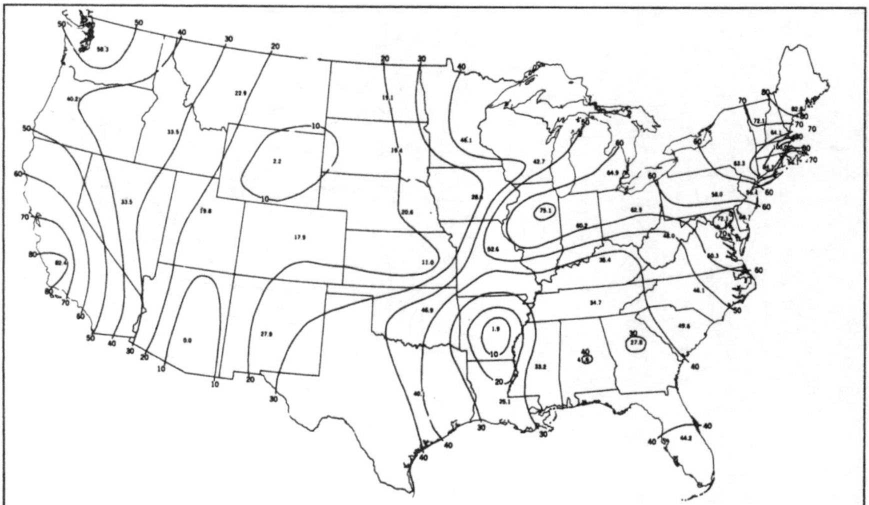

FIGURE 11.8. *The view from Minnesota: the second dimension.*

second perception surfaces of Minnesota and Pennsylvania are moderately high in the West, dip to a low trough, rise again with considerable regularity eastwards from the Great Plains, and then finally dip again in the New England area. We might hypothesize, therefore, that there is a second dimension to the mental maps that illustrates a propensity to perceive and evaluate the geographic space of the United States is a fairly consistent east-west direction. Thus, when the point of perception is on the "edge" of the rectangular space, as in California and to a lesser extent in Pennsylvania, an apparent, though spurious distance effect appears tenable. Only when the perception point moves to the center of the space, as in the case of Minnesota, must the distance hypothesis be discarded to be replaced by an east-west interpretation that is much less intellectually satisfying because it is difficult to relate to any other insights we have. We do know that east-westness in travel produces distinct psychological and physiological effects compared to north-south movement (Hauty and Adams, 1965), but this observation provides little confirmation that we are on the right track, and the hypothesis awaits much more study with larger and more numerous samples. For example, if one looks at the map of Alabama (fig. 11.9) through reasonably charitable eyes, some confirmation of the east-west hypothesis is obtained. Moving west to east, the high Pacific ridge dips to a north-south trending trough at the 110 meridian followed by a rise in the

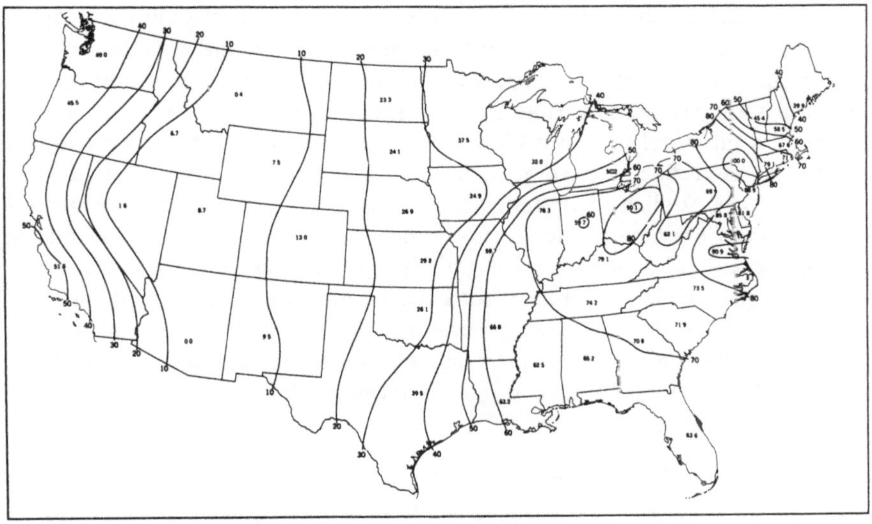

FIGURE 11.9. *The view from Alabama: the second dimension.*

eastern part of the country. But severe local anomalies such as Arkansas occur, and the surface appears equivocal in its ability to support or deny the rather clear configurations of the other three examples.

Europe Looks at Europe

In these days of newly emerging nations in Asia and Africa, we are apt to forget that it is really Europe that has experienced the most shattering political change of the twentieth century. Smashed by the First World War and beset afterwards by violent revolution, catastrophic inflation, and economic depression, the stable and ordered continent of the early years of the 20th. century is now but an historical curiosity to the last three generations. Laid waste by yet another world war in which civilian casualties alone were counted in the millions and divided afterwards by ideologies that confront one another across a line drawn through her very heart, Europe is still in the process of settling down to a new, though hopefully a dynamic, equilibrium.

Faced for the first time with two powers larger than any that they have experienced before, the countries of Europe are seriously considering a degree of cooperation and unification that would have been unthinkable a few decades ago. Military alliances, steel communities and joint atomic facilities, common aeronautical projects and Common Markets—all are signs of an overall process of drawing together. True, the process is not altogether smooth, and there are backward steps as well as forward gains, but as economic cooperation leads towards some degree of political integration, it is pertinent to enquire about the mental images and preferences that Europeans have of other countries in the larger European community. For in the last analysis, the barriers to unification are mental, and where a government is ultimately responsible to an electorate the mental images that a people has are reflected to some degree in the policies towards other nations.

It would be foolish to pretend that the space preferences for residential desirability of university students are indicative of the overall mental images of one nation held by another. Nevertheless, by posing the question in terms of residential choice, we may record the mental maps of a small, but important group in the post-war generation—a generation, let it not be forgotten, that has known only an uneasy peace broken by sporadic outbursts that could all too easily have triggered Armageddon. Such a generation is likely to give thoughtful answers.

EUROPE FROM DIFFERENT VIEWPOINTS

The sample from Sweden displays the highest degree of perceptual homogeneity with seventy-two percent of the variance collapsing upon the first component (fig. 11.10). On this general, overall scale, Sweden is by far the most preferred, followed by Switzerland, Norway, Denmark, and the United

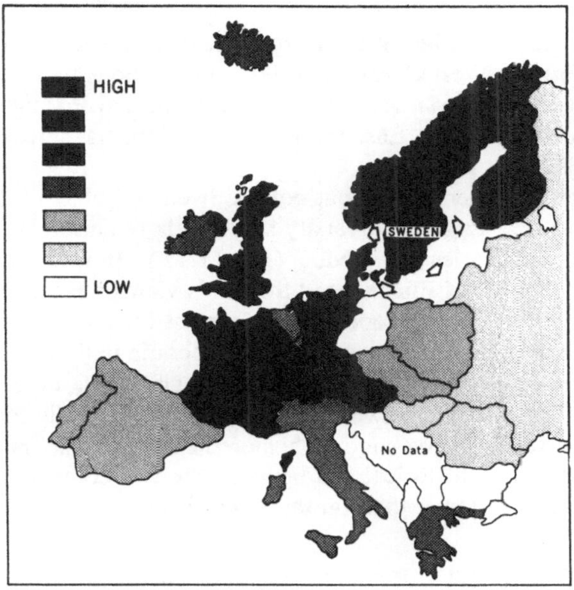

FIGURE 11.10. *Europe viewed from Sweden: the first dimension.*

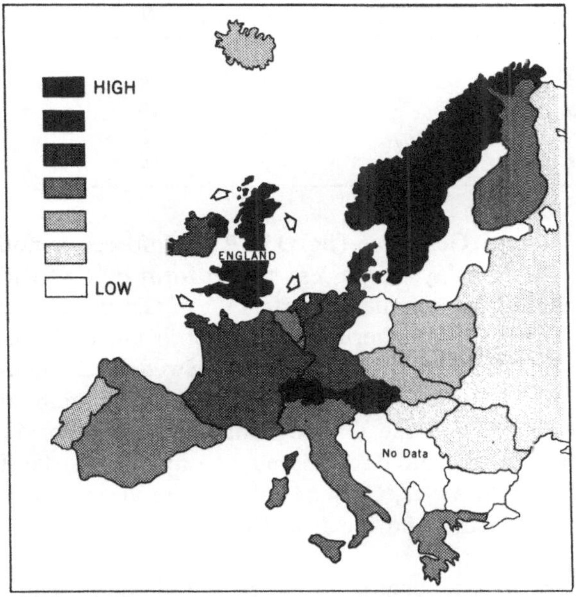

FIGURE 11.11. *Europe viewed from England: the first dimension.*

Kingdom in a tight cluster. Other West European, democratic nations appear next, such as France, West Germany, Italy, and the Benelux countries, followed by those with dictatorial regimes such as Spain and Portugal. The Eastern bloc is scored low, with East Germany and Albania perceived as the least desirable of all.

While not quite as homogeneous in outlook as Sweden (table 11.2), the overall mental map of the English university students is remarkably similar to that of their contemporaries in Sweden (fig. 11.11). Indeed, with the exception of England's view of Finland and Sweden's view of Eire, students in these countries seem to view Europe through almost identical eyes. Both groups display a strong preference for western, democratic nations and tend to shun authoritarian governments able to exert a high degree of coercion upon their citizens. The contrast between east and west is virtually identical in every respect. Portugal scores slightly higher on the English map, indicative perhaps of the long historical ties between the two countries, while Iceland is placed somewhat lower than on the Swedish scale.

TABLE 11.2. *Variation accounted for by components I and II*

Country	*Percentage of Variance Extracted* *I*	*II*
Sweden	72	6
United Kingdom	66	8
Germany	56	10
Italy	55	8
France	45	15

The view from Western Germany (fig. 11.12) reinforces a notion of cultural affinity that was displayed in a less severe form by Sweden's preference for the Scandinavian countries. After West Germany itself, the German-speaking countries of Switzerland and Austria are preferred over all others, with such linguistically similar nations as Sweden, Denmark, and the Netherlands following closely. Significantly, Belgium, Luxembourg and France are preferred next, and the United Kingdom scores considerably lower on the German map than does Germany on that of the United Kingdom! On the whole, however, Germany shares the viewpoint of England and Sweden for the West, and the Iron Curtain is again a most vivid dividing line.

Similar to Germany in the degree of perceptual homogeneity (table 11.2), but displaying a mental map that is just as distinctly individualistic (fig. 11.13), Italian students have a view of Europe that places the Catholic and Latin countries of France and Spain much higher than any other. Greece, too, with its sunshine and blue waters, is perceived much more

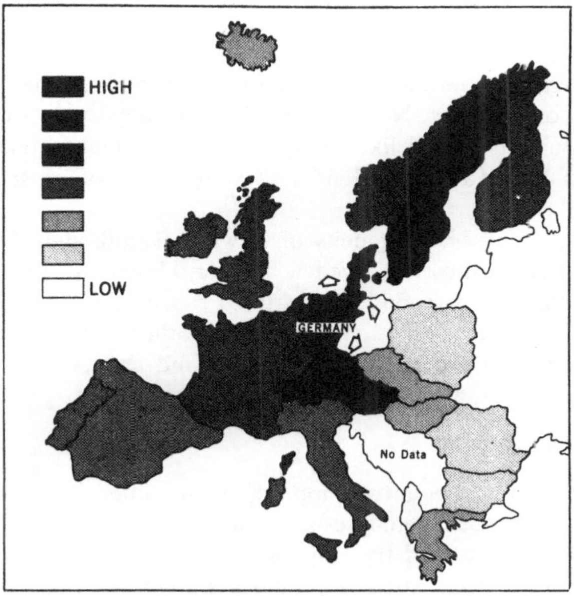

FIGURE 11.12. *Europe viewed from Germany: the first dimension.*

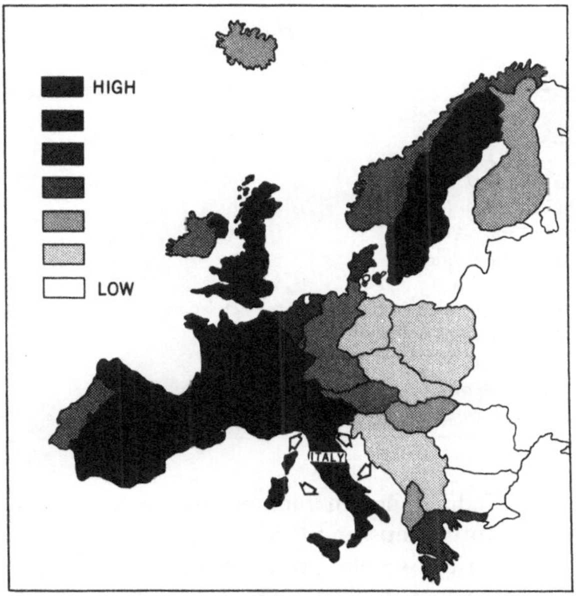

FIGURE 11.13. *Europe viewed from Italy: the first dimension.*

favorably, while Western Germany is given a lower rank in the Italian sample than by any other group. Nearby Albania, normally the last country in Europe where students would like to live, receives a higher rank than usual, close behind Finland and Iceland and well ahead of most of the Eastern bloc.

France possesses the least homogeneity of viewpoints, although the dominant one is very close to those of Sweden and the United Kingdom (fig. 11.14). However, there is a second view from France (fig. 11.15), producing a larger second dimension (15%) than any other, which places the U.S.S.R. and Yugoslavia on a par with the homeland and eschews Scandinavia, the Low Countries, and perfidious Albion!

OVERALL VIEWS OF EUROPE

We can consider the scores of countries upon the main dimensions extracted by the five principle components analyses as values on new variables that are a result of linear combinations of the original rank order lists. These variables display some interesting summary interrelationships. For example, while the dominant views of all five samples are closely related (table 11.3), Italy meshes less strongly than the other four, agreeing least of all with Germany.

TABLE 11.3. *Correlations of component I scores*

Country Pair	*Similarity of Viewpoint*
Sweden-U.K.	.94
U.K.-Germany	.90
Germany-Sweden	.90
Sweden-France	.89
France-U.K.	.88
Italy-France	.81
Germany-France	.79
Italy-U.K.	.73
Italy-Sweden	.69
Italy-Germany	.62

Given the problem of indifference in ranking, the second components, which do not produce uniform maps, must be used with caution. Relationships between the scores on this second group of artificial variables (table 11.4) are not always strong, but the directions of the signs are intriguing: all are positive with the exception of Germany. Whatever the second viewpoint from Germany represents, it relates slightly to the second from Italy, and inversely with all the others. Favorably evaluating the surrounding

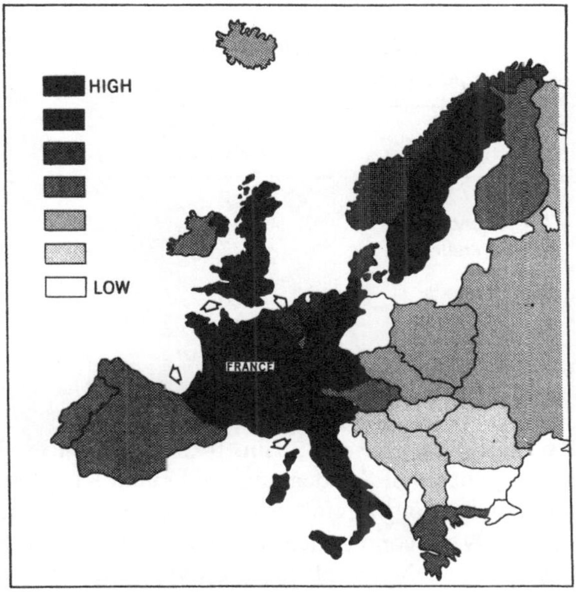

FIGURE 11.14. *Europe viewed from France: the first dimension.*

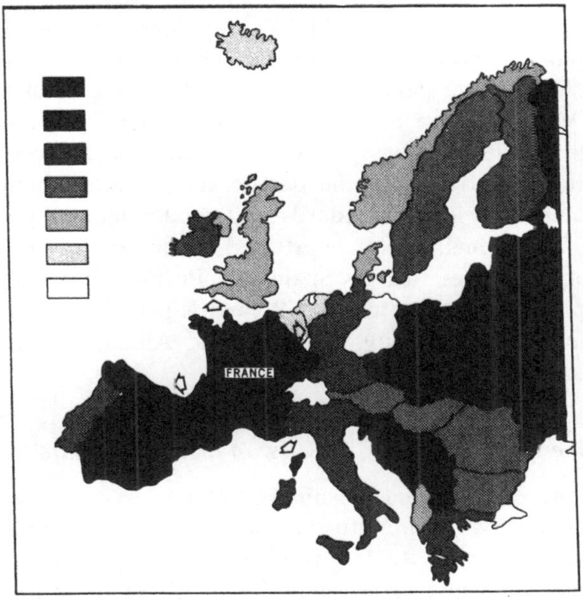

FIGURE 11.15. *Europe viewed from France: the second dimension.*

TABLE 11.4. *Correlations of component II scores*

Country Pair	*Similarity of Viewpoint*
Sweden-U.K.	.60
U.K.-France	.46
Italy-Sweden	.32
Italy-U.K.	.09
Germany-U.K.	—.06
Germany-Sweden	—.14
Germany-France	—.17
Germany-Italy	.16

neighbors (fig. 11.16), including Fascist Spain, this dimension is virtually unrelated to the first components of Sweden and the United Kingdom, components that displayed such strong preferences for democratic institutions (table 11.5). Similarly, Italy's second dimension relates inversely to Sweden I and the United Kingdom I, although all these should be interpreted with caution since both Germany II and Italy II have high variation remaining on the diagonal of the residual matrix, indicating that low communalities would be appropriate and that the specificity of their variation is high.

A second, or higher principal components analysis provides us with an overall scale for all the samples (fig. 11.17). Summarizing the forty-five percent of the variation, it may be interpreted as a complex scale that seems to reflect both the level of the standard of living and the degree of political authoritarianism. Very high are Switzerland and Sweden, both havens of neutrality with high standards of living and liberal democratic governments. This cluster is followed by most of the democratic socialist countries of Western Europe with high living standards and predominantly protestant church affiliation. The remainder of Western Europe scatters down the scale, with dictatorial regimes such as Spain and Portugal coming last. A fairly wide gap separates the Eastern bloc: Poland is perceived as the most liberal of this group. East Germany, Bulgaria, and Albania are the least desired countries for residential purposes.

Government Assignment: The View of an African Elite

In most of the countries of Africa the number of university students and graduates is still extremely small. With the exception of a few in business and private law practice, most enter government service in a variety of diplomatic, administrative, and teaching positions where the pace of advancement can be very rapid indeed. The result is that positions of considerable responsibility are often held by men and women only a few years away from graduation. When such decision-making power is in the hands of a fairly small elite group, the view they hold of their own nation becomes

TABLE 11.5. *Correlations across components*

Country Pair	*Similarity of Viewpoint*
Sweden I-Germany II	—.23
U.K. I-Germany II	—.13
Sweden I-Italy II	—.52
U.K. I-Italy II	—.43

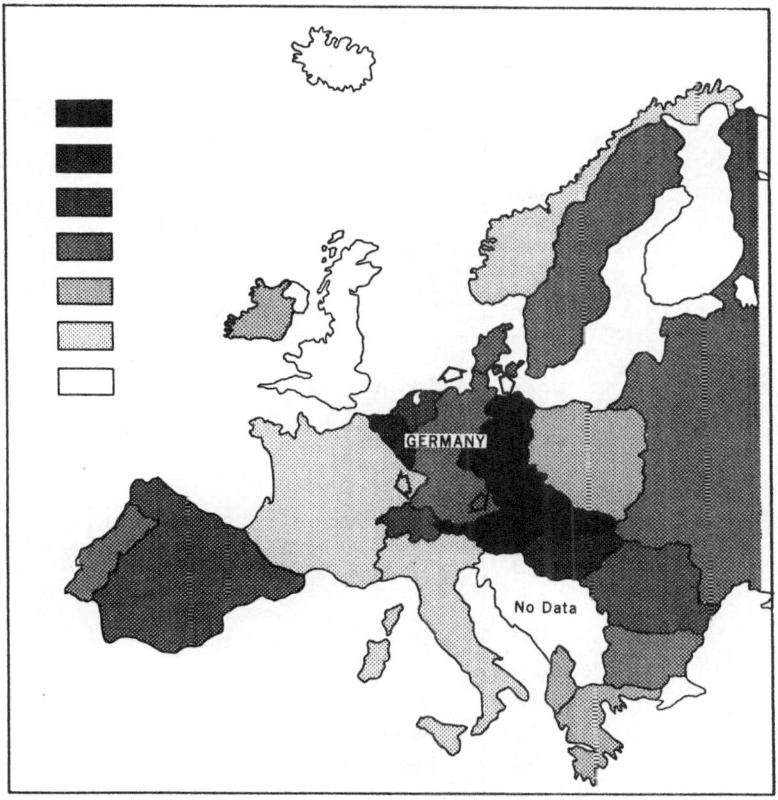

FIGURE 11.16. *Europe viewed from Germany: the second dimension.*

of more than passing interest. For example, investment must be assigned spatially, as well as sectorally, and the overall images held by such a group may influence the allocation of the meagre development funds for which there is such intense competition.

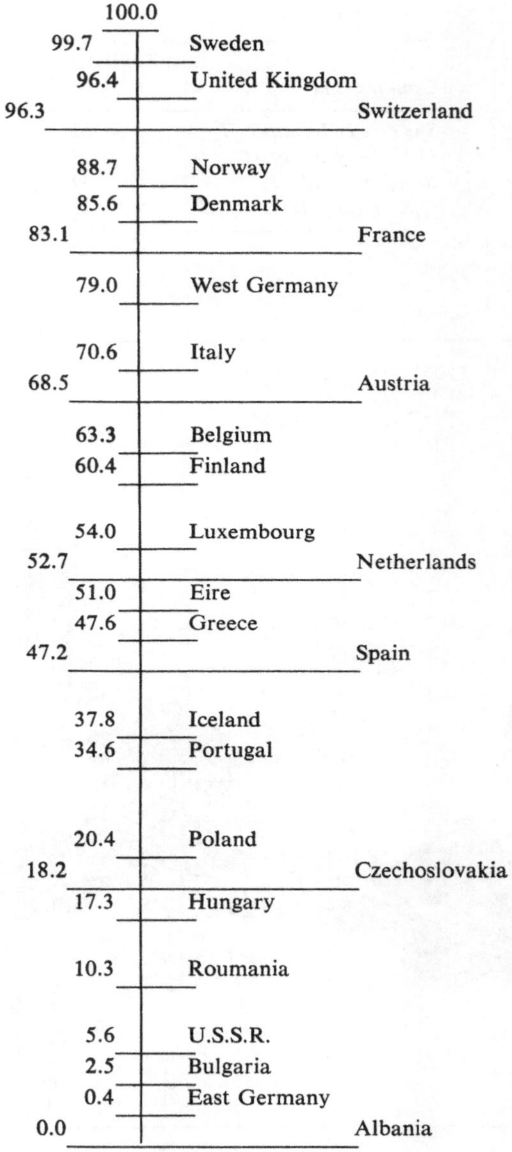

FIGURE 11.17. *Space preferences in Europe: the overall view.*

GHANA

In sampling the mental maps of Ghanaian and Nigerian students, the question was posed in terms of the residential desirability of districts given complete freedom of choice of assignment in some type of government service. In Ghana (fig. 11.18), there is a very high degree of agreement of spatial perception, with 67.4% of the total variation collapsing upon the first component. Generally the coastal districts containing the major urban centers have high scores, and the preferences decline fairly regularly away from the southern "core" of the country towards the north. On successive components the drop in the explained variation is marked. After the variation in the overall mental map is extracted, the remaining components appear to be

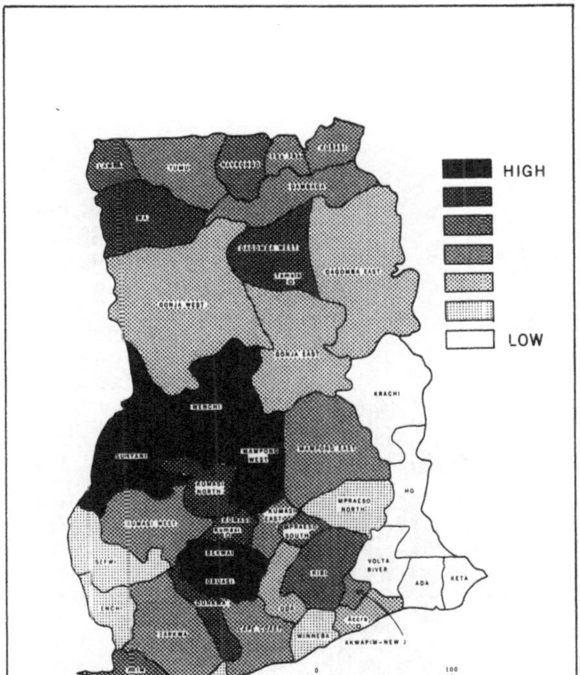

FIGURE 11.18. *The view from Ghana: the most general viewpoint.*

small regional effects that contrast one traditional area against another. For example, Component II, extracting 7.5% of the variation (fig. 11.19), contrasts vividly the eastern Ewe area with the core area of Ashanti, and such an interpretation is bolstered by the close matching of contrasting signs in the factor loadings with places of birth and residence of the students. The third dimension highlights in a similar fashion a core area centered on the town of Koforidua north of Accra.

The degree of spatial regularity in the scores of the overall mental map (fig. 11.18) suggests that a fairly simple perception surface could describe the national pattern with some accuracy and so allow us to separate broad trends from local anomalies. If the perception scores are related to their geographical coordinates, a quadratic surface accounts for seventy-nine percent of the variation (fig. 11.20). It appears to be the "best" in the sense

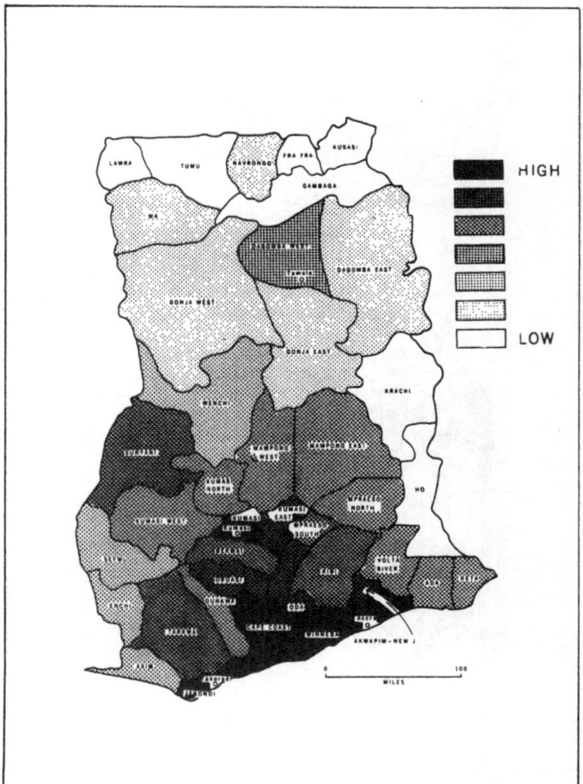

FIGURE 11.19. *The view from Ghana: regional contrasts.*

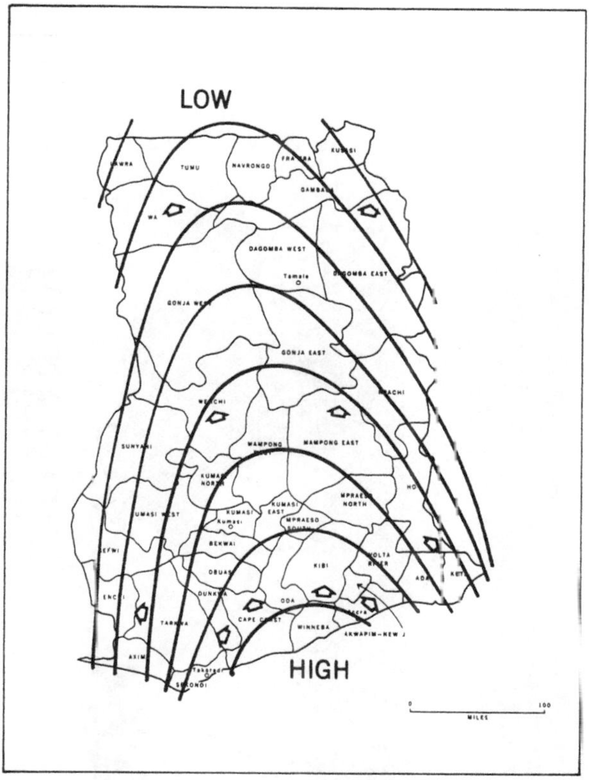

FIGURE 11.20. *Ghana: the quadratic perception surface.*

that it combines parsimony of equational terms with an ability to account for perception scores by their geographical location. For example, adding all the cubic terms only raises the sums of squares by three percent.

Having estimated the overall trend surface, the anomalies may now be interpreted as the power of the urban centers to cast especially bright images in the minds of men (fig. 11.21). Accra, the capital of Ghana; Kumasi, the center of Ashanti and the principal inland town; Ho, the administrative center of the Volta region; and Tamale, the center of the north—all have images far brighter than the overall, national trend would predict. Even Kusasi, Navrongo, and Wa in the far north are above the surface, for they contain the main towns and administrative centers. Of particular interest is the image that the Sunyani district holds, due not only to the presence of a main town, but because this is an area that is exploding economically. Here traditional authority and modern higher education blend to spark an area whose dynamism is clearly perceived so that it appears much more attractive

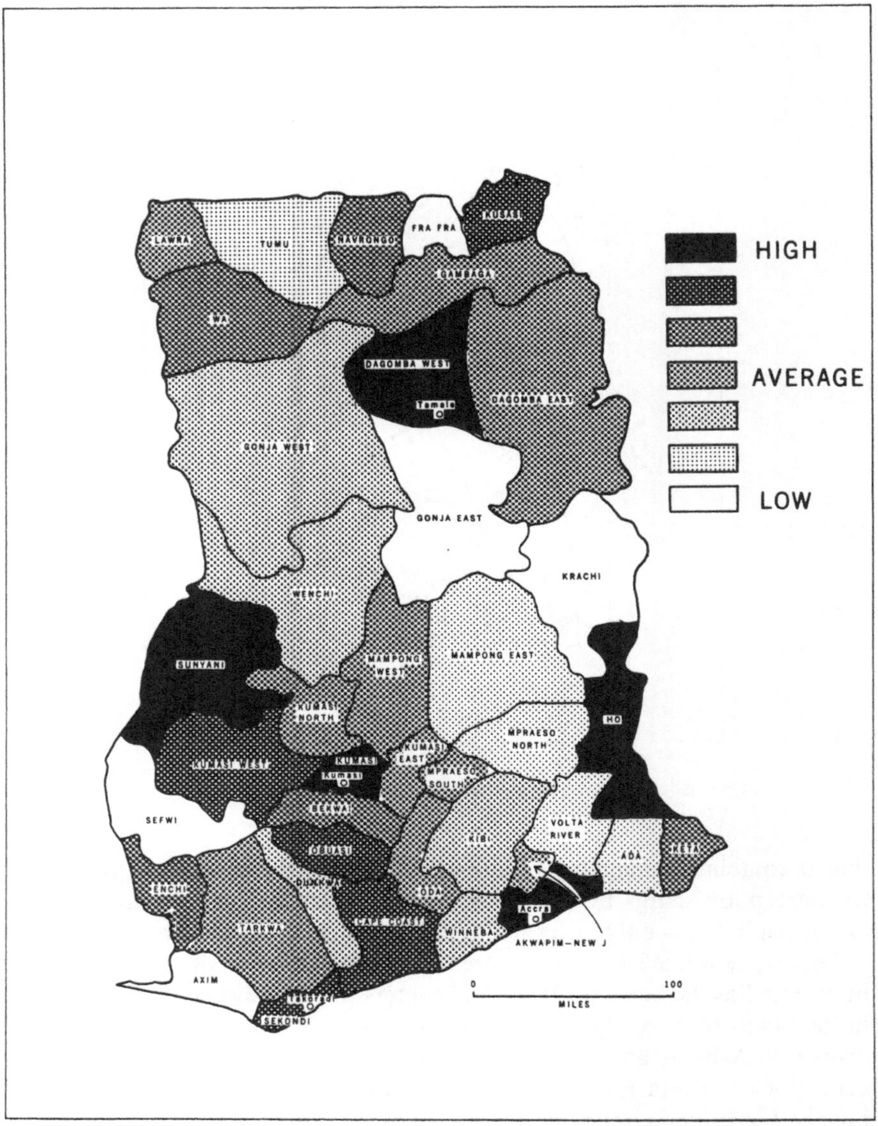

FIGURE 11.21. *Ghana: residuals for the quadratic perception surface.*

than its peripheral position away from the core would suggest. Similarly Enchi in the southwest is higher than one would predict due to recent economic developments in the area. It too partakes of the pioneer glow, perhaps because of the increased information people have about these areas from newspapers and news reports.

The negative residuals, those areas perceived as even less desirable than the overall trend would suggest, lie mainly in a zone trending from the southeast to the northwest. Many lie in the Barren Middle Zone, an area that for environmental reasons (tsetse fly and low, irregular rainfall) and historical reasons (devastated by slave raiding from the north and south), appears especially unattractive. Indeed, it is only now that population is trickling back to the area (Grove, 1963), settling along the major north-south roads whose original purpose was to provide administrative and economic links between the poorer north and the bountiful south.

NIGERIA

The overall viewpoint of university students in Nigeria is only slightly less homogeneous (62.5 percent) than that of Ghanaian students. For those who are aware of some of the less optimistic political prognostications that have been made in recent years, such agreement will come as a pleasant surprise. While Nigeria is assumed to face divisive forces along traditional regional lines between the North, Southeast, and Southwest, the strong agreement in the space preferences of university students has considerable implications for forging national unity. Despite different backgrounds and home residences, nearly all the sample loaded highly on this first, general dimension, indicating that each agreed to a marked degree about this overall mental map (fig. 11.22). The perception surface, however, is not so simply described as in the case of Ghana. Two district cores appear; one in a band stretching over most of the southern portion of the country from the Yoruba east to the Ibo west, the other a perceptual peak in the north centered on the Jos, Zaria, and Kano districts. While the overall linear trend is clearly from the southwest to the northeast, the variation accounted for by even the quadratic surface (52 percent) is considerably lower than in the case of Ghana. One is reminded vividly of Whitten's examination of the multiple intrusion hypothesis in the Lacorne granitic massif (Whitten, 1965), where two cubic surfaces were required to raise the sums of squares significantly. Here there is little doubt that a similar division of the area along the Barren Middle Zone, an area that is quite marked on the mental map, would provide a much more accurate description of the overall surface. Nigeria, at least in the minds of this small elite group, seems to be divided into two quite desirable parts—the Northern core and the Southern band composed of a blend of the Eastern, Western, and Benin regions.

Away from the Northern core perceptual scores drop sharply, and much of the northeastern part of the country is not regarded at all favorably. Generally there is some element of "peripheralism" as in the Ghanaian map, and it is striking the way in which both of the perception surfaces closely reflect the pattern of road density (Gould, 1960; Taaffe, Morrill, and

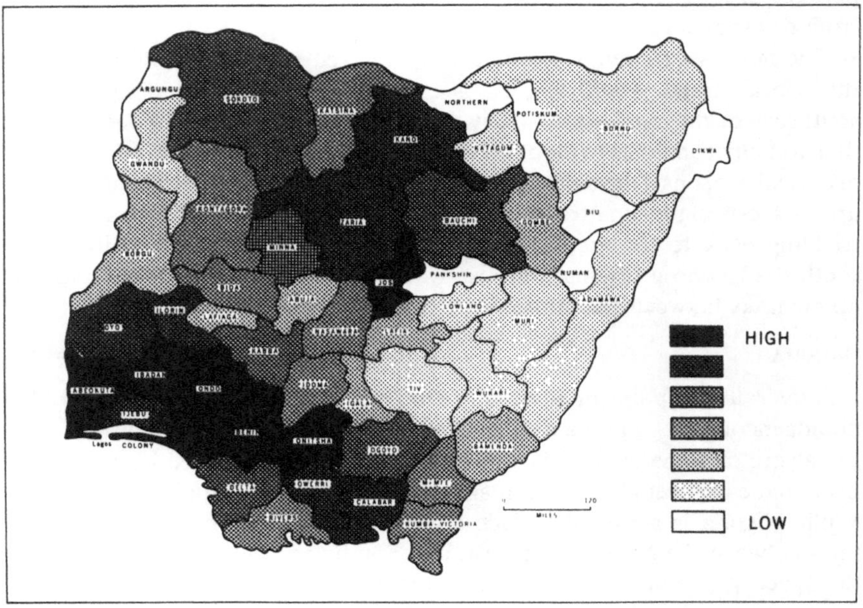

FIGURE 11.22. *The view from Nigeria: the most general viewpoint.*

Gould, 1963). Thus we may have some confirmation that road density is a useful surrogate measure not only of accessibility, but of those aspects of the modernization and development process that seem to be closely related to this slippery, but useful concept.

The second component, like the subsidiary dimensions of Ghana, accounts for only five percent of the variation and is a scale upon which small regional effects are contrasted. Compared to the strength of the first dimension, these do not appear to be strong or important, and they support the idea that for Nigerian university students petty regionalism has long been put aside.

On the Reconstruction of Perception Surfaces

On the assumption that people's actions in an area may be partially related to their perception of the space and the differential evaluations they place upon various portions of it, it is possible that by working backwards we can make some rough reconstructions of the mental images held by men long ago. For example, in a recent study of the process of historical settlement in western New York state just after the revolution (Fuller, 1966), trend surface analyses up to the cubic were carried out in which dates of first settlement in an area—the time dimension—were related to the geographic locations, or the two space dimensions (Chorley and Haggett, 1965).

We might consider the even march of the isocrones defining the simplest, or linear, surface (fig. 11.23) as indicative of the waves of settlers that might have moved across the country from east to west *if* the area had been perceived as a uniform transport surface completely isotropic in all the opportunities it presented to settlers at that time. However, such an assumption is obviously not tenable. Roads and tracks were beginning to lace the area at this time, making it easier to travel in some directions than others, and the *information* people had about different portions of the space varied and was strengthened by differential feedback processes. While analogies may be dangerous, I agree wholeheartedly with Bauer that they "may play two roles: the scientific role of developing generalized knowledge and the practical role of illuminating other events" (Mazlish, 1965). Thus, in the same way that a submarine valley can distort an evenly spaced wave train (fig. 11.24), so the underlying surface of perception distorts the even waves of settlement over the land (Kinsman, 1965). Fitting the quadratic surface (fig. 11.25), which represents the next level of accurate description gained at the least expense of complexity, provides us with some notion of the ease of travel in certain directions, the information flowing back to the points of origin, and the way opportunities were perceived by the people at the time.

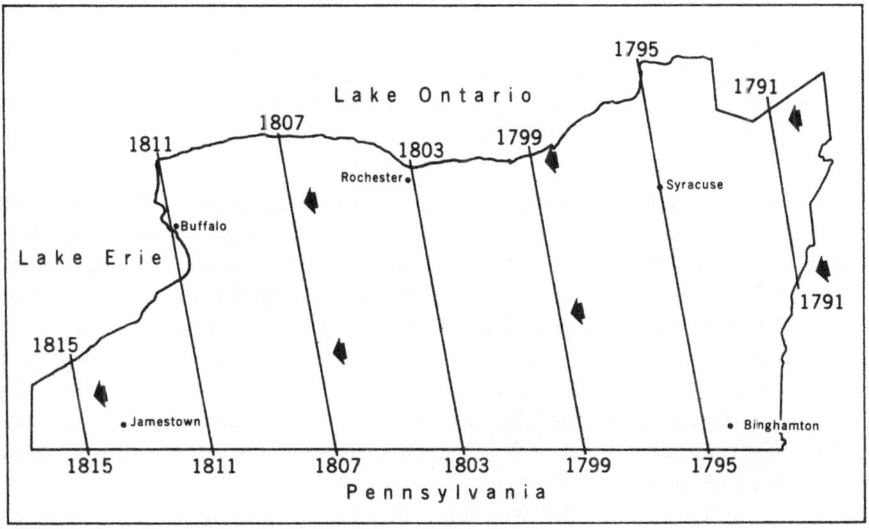

FIGURE 11.23. *Pioneer settlement waves in western New York State.*

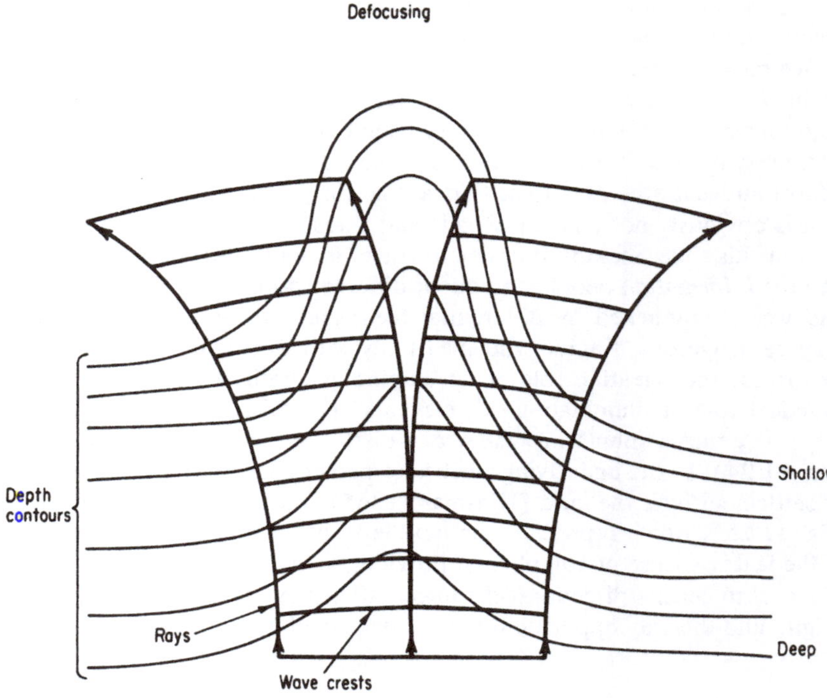

FIGURE 11.24. *The refraction of a wave train over a submarine valley. From Blair Kinsman,* Wind waves: their generation and propogation on the ocean surface, *(C) 1965. Reproduced by permission of Prentice-Hall Inc., Englewood Cliffs, New Jersey, and the author.*

The even settlement waves are pulled along the main route to the west, and the lags to the north and the south of this main corridor are marked. Describing the time and space relationships with the next most complex surface (fig. 11.26), the cubic indicates even more strongly the way in which the Lake Ontario plain was perceived as a less desirable area for settlement, for the time gradient is extremely steep to the north as the settlers bypassed it in their push westwards along the Lake Erie corridor to the new opportunities in Ohio. This was also an area of military activity: towns were frequently raided by the British in the early years of the nineteenth century. Similarly, the southwestern corner forms a pocket of late settlement in an area of rougher terrain that was filled in after the initial waves of settlers had pushed into the new lands of the west. Maps of residuals highlight the areas perceived as particularly attractive or repellent (Fuller, 1966).

To separate out the broad regional regularities and trends from the

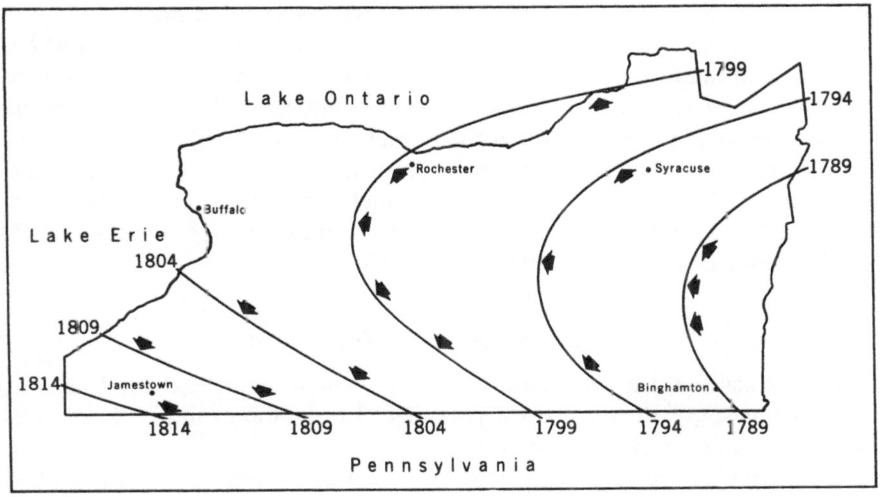

FIGURE 11.25. *Western New York State: the quadratic surface.*

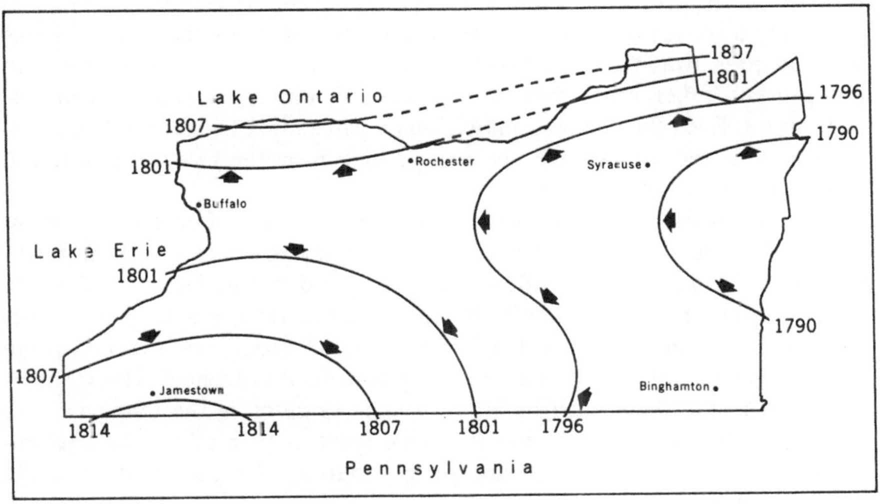

FIGURE 11.26. *Western New York State: the cubic surface.*

smaller local effects has always been a challenging task for the geographer. Where a dynamic spatial process such as pioneer settlement is going on, the use of trend surface analysis, combining space and time, may not only allow us to achieve such a goal in an efficient and objective manner, but to obtain in addition some insight into the mental images that men held at the time. If such a notion is valid, we may be able to examine the way in which such mental maps change through time, and so trace the line of inheritance for these images. Perhaps a series of careful content analyses may allow us to observe which areas maintain their brightness in the minds of men and which are quickly tarnished as new opportunities, new technologies, and new values change the very matrix in which they are evaluated and perceived (Budd and Thorp, 1963; North, Zaninovich and Zinnes, 1963; McClelland, 1961; Merritt, 1962; Aiken, 1968).

Some Implications of Mental Maps

What are some of the implications of these mental maps, and what lines of further investigation seem worth pursuing? I hope the examples have indicated that there may be an area of enquiry here that is not only geographically intriguing, but capable of crossing the line between pure and applied research. For perhaps the most obvious implications lie in the broad area of planning, whether this is undertaken by governments or individuals. Many locational decisions in industry are going to be influenced by the mental maps of a few key people. We can see this in the choices of many footloose industries in this country, while in England the image of the southeast is becoming a source of continuing frustration for planners trying to disperse new factories away from the London "magnet" to relieve congestion and to pump-prime other areas that are in need of additional employment opportunities. Even the channel tunnel, which will simply bolster the locational advantage of the southeast, is receiving criticism on the grounds that it will reinforce the pull of the area.

In much of the underdeveloped world the allocation of social investment is still of critical concern as many countries try to forge the basic infrastructure of transport, education, sanitation, and health facilities. Are the areas that are already "mentally bright" going to receive a large share because they are prominent in the minds of men? Would an awareness and self-knowledge of this tendency have any beneficial influence? The stricture "Unto them that hath shall be given" seems to describe the tendency of a system of allocation with strong feedback features to produce the agglomerations and clusters of goods and people that are the main feature of the urban revolution.

There are also implications for administrative planning. In the African countries particularly, the mental maps closely correspond to the accessible, modernized areas illuminated by the bright lights of the cities and towns. Yet one of the great needs in most of these countries is to get people, particularly teachers of all kinds, to go into the "bush" areas that are so disparagingly viewed. Are there not some implications here for incentive allowances that might be inversely related to the perceptual scores that various areas receive? Of course, this problem is not unique to Africa. Salaries for teachers in Alaska are incentively inspired beyond the difference in the cost of living, and the Soviet Union is using very high incentive pay to lure her people into the new and dynamic lands of Siberia.

In the area of migration, too, mental maps may shed some light on the gross and long-term movements of people. Thomlinson, for example, after trying to estimate the effect of many variables on migration in the United States, commented upon the high residual variation of the Pacific states (Thomlinson, 1961). Interestingly enough they are all part of a predominant ridge of desirability that is consistent across the mental maps of all the students sampled. Similarly, the areas of marked migrational loss in this country, the Great Plains in particular, are low troughs and sinkholes. The implications for depressed areas are obvious, and in some of the backward pockets of Appalachia it would be useful to know about the mental maps of the young and old. In England work is currently proceeding on the mental maps of pupils about to leave school in the hope that they will shed some light on the migrational streams of young people that are causing such concern to regional planners.

At the more academic level, the mental maps raise the question of the geographical implications of the information flows to which people are subjected. Many writers, across a range of disciplines and concerns, have commented upon the way in which viewpoints are molded by the available information. As Herbert Simon notes in a critique of some common clichés (Simon, 1965):

> Does a man live for months or years in a particular position in an organization, exposed to some streams of communication, shielded from others, without the most profound effects upon what he knows, believes, attends to, hopes, wishes, emphasizes, fears and proposes?

What are the flows of information that form and mold the surfaces of mental maps? All other things being equal, do they change in content and intensity as one moves up the ladder of central places to the critical nodes of connectivity in an interurban network? Saint Paul's migration to Rome might well have been influenced by his mental map of the geographic space that comprised his "world." Surely his demonstrated awareness of the relationships between location, information, and space might allow modern geographers to claim him as one of their own? After all, they claim the best of everything else!

In western New York, it was noted that differential flows of information may have had a profound influence upon the rate and direction of pioneer settlement. At a later time, and a little further west, Cochran has described the psychological effect of the railway in altering geographic horizons, and the way the "big city" newspapers raised the information level of the rural population and altered their consciousness of time and space (Cochran, 1965).

Finally, there is the question of the information available to one generation as it is filtered through the minds of the last. To what extent do we inherit our mental maps? It would be interesting to sample the geographic images in successive generations to see what significant changes exist between them. To what extent, for example, is the bright image of the Jos district in Nigeria due to it being the traditional "local leave" resort for the European population in colonial times? How closely would the mental maps of district officers in the 1920's and 1930's match those of the present? As Happold notes in a different context (Happold, 1966): "One cannot stress too strongly the extent to which our world view has been conditioned by our mental history and development." Using the notion of the positions of neutral points as indices of parochialism, we may be able to *measure,* though crudely to be sure (Guttman, 1966), the changes in mental images from one generation to the next. As Mencius noted more than two millenia ago (Kelley, 1935):

> By weighing, we know what things are light, and what heavy. By measuring, we know what things are long, and what short. The relations of all things may thus be determined, and it is of the greatest importance to measure the motions of the mind. I beg your Majesty to measure it.

Postscriptum 1970

This discussion paper, originally given as a talk to the Michigan Inter-University Community of Mathematical Geographers in 1965, appears to have played a modest catalytic role in the analysis of geographic perception and residential desirability. A number of reviews have appeared since then (Doherty, 1968a; Downs, 1967, 1968; White, 1967b; Bordessa, 1969; Goodey, 1968b, 1970), and some comparable studies have been completed at a variety of scales. Some have replicated the original study at roughly the same, national scale in the United States (Aangeenbrug, 1968; Goodey, 1968a), the United Kingdom (White, 1967a; Cole, Falconer, Brandwood, Simmonds, and Young, 1968), and Nigeria (Ola, 1968), while others have concentrated upon the perception of gross urban areas in the United States (Doherty, 1968b) and New Zealand (Johnston, 1970). Similar studies have been made at the intraurban level, investigating the residential desirability of various parts of a city (Eyles, 1968; Johnston, 1968; Peterson, 1967).

However, many crucial questions remain to be answered, or even properly posed. Perhaps one of the most difficult is the role of the information that people receive and then structure into preference surfaces or mental maps. In Nigeria, for example, it appears that much of the information possessed by children or young adults comes from relatives who are linked together in very extended family structures (Ola, 1968). Unfortunately, this does not say as much as one supposes. Most people have relatives located where most of the people of a region are, so we are really led back to the very simple idea that many people generate much information and vice-versa. Thus, to measure the intensity of information we may be able to use crude surrogates such as gross population, or some of the strongly interrelated modernization variables such as road density (Abler, Adams, and Gould, 1971). Some confirmation of this idea comes from Sweden (chapter 13), where the information surfaces of children are highly predictable from very simple gravity model notions. Furthermore, the space preferences appear to be related in Sweden, Nigeria, and the United States (Dutton, 1970), to information that is, in turn, strongly related to the population of the generating area and the distance from the perception point. The latter variable may not be expressed as a simple function, and a number of workers in a variety of fields have investigated various aspects of the effects of distance in geographic perception studies (Rushton, 1969a,b; Dornic, 1967; Ekman and Bratfisch, 1965; Reynolds and McNulty, 1968; Stea, 1969).

Work with children has also indicated that there are some intriguing questions of learning interwoven with these more general geographic problems. There is some tentative evidence that the degree of agreement in a group of children about residential desirability rises in a very regular manner described by a simple logistic or learning curve (Gould and Ola, 1970). If we think in terms of a signal/noise ratio, then the "spatial signal," or preference, gets stronger and stronger as the children get older, until it levels off in the early twenties. Interestingly, groups of children in the magistral of a region, literally "in the center of things" where the information flows might be expected to be more intense, seem to have stronger signals than those at the periphery. But this possibility has to be investigated much more carefully.

Information also influences the spatial discrimination of a group of people. This paper raised the possibility in the case of students from Alabama viewing the South—an area homogenized by the northern students—and confirmation comes from Britain (Gould and White, 1968) and the United States (chap. 10). The decay of information and the ability to discriminate with distance is so marked that we can often decompose a residential perception surface into two parts: a national surface shared by all the people in a region or country, and a strongly peaked dome of local desirability centered on the perception point.

Work is also proceeding on some of the practical implications of mental maps. A study of Tanzanian university students (Gould, 1970b), confirmed that it may be useful to smooth perception surfaces with salary differentials to make the assignment of scarce personnel easier in many underdeveloped areas. In Sweden a large study is investigating the links between information, perception, and the decision to migrate to see if perception is translated into a form of action that is, in turn, intimately linked to the urbanization process itself. The relationships may be difficult to untangle, for differences in location, age, sex, and class may all distort a simple conceptual chain. But while there is still much to be done, there is little doubt today that "geography can be mixed up with psychology"!

IV

The Development of Spatial Cognition

Introduction

As indicated in the other sections of this book, adults form cognitive maps and images of spatially extensive environments, orient themselves in a variety of such environments, execute judgmental preferences, and form impressions concerning distance, boundedness, and other metric characteristics of spatial environments. Clearly, there is no Zeus' forehead of geography from which such cognitions emerge full-blown: they are the product of the collated experience of the mobile human being, the result of development *and* learning.

The terminological problems associated with studies of the development of spatial cognitions have yet to be resolved. For the reader who is not entirely happy with the distinctions among spatial perception, spatial orientation, spatial cognition, cognitive representation, and cognitive mapping (see chap. 1), Hart and Moore (chap. 14) return to these topics in a developmental context. They also confront the problem of defining development and distinguishing it from other allied processes.

This definitional question is but one of the problems in understanding the process of cognitive development. Studies of cognitive development have emerged from general developmental psychology, a field which includes both cognitive and noncognitive viewpoints. The noncognitive position is exemplified by the operant school of psychology, which had its genesis in Skinner's first major publication (1938). Skinner discussed environmental influences upon development in his controversial *Walden Two* (1948), but this had no significant effect upon later operant views of development (e.g.

Bijou and Baer, 1961), which have little to say about the development of environmental cognition. A few members of other disciplines who have been influenced by the operant view have adopted positions on some nondevelopmental aspects of spatial cognition, notably in environmental design (e.g. Studer, 1966; Studer and Stea, 1966). However, while the operant approach has stressed the environmental *control* of behavior, it has overlooked the *spatial* aspects of the environment.

The cognitive viewpoint on development is not a unitary school of thought; rather, there are three major approaches represented in the writings of Bruner and his associates (Bruner, 1964; Bruner, Goodnow, and Austin, 1962; Bruner, Oliver, and Greenfield, 1966); Piaget (Piaget and Inhelder, 1967); and Werner, Wapner, and Kaplan (Werner, 1948, 1957; Werner and Wapner, 1956; Werner and Kaplan, 1963). Amalgamations from a spatial viewpoint are exemplified by Rusch's (1970) combination of Brunerian and Piagetian terminology and stages, and by Hart and Moore's inclusion of Werner's developmental theory (chap. 14). Concerning the formation of images, Rusch (1970, p. 60) comments on his "imaginal" period, which begins somewhere between eighteen months and three years:

> Differentiation of habitual action leads to the representation of that action by mental images. The child begins gathering images of complete patterns of action . . . which . . . can be thought about, contemplated, examined independently of the actions themselves. But the images tend to remain literal or iconic, composed of surface cues and with little recognition of underlying structure.
>
> Gradually the child begins to recognize invariances (that which is constant) in the objects about him and in his images of objects and patterns of action. These invariances soon become attached to the language forms he has been learning since his second year. Linguistic symbols are used to stand for or represent actions and images.

At a later stage, the developing child's verbal ability becomes sophisticated, and as in the adult world which he is expected to emulate, he is increasingly reinforced for exercising his verbal prowess (often to the exclusion of other symbolic skills). Concerning this verbal emphasis, certain of the ideas raised by Stea and Blaut (chap. 3) are reinforced by Bower (1970, p. 510) who claims:

> A plausible hypothesis is that very young children rely initially on sensory (and possibly motoric) imagery for representing the world, but this method becomes superseded and pushed aside by the child's developing linguistic competence. According to that hypothesis, linguistic encoding comes to be preferred because it frees cognition from the immediate sensory impression and concreteness of experience, enabling more abstract groupings, concepts, and relations to be used in structuring and communicating about the diversity in direct experience. On this account, then, the course of development in

> many cases leads to the gradual withering away of imaginal processes, which die from neglect and disuse. Visual impressions are no longer remembered in their full, vivid richness, but rather become conventionalized in terms of conceptual stereotypes.

It is notable that none of the major cognitive development theorists have dealt extensively with cognitions of spaces too large to be perceived. That their views are relevant to this topic, however, is attested by the number of researchers in large-scale spatial cognition who acknowledge their debt to the intellectual giants of developmental theory. Bruner has had a strong effect upon Appleyard and other members of the design profession (see Carr, 1970; Rusch, 1970). Outside of psychology and educational theory, interpretations (and often gross simplifications) of Piaget have influenced geographers, especially those concerned with geographic education (Almy, 1967; Eliot, 1970; Miller, 1967; Towler and Nelson, 1968). While Piaget has dealt less with the question of the cognition of macro-space than of micro-space, his techniques and observations have influenced many writers in the area of cognitive mapping.

As Hart and Moore state, the theoretical framework which guided their literature review stems from the Werner-Piaget tradition. Such a tradition implies the presence of distinct developmental stages or a developmental order in the emergence of spatial cognition:

> The terms *fundamental spatial cognition* and *macro-spatial cognition* are introduced . . . to distinguish between two varieties of spatial cognition: on the one hand, the development of fundamental concepts of space and, on the other, the further differentiation and elaboration of these concepts into the development and representation of large scale environments.

Hart and Moore state that the above distinction "should not be taken to imply that these are necessarily different entities or follow different psychological laws and processes." Their indication that the development of these two forms of spatial cognition "intermesh" accords with the findings of Stea and Blaut (chaps. 3 and 12).

We are still faced with the question: Do the fundamental spatial cognitions progress from a *micro*-scale to a *macro*-scale? Suggestions that this is not the case are provided by Gibson (1970, p. 101) who cites evidence from Bower (1966) in discussing two modes of perceiving: the perception of space and events in space; and the perception of permanent objects:

> (These) can be approached and examined closely . . . there is reason to think that in their phylogenetic development there is a considerable difference between them. Localizing oneself in the spatial layout or monitoring events going on in the space around one seem to develop earlier and to be neurologically more primitive than fine-grain identification of objects or patterns . . . (which) is the later achievement; its development continues over a long

> time; learning plays a prominent role in it as compared with perceiving the spatial layout and events; and we can expect to find more striking phylogenetic differences.

Here, we are taking evidence concerning environmental perception and extrapolating it to environmental cognition, since the "perception" of *macro*-space and events in *macro*-space, a common and adaptive achievement in many organisms along the phylogenetic scale, is in fact *cognition* of the environment.

At this point we choose not to return to the question, "*How* does environmental cognition develop?" but to consider instead, "Cognition of *what*?" It is often taken for granted that the developing child tends toward, and rightfully *ought* to tend toward, the "achievement" of "standard" Euclidean and Newtonian concepts of the world. But it is often forgotten that the Euclidean and Newtonian views are effectively *models* of real world geometries and physical systems rather than *givens*. Because these represent the ways in which we have been taught to think, we impose criteria derived from our views of the world upon the developing child, referring to his attainment of adult models as development. This may be perfectly legitimate, but certain external criteria are lacking: we do not know for sure that the geometric model of the world we are imposing upon the child is the only "true" model. As Bruner (1966, p. 320) argues:

> Our knowledge of the world is based on a constructed model of reality, a model that can only partially and intermittently be tested against input . . . our models develop as a function of the uses to which they are put first by the culture and then by any of its members who must bend knowledge to their own uses.

It may be that the world view toward which the developing child is tending is like that of the children in Lewis Padgett's classic tale "Mimsey were the borogroves"—an outrageous, multidimensional universe.

The key to understanding the development of spatial cognition lies in the answers to the proverbial four questions:

1. What is "fundamental"?
2. Does that which some consider fundamental necessarily occur earlier in a developmental sequence?
3. Is the development of spatial cognition a unitary sequence, or a combination of parallel intermeshed sequences as Hart and Moore suggest?
4. If it is not a unitary sequence or a combination of sequences, what then?

Unlike the Passover Seder, there is no youngest son ready to provide the answers.

Evidence of development in spatial cognition is not confined to the sources cited in the preceding paragraphs. The Stanford-Binet Intelligence Tests have long utilized a simple map test as a developmental index. Interestingly, the map seems to duplicate the test constructor's (generally correct) concept of a child's map, for while the roads are correctly depicted in the plane of the paper, the facades of the houses are also rotated into the plane of the paper (see Ladd, 1970, for examples and a discussion of this phenomenon.)

Children continue to develop into early adolescence, and Gould (chap. 13) treats a neglected aspect of spatial cognition in a study of adolescent Swedish school children in which he traces, and models, the change in the *amount* of spatial knowledge as a function of increasing age. Ladd (1970) had earlier studied neighborhood image maps produced by black adolescents in a Boston school. These later stages in development, which must be looked at cross-culturally and cross-nationally, are still very much open ground. For example, there is anecdotal and other evidence concerning the territorial and home-range behavior of teenagers and teenage gangs (Whyte, 1943; Goldberg, 1969; Tindal, 1971; Lyman and Scott, 1967), but little about the ways in which they cognize and cognitively subdivide their spaces.

Thus far we have said little about learning *per se* because little has been said about spatial learning generally. There have been a few exceptions (Golledge, 1967, 1969; Stea, 1968, 1969), but much of the effort has been directed to a reinterpretation of earlier theoretical formulations and empirical results in learning from a *spatial* point of view. Geographers have been influenced by the ideas of Tolman (chap. 2).

The papers in this section range widely: micro-, meso-, and macro-spatial cognitive maps, derived by a variety of techniques, are presented for birth through adolescence, but the selection is intended to be representative rather than exhaustive. Through all of the papers runs a common thread: spatial cognition is of fundamental importance in the developmental process and its neglect by traditional theoretical and educational systems is unfortunate. Arnheim (1967, 1969) has addressed himself to this issue, and as Bower (1970, p. 510) so succinctly puts it in speaking of imagery:

> In light of the research story told previously, that visual imagery improves memory, we may rue the general cultural deemphasis and decline of mental imagery and ask for research on techniques for developing imagery skills in adults. To end on a practical note, our prescription to the adult in approaching a new learning task is for him to become as a child again, to tap the wellsprings of his suppressed imaginative talents that have lain buried under years of linguistic development.

12

Some Preliminary Observations on Spatial Learning in School Children

DAVID STEA AND JAMES M. BLAUT

The theory and data described in the following pages emerged from an observation made early in 1967. The observation was an outgrowth of a proposal to teach cartography to children via the prior teaching of aerial photographs. However, preliminary testing of 6-year-old Puerto Rican children indicated that, contrary to all expectations, they could already read black and white vertical aerial photographs with very little difficulty. This surprising result led to the initiation of a cross-cultural and developmental study of spatial (environmental) learning, involving children in both Puerto Rico and Southeastern New England.

What our Puerto Rican subjects achieved is reported here, but the results are not specific to this particular group. We argue, in fact, that what we shall term "place learning" occurs quite generally without formal schooling, and that little is known about the cognition of large scale environments.

Theory

Place learning is a special case of learning in general. We take as assumed the general model of learning, and for each of its propositional variables we stipulate certain values (or limitations) which relate to place learning as such. The variables in general learning may be listed as follows: first, a stimulus or complex of stimuli that set the stage for learning or make it possible for learning to occur; second, an organism, in this case a human

The research reported here was supported by U.S. Office of Education Grant No. 4493. A slightly different version of this paper is forthcoming under the title: "Some Preliminary Observations on Spatial Learning in Puerto Rican School Children" in the *Proceedings of the Second Conference on the Family in the Caribbean* to be published by the University of Puerto Rico Institute of Caribbean Studies.

child, who is to learn to make some response to the stimuli; third, an inferred cognitive process within the central nervous system that encodes and stores the experience and recombines its elements in line with previous experience already in storage—a process that is hinted by such words as "memory" and "thinking"; fourth, a response that is to be learned or to be attached to the stimuli; fifth, a reward or reinforcement for the behavior (that makes the learning "satisfying" in some way); sixth, some criterion of "learning" allowing us to say that learning has in fact occurred.

"Place," in a psychological sense, is a special case of the stimulus situation. It has two defining attributes: (1) its scale is large relative to the learning organism, and in general too large to be perceived as a whole at one time; (2) its level of complexity extends to all forms of perceivable objects or events in a given space, and not simply to one category of object or event or person.

Between the external, "objectively observable" world of the child and his manipulation of, or responses to, this world—his environmental behavior—is a nebulous intervening entity. This has its locus, presumably, somewhere in the "black box" of the central nervous system. It acts to integrate what he presently perceives with his environmental memory and current plans. It is an evaluative and purposeful entity, and, in line with the term coined by Tolman, we shall call it a "cognitive map." A cognitive map enables a child to predict the environment which is too large to be perceived at once, and to establish a matrix of environmental experience into which a new experience can be integrated.

We cannot tap such cognitive maps directly; we can only infer that they *must* exist from certain actions of the child. Our earlier studies show that he can draw a map of the route from home to school, or even of his neighborhood; he can use toys to model his community or a reasonable facsimile thereof; he can tell the locations of objects he cannot see from those he can see, though never having traversed a straight-line path between them; he can immediately "pick out" new routes from home to a goal on the basis of old routes once traveled (Blaut, McCleary, and Blaut, 1970; Blaut and Stea, 1971; Muir, 1970; chap. 3). Theories of learning based upon simple stimulus-response connections fail to account for such integrative human behavior.

There seems to be a conflict or tension between the experimental psychologists' tendency to view learning (especially in the rat) as a value-free process (except for positive or negative values attached to reinforcements and to stimuli which become secondary reinforcers), and the opposite tendency on the part of so many educational psychologists to view learning as principally a matter of socialization and, thus, of valence. Place learning lies somewhere between the two. No doubt every higher organism attaches valences to environmental features; hence, we must build into our model capacity for taking account of values. At the same time, we are not dealing with social learning *per se.* A second very fundamental property of the place-learning model is postulated as follows: a significantly high proportion of place learning experiences make use of small-scale surrogates or simulations of the environmental situation such as pictorial imagery (as in television, or aerial photographs), toys, and environmental games ("territory," blind man's bluff, football). The rationale, we assume, is the difficulty a child faces in perceiving and manipulating events at what we have called "an environmental scale" and the need to deal with these events in a surrogate, simulated, or small-scale form (it may even be true that this form of simulation is the earliest use of "model thinking" and therefore true "scientific" reasoning in the child).

The reward for place-learning is self-evident in the phylogenesis of the human species; to survive, the human organism and its prehominid ancestors have had to learn how to deal with, and move about in, a spatial environment (see chap. 4), and this learning is certainly repeated ontogenetically in the development of every child. The reinforcements for environmental learning are to some extent rather different from those for social learning and small object learning: stubbing one's toe is often sufficiently punishing for stepping in the wrong place or in a wrong way; in larger-scale movements, getting lost can be an even worse punishment for not knowing where you're going.

We can affirm that environmental learning has taken place in a child if, prior to a presumptive learning experience, the child has shown himself incapable of performing an action which the adult world, as jury, considers to be appropriate environmental behavior, but if after the learning experience the child can, indeed, perform that action. Scientifically, however, we search for criterial actions which can be generated in an experimental situation designed to simulate something in the real environment as tests of environmental learning achievements. Four such forms of criterial action have been selected for our experiments: first, the ability to form and represent a cognitive map of the geographic environment; second, the ability to solve navigational problems using the representation; third, the ability to develop new mapping skills very rapidly with minimal instruction; fourth, more generalized environmental behavior as manifested in toy play.

In the following sections of this paper, we summarize the evidence from these experiments based upon two of the above criterial actions and what we believe to be their educational implications.

Empirical Data

In Autumn, 1968, we tested kindergarten children from four Puerto Rican communities in aerial photo interpretation. A vertical photograph of each community at a scale of 1:5,000, centering on the school attended by the children being tested, was taken. The communities were representative of four strata of Puerto Rican society: urban middle class, urban lower class, coastal plantation, and mountain peasant. Our object was to discern to what extent the children in these communities could read aerial photographs, and what differences in photo-interpretation ability existed among them. Thirty five-year-old children were tested in each community, each child being asked to identify (name or point to) whatever they saw in the photographs, to supply a Gestalt name for one preselected area within each photograph, and, finally, to name the photograph itself.

The results of the kindergarten tests in Puerto Rico indicated the following: first, taking a "composite" view of each community studied, table 12.1 shows that all except the urban lower-class children produced about the same number of total responses; the latter emitted fewer. However, the urban middle-class group produced substantially more correct responses than the others, who were about equal in this respect. Interestingly, the urban children gave markedly fewer incorrect responses than did the rural groups.

TABLE 12.1. *Responses to aerial photographs by community*

	Urban Middle Class	*Coastal Plantation*	*Mountain Peasant*	*Urban Lower Class*
Mean number correct identifications per picture	3.04	2.16	2.01	1.93
Mean number incorrect identifications per picture	0.69	1.42	1.51	0.80
Mean total identifications	3.73	3.58	3.52	2.73

There was no indication that children tended to make more correct identifications in photographs of their own communities, nor were rural children better able to identify rural communities and urban children, urban communities. In fact, when photos to be identified are considered separately, the only generalization that can be made is that the fewest correct identifications were made on the photo of the flat, regular plantation community.

Of the specific features identified, houses were most frequently pointed out, followed closely by motor vehicles, roads and streets, and, with somewhat less frequency, trees. The ordering, however, was different for each group. Roads were most significant in the urban-middle and plantation communities, vehicles in the two urban groups. Both houses and trees were most frequently recognized by urban middle-class children.

Thus, it appears that urban middle-class children perform best in aerial photo interpretation, but that the two groups of urban children we tested were less prone to "take a chance" on a wrong response. What most interests us, however, is the fact that children are able to perform this identification in the absence of any formal schooling; further, flying experience was fairly negligible for all groups. However, economically "better off" children could be expected to have more access to two aids hypothesized to be essential in environmental learning of the type tested here: television and "environmental toys" (model houses, vehicles, foliage). Both of these enable the child to view a model or small-scale world from other than his usual vantage point. Television frequently presents oblique, black-and-white aerial photographs of geographical areas, both familiar and unfamiliar, reduced to two dimensions, while a child playing with environmental toys typically places himself above the toys with which he is occupied. Altitude above ground should also be important, but its influence has yet to be evaluated. Presumably, children who start life on the upper floors of high-rise apartment buildings and in mountain-top communities have substantially greater opportunity for aerial views of their surrounding world.

One fairly obvious question emerging from the study of kindergarten children is: what happens to place learning, as measured by aerial photograph identification, as a child progresses through the grades? To assess this, a study was initiated in Puerto Rico during the summer of 1969 which tested children from the second, fourth, and sixth grades of the same schools involved in the previous study and on the same four photographs.

A total of seventy-six students from the mountain peasant community (Corozal) and ninety students from the urban middle-class community (Villa Nevarez, San Juan) were tested. The results of this test were combined, in part, with those from the kindergarten children of the previous study to yield figure 12.1.

Several results are suggested. First, the most marked increase in correct identifications and the most marked decrease in incorrect identifications appears to occur between grades "K" and "II". Total identifications and correct identifications for both groups show little improvement beyond grade II. Taking the data as a whole, one result is strongly indicated: the sort of environmental learning tapped by aerial photo identification seems to taper off beyond grade II, somewhat *before* the (often erroneous) interpretations of more orthodox theories claim that a child is ready to commence such learning.

One possible counter-argument involves a so-called "ceiling effect." That is, it may be that children seem to reach an asymptote because there really is no more to identify. To refute this argument, we have assembled the data of table 12.2 by taking the largest number of correct identifications per photo and aggregating these to derive a mean value for each grade level in each community. If the graphed means were in fact ceiling values, we would

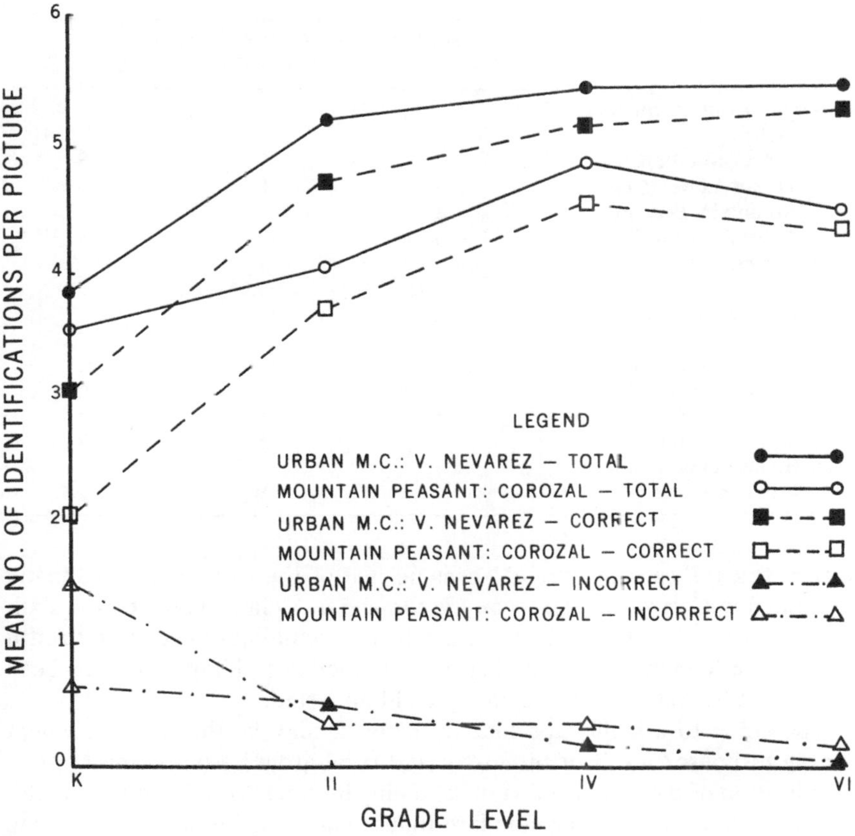

FIGURE 12.1. *Total correct and incorrect identifications as a function of grade level, 1968 and 1969 studies.*

expect that there would be very little dispersion about these means; that the means of the maximum scores would be very close to the overall means. This is clearly not the case. If we consider the overall scores of the best child in each group, he is clearly above the overall mean, but only in one case did a single child produce all the high scores. In no case does the total number of identifications—based, perhaps, upon perceived possibilities—approach the maximum allowed by the scoring system. Fifteen categories of possible response had been provided, with provision for as many as nine more to be added if the responses received warranted such addition. Thus, it would have been theoretically possible for a student to give twenty or more correct responses.

As with the kindergarten children, all students most frequently identified

TABLE 12.2. *Highest scores by grade (second, fourth and sixth) and community*

	Highest Score (Correct Identifications)		
	Grade Two	*Grade Four*	*Grade Six*
Mountain Peasant Community			
Photograph:			
Urban Middle Class	7	7	8
Urban Lower Class	6	8	6
Mountain Peasant	8	9	8
Coastal Plantation	6		8
Mean Highest Score, All Photographs	6.75	7.75	7.5
Urban Middle Class Community			
Photograph:			
Urban Middle Class	10	10	11
Urban Lower Class	10	10	10
Mountain Peasant	8	8	8
Coastal Plantation	10	8	9
Mean Highest Score, All Photographs	9.5	9.0	9.5

houses. But the category "road" became one of the three most frequently mentioned in Corozal by the second grade and in Villa Nevarez by the sixth grade. By the sixth grade, the generic category "buildings" had risen to fifth place in the identification ordering of Villa Nevarez children, having been pointed out by sixty-two of the ninety children tested.

Proceeding to a highly speculative realm, it may be that the equivalent "wrong response" levels of the urban and rural groups from grade II on is an indication of the effectiveness of schooling in "socializing" the rural child, in molding him to the urban standard that "risk-taking" is somehow a thing to be avoided. Further research would be needed to support or refute this contention.

We attempted to correlate ability to read aerial photos in the Puerto Rican communities with toy play behavior. A standard kit of environmental toys consisting of low- and high-rise buildings, cars, trucks, trees, and parts of a train, was presented to these children and they were asked to "play with the toys" on a large sheet of paper. When they had finished, or after fifteen minutes had passed, the positions of the various objects on the paper were traced. While the data have still to be fully analyzed, it is evident that in the five-year-old bracket there are some children who model communities with environmental toys while others use some noncommunity basis for arrangement. Some use a Gestalt (e.g., a town) to label what they have produced, while others call their product only by the elements incorporated. More detailed analysis should tell us something about the kinds of forms produced, their relative degree of order, and, when the testing is extended to lower ages, when community modeling first appears.

Although the data have still to be statistically analyzed and scrutinized in detail, it is clear that some children can read aerial photographs at the age of five, can trace "maps" from these photos at a slightly later age and use these maps to draw trip routes (Blaut, McCleary, and Blaut, 1970), and can use aerial photos to learn map reading as a symbolic language (Muir and Blaut, 1970). (Two four-year-olds did remarkably well on the identification task with poor aerial photos, but further corroborating evidence at that age level is needed.) These facts run contrary to those theories in psychology and education that tend to ignore learning that occurs in the absence of formal instruction during the preschool years. Moreover, our results suggest that the formation of "cognitive" or "mental" maps begins at quite an early age, and finally, that *schemata* or cognitive organizations of the child's world, and indeed, the world of any organism that requires the representation of the world as it would be if viewed from above, are fundamentally adaptive.

Possible Implications of our Findings for Education

Spatial learning has been neglected as much by educators as by psychologists engaged in the study of learning. Primary and secondary school geography classes still spend an overwhelming amount of time on the nonspatial aspects of geography; cartography as such is often not taught at all, and maps serve as an (often, as they are used, dispensible) adjunct to other aspects of study —when they are not being used merely as coloring exercises. The emphasis is still upon information conveyed by our major mode of communication: the printed word.

Much has been written and much much more spoken about tremendous difficulties encountered by children—especially "disadvantaged children"— in learning to read. Let us consider what is involved in reading, one form of communication via a system of abstract symbols. A letter is a nonideographic symbol. Apparently arbitrary combinations of these abstract symbols make up words, some of which have referents in concrete objects ("horse") or observable actions ("trotting"). Centuries of usage have caused these symbol aggregates—first, second, sometimes third order abstractions—to be combined in ways that seem arbitrary to the young child, and only somewhat related to his spoken usage. Only in some cases does a one-to-one correspondence between the word and the object exist, and even then the symbol is so complex in itself (how many squiggles are there in "horse"?) and so difficult to distinguish from other symbols that we ought not to be surprised that the child experiences great difficulty.

If we view the "reading problem" as a member of a more general problem group—learning languages in graphic form, with language defined as generally as possible—then we may be prompted to ask: Is there some notational system, some language, which is easier to learn than that used in reading, that will teach the child the principle of abstract symbols and their combination by building upon things he already knows?

Based upon the research findings outlined, we feel that the answer to the above question is "yes" and we further suggest that the usual order of teaching—reading first, then map reading in the later grades—is backwards. For if a child, in the absence of any formal instruction, can correctly identify objects in a vertical aerial photograph at age four, and if this enables him to read and interpret the symbols on a map (two-dimensional graphic symbol system) after only one month's instruction at age six, then we already have what we are seeking.

But if this "appreciation," if we may call it that, of the spatial relation among entities is initially there to build upon, it seems to be ephemeral. Replaced by other modes of understanding, it apparently decays as the child "advances" by other standards of the educational system. If sixth-graders are no more adept at aerial photo interpretation than are first-graders, then adults, too, may be no better. While we will not go so far as to suggest this is why adults find maps "so confusing," we do suggest that early instruction in cartography may aid the learning of languages in general. Further study is required to bear out the latter contention. But learning cartography as a multi-dimensional language may be of value in its own right, presenting the child with an alternative to the rigidity of linearly associated symbols.

13

The Black Boxes of Jönköping: Spatial Information and Preference

PETER R. GOULD

The Chain to Urbanization

One of the most striking aspects of the population pattern in many developed countries is the extraordinary unevenness of the distribution. Intense agglomerations of people form sharp peaks on the population surface, these summits being approached from all directions at exponential rates. In most countries, the urbanization process has run virtually unchecked so that today many of the urban mushrooms are coalescing into continuous strips to form the megalopolises of tomorrow. High densities, overcrowding, air pollution, chronic problems of waste disposal, and virtually unmanageable rates of sensory stimulation are only a few items in a long and wearyingly familiar list that characterizes much of human life in these areas. The results of such cumulative environmental stress can be seen in the glazed eyes of numbed commuters, the high rates of mental illness, burgeoning levels of crime, and appalling rates of drug addiction.

The apparently chronic levels of urbanization are the latest manifestation of a long and circularly causal chain stretching back more than a century. Much of the growth came from large numbers of rural migrants flowing to the cities. Often they followed the same sort of stepping-stone route—from village to provincial town to big city— that characterizes population movements in the underdeveloped world today. For example, many young men and women in Africa perceived the towns and cities as highly desirable places of residence where new opportunities will lead to a socially more exciting, economically less uncertain, life. Cities are in the swing of things; the rural areas are disparagingly described as "the bush."

The perception and evaluation of residential desirability varies markedly from place to place, and may be described by perception surfaces, or mental maps, characterizing the shared outlook of a group of people at a particular location. But the way in which such surfaces are formed and molded has been much more difficult to specify. Presumably people's perception of various places is influenced by the spatial information they have, but this only pushes the basic question back as we try to specify such information in a reasonably plausible manner. And even if we succeed in this task, the aggravating scientific child in us still whines a plaintive "Why?" Why does the spatial information available to a group of people at a particular place vary? What lies behind the *information* that leads, perhaps through *perception* and *migration*, to *urbanization*? Perhaps the following Swedish data allow us a modest speculation about the answers to such questions.

The Children of Jönköping

As a small part of a much larger study investigating residential perception surfaces in Sweden, the preferences of many children were sampled at Jönköping, a town in south-central Sweden lying at one end of Lake Vätter and between the three main urban focuses of Stockholm, Göteborg, and Malmö (fig. 13.1). The children worked directly upon large maps showing the seventy A-level, or functional regions, with major roads, towns, and waterbodies shown as a light blue underprinting to give some general orientation. Preferences were recorded on the maps as ranked values from one to seventy, and from a basic data matrix of seventy regions $\times$ N children, scores on the principal component of the correlation matrix were calculated for each of the areas. These values were plotted and used as interpolation points to construct a perception surface lying over the map.

The youngest children, only seven-and-a-half years old, were unable to give their preferences. Unfamiliar with maps, with little information to guide them, their preferences hardly seemed to have crystalized out of the spatial flux. But for those nine-and-a-half years old (fig. 13.1), a collective image has already formed, although the degree of agreement in the group is still quite low. If we think of the overall image shared by a group of children as a signal subject to the random noise of ignorance and indifference, then the signal-to-noise ratio is only 23.4 on a scale ranging from one to one hundred. In fact, with the samples of forty used here, we might expect a value of 7.1 to occur if each child had a completely random set of preferences with respect to all the others. Since the variance of the distribution of expected values is very small, we have strong evidence that we have a weak, though quite significant, preference signal.

The most striking aspect of this mental map is the strong local dome of desirability that stretches from the perception point at Jönköping, the peak of one hundred on the perception surface, westwards to the city of Göteborg. The isopercepts decline in all directions from this ridge of local desirability, rising to a definite peak again at the extreme southern tip of Sweden around the urban center of Malmö and its neighbor Hälsingborg. The capital, Stockholm, is not a peak, but lies on a plateau of average

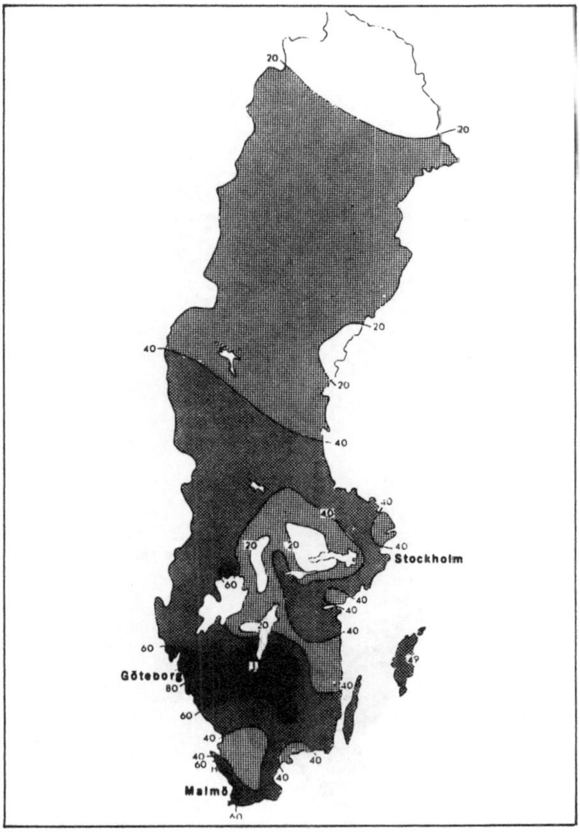

FIGURE 13.1.

values broken by deep perceptual sink-holes in middle Sweden. If we go back to the original, raw data we can see that Stockholm suffers from a split image—it is either rated very high or very low—so its perceptual score is only moderate for the overall, group viewpoint. Many of the other regions are also in this middle range, from forty to sixty, indicating that preferences are not strong. We have a lot of indifference, ignorance, and randomness in this surface as we might expect from the low level of overall agreement.

By the time children are eleven-and-a-half years old, however, spatial preferences begin to sharpen (fig. 13.2). The measure of agreement, or the signal-to-noise ratio, rises to 34.3, and a much crisper configuration is seen on the map. High preference values blanket the south, and there is a distinct north-south split as no northern region reaches a score of even twenty. The local dome of desirability, extending to a ridge at Göteborg, is joined by two other peaks, Halmstad to the southwest, and the island of Visby off the east

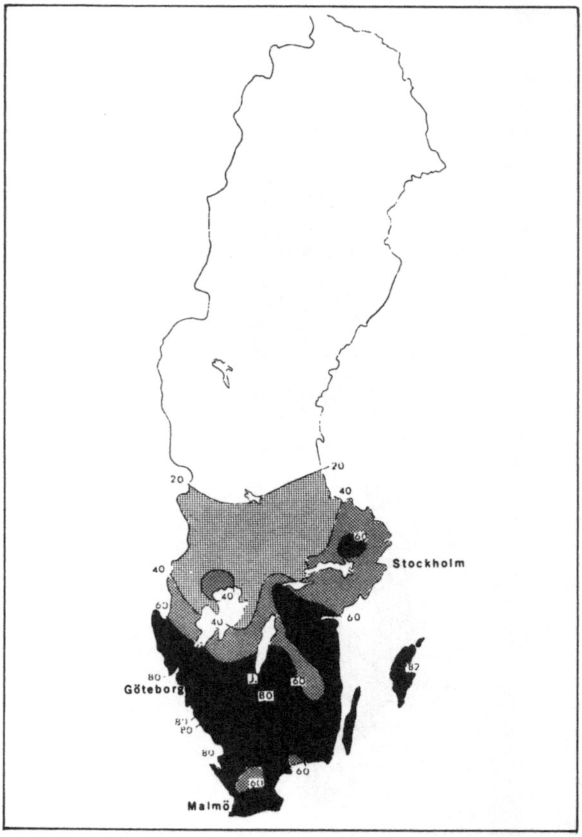

FIGURE 13.2.

coast. These are both vacation areas, and we might expect children with some holiday experience to view them very favorably. Stockholm still suffers from a split image and lies, once again, in the middle range of values. However, on the thirteen-and-a-half year old mental map (fig. 13.3), the capital finally emerges as a major peak, joining the local area of Jönköping, and the two other major urban centers of Göteborg and Malmö. While the distinct north-south split is still extremely strong, it is notable the way the older children are far more discriminating. Whereas the south was blanketed before by moderately high scores, now the southern portion of the perception surface is more convoluted as preferences are assigned with greater care. Not only do the major cities emerge more clearly as peaks, but the local preference ridge to Göteborg has been eroded, and Visby (nice for a holiday . . . but living cut off there all year?) has suffered the greatest decline of any region.

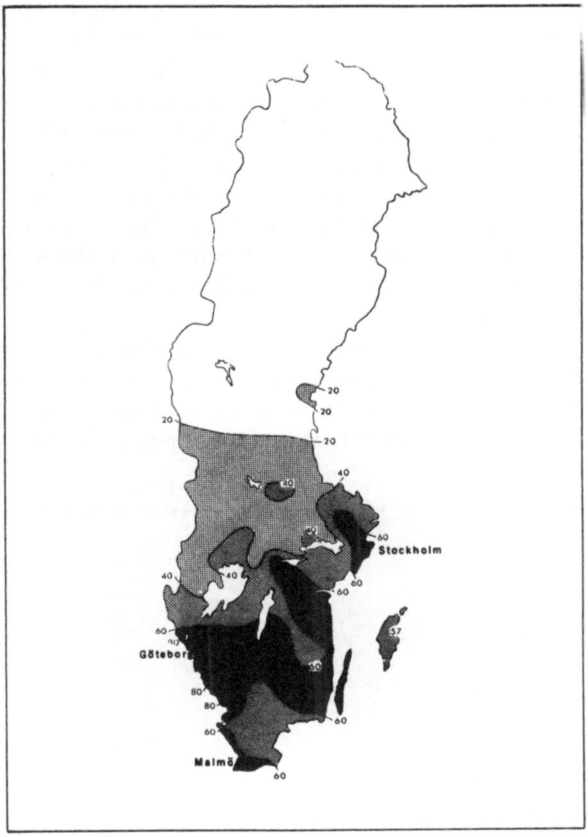

FIGURE 13.3.

Flows of Spatial Information

One of the very basic, yet most difficult questions to consider is the information that people have available to them as they try to think through their own, quite personal preferences. Spatial information presumably comes from reading, listening, radio and television, traveling, formal and informal teaching, and a host of impressions ranging from visits to grandmothers to brightly colored railway posters. While some progress has been made with polar bears, we obviously cannot hang automatic tape and light recorders around the necks of people from the day they are born to measure and analyze all the sense impressions about geographic space. We must be content with a much cruder, *surrogate* measure of information; in this case, the information stored in the little black boxes of Jönköping. As an experiment, the children were asked to write down, in exactly five minutes, all the names of villages, towns, and cities in Sweden they could remember. Individual lists were not expected to be very enlightening, but by requiring the place information in the black boxes to be spilled without warning, and under time pressure, it was hoped that some crude, but meaningful aggregations could be made. The individual lists from a group of children were simply recorded on a map of the seventy regions, and the raw counts used as measures of the information available. No attempt was made to weight the places by their positions in individual lists, but there are some very crude indications that the children mentally slide down the urban hierarchy as they try to recall names. But this question is far from being properly pinned down at the moment.

From the raw counts generated by groups of children of different ages, we can construct "information surfaces" lying over Sweden (fig. 13.4). Note that the intervals between the isolines defining the surfaces are exponential, each value being three times that of the next lowest. All the maps

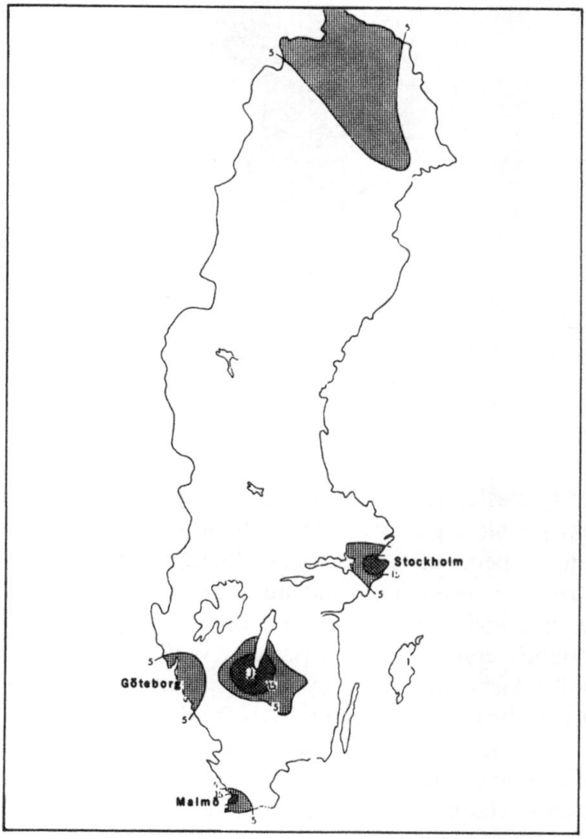

FIGURE 13.4.

are directly comparable on an absolute basis since the same number of children (forty) contributed to each. The seven-and-a-half year old children have a rather low information surface, with the highest point directly over the local area itself. From Jönköping the surface declines in all directions, rising only around the three major nodes of Stockholm, Göteborg, and Malmö. The rest of the information surface is very low, consisting mainly of "grandmother noise" between the strong signals from Sweden's principal cities. A small rise in the far north at Kiruna reminds us that the sample was taken during the time of the big strike in the iron mines. This was a major, if somewhat embarrassing event in the country's excellent record of labor-management relations, and it was recorded at length over both the national television stations. Even seven-and-a-half year olds were conscious of the name at the time.

For the nine-and-a-half year old (fig. 13.5), the information surface has risen considerably, and we should recall that this is the first age for which

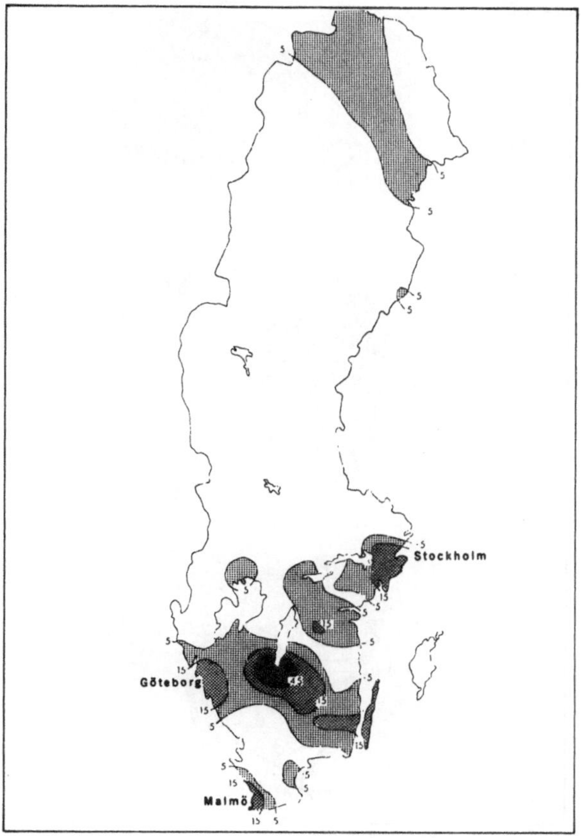

FIGURE 13.5.

we are able to construct a preference surface for children. Much more information is available, and the local dome has more than doubled in height. While there is still a decline in information away from the perception point, the local effect reaches out strongly along a corridor to the east to include the narrow island of Öland, perhaps as a result of holiday trips to this famous resort area. The major urban areas are still prominent peaks of information, and between Jönköping and Stockholm another hump of information is forming around the major towns of Linköping, Norrköping, and Örebro. Elsewhere the surface is very low, and a few of the regions are still at zero level—no child in the group has recalled *any* spatial information about them. In the north, the effect of the Kiruna strike is clear, but the black boxes are also spilling forth a little information about some of the northern towns like Luleå, Piteå, and Umeå. These form small informational blisters in the otherwise low-lying surface.

For the eleven-and-a-half year old (fig. 13.6), some of the interstitial

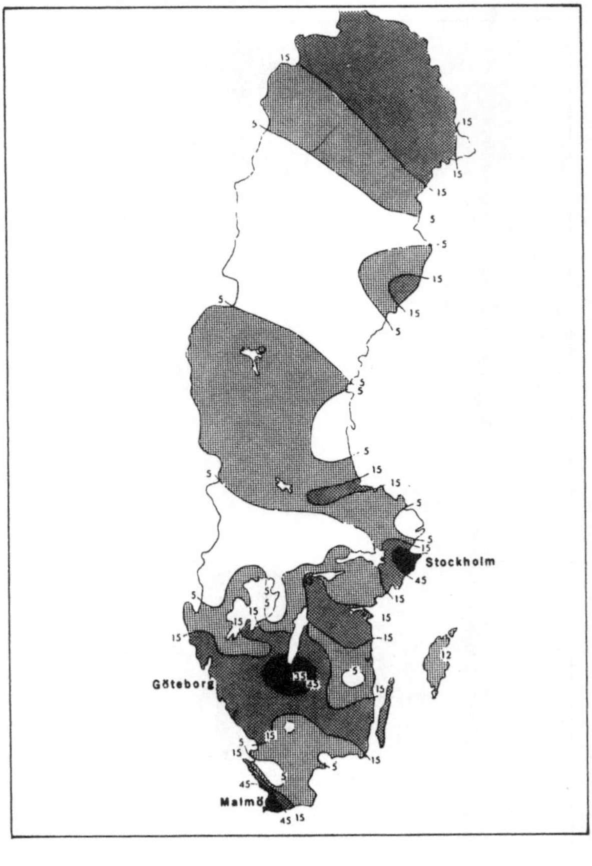

FIGURE 13.6.

areas are beginning to fill in with information, and no region remains at the zero, unrecalled level. The local information peak is extremely pronounced, and Stockholm and Malmö have also broken through to the next isoline. In the north, Umeå stands out as an island of information, while the area around Kiruna and the coastal town of Lulea is at a similar level. But the surface is not rising evenly as the children get older, and deep pits of extremely sparse information characterize the south, while middle Sweden, directly west of Stockholm, is an informational desert. Modest increases over the previous surface occur somewhat further north in an area famous for its old folk traditions in mid-summer, and for its skiing in the winter.

Finally, for the child thirteen-and-a-half years old (fig. 13.7), the local dome seems to be reaching a saturation level, but areas immediately around it in the south are still filling in with information. Göteborg breaks through to the second level to form the fourth distinct peak, but middle Sweden

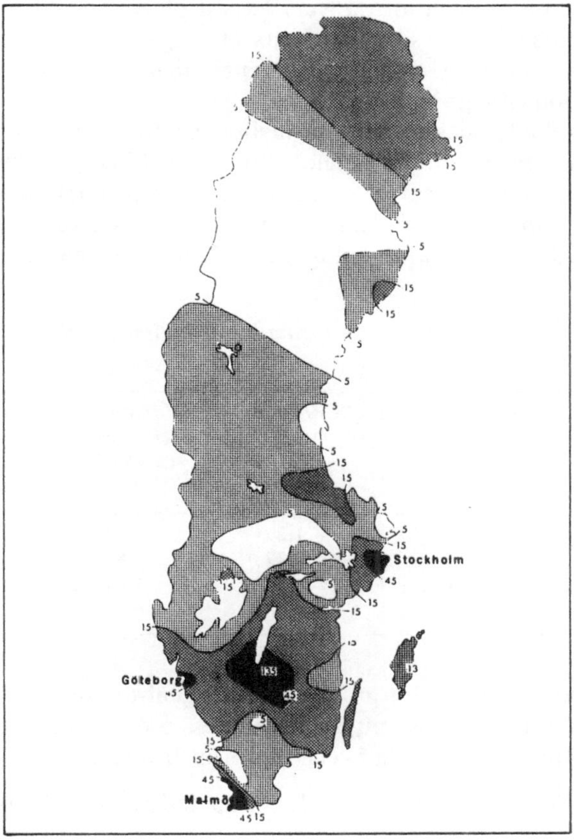

FIGURE 13.7.

remains a virtual blank. In the north, the surface has hardly changed from the younger group's viewpoint.

How can we account for the growth of these information surfaces—surfaces that develop in such a regular manner that we can almost use the first series of maps as a crude model to predict the next in the sequence? Perhaps the black boxes of Jönköping are small information receivers picking up and storing spatial information transmitted by the various regions of Sweden. People, of course, generate information, and we might expect regions with large numbers of people to transmit more information than those with just a few. On the other hand, regions far away from Jönköping might have weaker signals than those nearby. Presumably information is received and stored after a signal reaches a certain threshold value; that is, it becomes sufficiently strong in the cumulative sense. This could occur in two ways: either by children being subject to very strong signals for a short time, or by receiving weaker signals over a longer period. Young children seem to store only the strongest signals of information from small places close to them, or the very largest places far away (fig. 13.4). Older children, subject to six more years of "regional transmission," store the cumulative impact of many weak signals that never appear to be received by the youngest group (fig. 13.7).

We can test the basic idea that information is a function of the population of a generating region and its distance away from Jönköping by casting the problem in the familiar gravity model form. In this simple form, with the variables transformed to logarithms, all the information surfaces are highly predictable (table 13.1), with multiple correlation coefficients around 0.74.

TABLE 13.1. *Values for the information equation for the children of Jönköping*

Age	*Multiple Correlation Coefficient*	*Constant* $a_{1.23}$	*Population Regression Coefficient* $b_{12.3}$	*Distance Regression Coefficient* $b_{13.2}$
7½	0.64	−1.27	0.73	−.31
9½	0.74	−0.87	0.86	−.54
11½	0.73	−1.19	1.01	−.40
13½	0.74	−0.60	0.84	−.41

The surface for the youngest group, whose signal-to-noise ratio we might expect to be a little lower, is only somewhat less predictable.

Taking the thirteen-and-a-half year olds as an example, we can write:

$$\text{Log Information} = -0.60 + 0.84 \text{ Log Population} - 0.41 \text{ Log Distance.}$$

Of course, we would not expect the equation to predict the *exact* amount of information for every area, but by substituting each region's population and distance into the expression we can examine the major exceptions, or residuals, to the overall relationship. Positive residuals, where children have considerably more information about a region than we would expect simply from its population and distance from Jönköping, are found immediately around the local area and at the holiday centers on the south coast. The northern regions are also positive anomalies, perhaps because of the great attention given to the strike in the Kiruna mines. Negative residuals lie in middle Sweden; that is, the equation *over*-predicts the information that the children actually have.

Information and Preference

The relationship between the information and preference surfaces is a very difficult one to untangle because it is obvious that information can be both negative and positive. Information and preference definitely move together ($r_s = 0.54$), but this naive expression hides more than it discloses. For six regions in the north, the children have quite a lot of information, but they still record very low preferences. Conversely, the children have little information about three of the sinkholes in the south, but these appear to acquire an aura of desirability by their southern location near regions of very high preference. If these nine (six in the north, three in the south) anomalous regions are removed from the analysis, the overall relationship between the information and preference surfaces strengthens considerably to 0.75. But such an analysis, even if it can be so dignified, is very crude and unsatisfactory. The problem of negative and positive information is a very difficult one, and a firm grip on it has not been achieved by the very simple measures of information used here.

14

The Development of Spatial Cognition: A Review

ROGER A. HART AND GARY T. MOORE

This paper is a review of current knowledge and a guide to the literature on the development of spatial cognition. We are interested in how people develop a knowledge of space. First, how do the fundamental concepts of space develop in the child? And, second, how do we come to know and represent in cognition the everyday physical environment?

In considering the child's conception of space, a number of questions arise. Is space given innately, is it built up empirically from sensations, or is it some interaction of the two? Is there an order of succession or sequence in the child's understanding of space? What is the role of experience in the understanding of space; most particularly, what are the relative roles of action, perception, imagery, and language? Further questions arise in considering how the everyday physical environment is known. How do children and adults develop a knowledge of the spatial entities and relations of large-scale environments like neighborhoods and cities? What systems of reference do they use? These questions pose a set of issues sufficiently im-

The authors are respectively Fulbright Scholars in Geography and Canada Council Fellows in Psychology at Clark University. Grateful acknowledgement is made for support in the preparation of this review by the Place Perception Project at Clark University, supported in part by the U.S. Office of Education (Grant #4493). The authors would like to thank the following people who have offered comments and suggestions on various parts of this work: James Blaut, Leonard Cirillo, Roger Downs, John Gittins, Bernard Kaplan, Leonard Mark, Richard Melito, David Seamon, David Stea, Jacques Voneche, Denis Wood, and Victor Zinn, plus members of the Environmental Cognition Workshop, Norman Carpenter who prepared the illustrations, and Gloria Graves, Phyllis Mosczynski, and Pat O'Donnell who prepared the manuscript. The review covers all literature to May 1971 located by thorough searches of *Psychological Abstracts* (1927–1972) and *Current Geographical Publications* (1938–1972) and from suggestions offered by our colleagues. A longer working-paper version of this chapter including studies excluded from this review due to space limitations is available from the authors.

portant to environmental design and planning and to environmental education to warrant special consideration.

Thus, this review is concerned with the development of cognitive representations of space, beginning with the child's first spatial concepts and advancing toward the adult's cognitive representations of large-scale environments. As will be seen more clearly in the next section, our focus is on the major empirically-based theories and related findings on the structure or form of space as seen from a Wernerian-Piagetian structural-developmental point of view. In systematically reviewing the literature back to the early part of the century, we have been impressed by, and tried to show, the fact that most of the data on the development of spatial cognition, including that on large-scale environments, can be comfortably explained within this theoretical point of view. Much of the work done within a strictly behavioristic framework is fragmentary and does not purport to trace comprehensively the developmental sequence. However, this data is discussed and compared with other more developmental work. Most of the work from the psychoanalytic, anthropological, and philosophical points of view addresses a somewhat different question—the content or meaning assigned to space—and is not reviewed here.

The review is in three sections. First, an attempt is made to provide some definitions and conceptual distinctions as a base for the field. The second section covers development of the fundamental concepts of space, starting with the epistemological problem of space and the resolutions of Déscartes, Berkeley, Kant, and Cassirer. Out of this tradition come Werner's and Piaget's empirically based theories of the development of the child's conception of space. Because Piaget has provided the most comprehensive explanation of the development of spatial cognition, an attempt is made to summarize his lengthy volumes and related papers. Piaget's theory has been the subject of many replicative and follow-up studies; these are briefly discussed with special attention to the work of Laurendeau and Pinard. Also, a survey is made of the different modes of representation which may be utilized in spatial cognition. The third section focuses on empirical research on the development of cognition of the larger physical environment. Following a discussion of the ontogenetic development of both geographic orientation and topographical representations and the systems of reference they utilize, we lead into the scant literature which deals with the microgenetic development of representations of new or unfamiliar environments. Parallels and discontinuities with the theories of Piaget and Werner are discussed throughout as a means both of providing coherence to otherwise unrelated studies and of suggesting avenues of theoretical explanation.

I. *Definitions and Conceptual Distinctions*

To provide a base for the field of spatial cognition, certain definitions and conceptual distinctions will be reviewed. First, we must clarify the terms *spatial cognition* and *cognitive representation* which, as used by the leading developmental psychologists, encompass the more specific terms *cognitive mapping* and *cognitive* or *mental maps* used both in the environmental behavior literature (Downs, 1968; Stea and Downs, 1970; Blaut, McCleary, and Blaut, 1970) and elsewhere in this volume (chap. 11). For instance, the Russian psychologist Shemyakin (1962, p. 190) defines "spatial representation" as a "reflection of space in the minds of men . . . a reflex activity of the brain [which] creates for men a unified system of space." Piaget and Inhelder (1967, pp. vii, 454) define "spatial concept" as "the fundamental idea of space," and "spatial representation" as the symbolic and internalized mental reflection of spatial action. Werner (1948, p. 167) also refers to a "spatial concept" as the fundamental idea of space, and "representation" as an "intellectualized version [of] reflective thought." Similarly, Laurendeau and Pinard (1970, pp. 13–14) define "spatial representation" as "an implicit action which is carried out in thought on the symbolized object . . . a mental reproduction [or] sketch of an object in thought."

Each of these theorists is referring to an internalized *cognitive* representation of space, as opposed to external representations such as children's drawings (see Arnheim, 1967) and the maps of cartographic psychophysics (Williams, 1958; Ekman, Lindman, and William-Olsson, 1961; Robinson, 1967; M. Wood, 1968; Dent, 1970). It will be helpful, therefore, to distinguish between *external* representations and *internal* or *cognitive* representations. We can only *infer* internal representations from external representations (e.g., drawings, maps, verbal reports, models) or from overt spatial behavior. Thus, external representations are only of interest in this review to the degree that they shed light on the development of the internal representation of space. In summary, *spatial cognition* is the knowledge and internal or cognitive representation of the structure, entities, and relations of space; in other words, the internalized reflection and reconstruction of space in thought.

The commonly used (and sometimes misused) terms *cognitive maps* and *cognitive mapping* imply map-like representations of geographic or other large-scale environments. As has been argued before (Blaut, 1969; Stea, 1969), it begs the question to suggest that spatial relations are necessarily represented in a cartographic form. Therefore, we prefer to use the more inclusive terms of developmental psychology—*spatial cognition* and *cognitive representation.* (It will be seen later that "cognitive mapping" is only one form of cognitive representation of large-scale environments). The terms *fundamental spatial cognition* and *macro-spatial cognition* are used, however, to distinguish temporarily between two varieties of spatial cognition: the development of fundamental concepts of space, and the further differentiation and elaboration of these concepts into the development and representation of large-scale environments. This distinction, however, should not be taken to imply that these are necessarily different entities or follow different psychological processes and laws. The distinction is used in this review simply to distinguish between two bodies of literature, one primarily from developmental psychologists concerned with understanding the development of basic spatial concepts, and the other from a wide range of geographers, urban planners, psychologists, anthropologists, and educators concerned with large-scale environments, but who, for the most part it seems, have paid little attention to the work of developmental psychology. This review, then, is a first step toward bringing these two bodies of literature together.

The review focuses primarily on the *form* or *structure* of space. The fascinating questions of the development of the *experience* of space (environmental experience), of *mythical conceptions* of space, and the *content* of space (the *meaning* and *values* assigned to spaces at different times and by different cultures), has received attention largely in the phenomenological, anthropological, and psychoanalytic literature, and is not reviewed here (see Hallowell, 1942, 1955; Cassirer, 1944, 1955; Werner, 1948; Eliade, 1959; Rasmussen, 1959; A. Wallace, 1961; Scully, 1962; Erikson, 1963; Romney and D'Andrade, 1964; Straus, 1963, 1966; Beck, 1967; Hall, 1966; Bachelard, 1969; Buttimer, 1969; Rusch, 1970; Wisner, 1970; Tyler, 1971; see also the reviews by Lynch, chap. 16, and Beck, 1964, pp. 3–33).

Another area of research closely related to the above is that of *environmental dispositions and preferences*. Unfortunately this latter area was entitled at one time "mental maps" (chap. 11) thus causing others to believe it a part of spatial cognition. This research deals with subjective or evaluative responses to environments, and only in some cases to maps or other external representations. Even in these latter cases, however, the external representations are constructed post-hoc by the experimenter and do not necessarily imply any parallel internal representation by the subject (see reviews by Craik, 1970a, 1970b).

An important distinction should be made between *spatial cognition* and *spatial perception*. Cognition, by lay definition held to include all the modes of knowing (perceiving, thinking, imagining, reasoning, judging, and remembering), would seem to include perception. Piaget and his followers (Piaget, 1963b, 1969; Laurendeau & Pinard, 1970) suggest that knowledge of the world includes two aspects: one of which is essentially *figurative*, related to the percepts or images of successive states or momentary configurations of the world by direct and immediate contact, and a second which is essentially *operative*, related to the operations which intervene between successive states and by which the subject transforms parts of the world into reconstructable patterns or schemas. Visual perception is only one form of figurative knowing, while cognition (or intelligence) is based on the operative mode. Another closely allied view, that of Werner and his followers (Werner, 1948; Wapner and Werner, 1957; Wapner, 1969), treats perception as a subsystem of cognition. In this view, knowledge about the world may be constructed by many means, perceptual judgments being only one. As development proceeds, perception becames subordinated to higher mental processes. Cognitive structures available to the organism influence perceptual selectivity which leads to a reconstruction of the world through selected fields of attention. Perception is thus both a subsystem of cognition and a function of cognition (see Langer, 1969, pp. 148–156, for a review of these two positions.) Spatial perception and spatial cognition, therefore, are two separate but reciprocating processes; the precise relationship will be discussed later (see reviews of spatial perception by Wohlwill, 1960; Howard and Templeton, 1966).

An additional distinction should be made between *spatial cognition* and *spatial orientation.* Spatial orientation, or geographic orientation as used in this review, refers to the way an individual determines his location in the environment. It utilizes a topographical cognitive representation which is related by a reference system to the environment. Geographical orientation is discussed to the extent that it sheds light on the more general development of topographical representations (see reviews by Shemyakin, 1962, and Howard and Templeton, 1966).

THE CONCEPT OF DEVELOPMENT

Finally, we must distinguish between *development*, *learning*, and related concepts, and indicate the value as we see it of a structural-developmental approach to spatial cognition. Four concepts have been associated with transformations in an organism: growth, maturation, learning, and development (Harris, 1957). Briefly, *growth* is a nonspecific term generally implying any form of accretion, while *maturation* is more specific, implying physiological growth. It is more difficult to distinguish learning from development, a traditional controversy. *Learning* involves quantitative changes in the reception and retention of information or subject matter. It refers to the situation in which information is presented to the individual who changes through reacting to it and corrects initial attempts in response to indications about his prior successes (Ausubel, 1963; Hilgard and Bower, 1966; Deese and Hulse, 1967). On the other hand, *development* implies qualitative changes in the organization of behavior. Most often it refers to the situation where the individual changes as a function of interaction between current organization and discrepancy with the environment (see Werner, 1948; Piaget, 1963a,b,c; a similar distinction is made by Flavell and Wohlwill, 1969). The basic aims of the study of development are twofold: to describe the characteristic pattern of each level of organization, and to explain the relationship and transformation between these levels (Werner, 1948; Werner and Kaplan, 1963; Piaget, 1963b; Langer, 1969). Thus, theories of development are concerned with qualitative changes in structural organization, whereas theories of learning are concerned with quantitative changes in the incorporation of specific information into structures (see White, 1970, for an excellent intellectual history of this issue).

Inherent in the structural-developmental point of view are some basic theoretical principles. White (1970) suggests that the broadest contemporary formulation of the position is the *comparative organismic-developmental theory* of Werner and his two principal colleagues at Clark University, Kaplan and Wapner (Werner, 1948, 1957; Wapner and Werner, 1957; Werner and Kaplan, 1963; Kaplan, 1966a, 1966b, 1967, in press; Wapner, 1969; Wapner, Cirillo, and Baker, 1971; see Langer, 1969, for a comparison with Piagetian theory).

One of the most important contributions of this group was to render the concept of development context-free so that it is not limited to processes unfolding over time or in relation to age, but is conceived as an *ideal of natural order* (Toulmin, 1961). In this way development becomes a manner of looking at and conceptualizing phenomena (Kaplan, 1966a, 1967). Implicit in the developmental point of view, and sometimes in tension with it, is the *organismic assumption* (Kaplan, in press). A human being is an organized unity in which the laws of the whole govern the functioning of the parts. Thus, human behavior is only understood in terms of, and in relation to, this underlying organization. The organismic focus on the description of structural organizations together with the developmental focus on the transformations between different levels of organization leads to the organismic-developmental perspective (see Moore, 1970, for a comparison of the assumptions of organismic-developmental and general systems theories).

Much of the richness of a developmental approach is contained in the *comparative analysis* of different forms of development. As well as *ontogenesis* (individual development over the life-span), other forms of development may be studied, including *microgenesis* (short-term individual change), *pathogenesis* (psychopathological change), *ethnogenesis* (cultural change), *historicogenesis* (historical change), and *phylogenesis* (evolution of species). The final step for a comparative organismic-developmental approach is to compare the results from each behavioral domain and to derive developmental laws applicable to mental life as a whole (Werner, 1948; Langer, 1970).

In summary, then, the values of the structural-developmental approach are twofold: first, only by understanding the development of a phenomenon can we fully understand and scientifically explain its mature form, or as Werner (1948, p. 5) has stated: "Complementary to [the discovery of the structural pattern] is the task of ordering the genetic relationships between particular levels." And Piaget (1970, p. 12): "Were it not for the idea of transformations, structures would lose all explanatory import, since they would collapse into static forms." Second, as any organism or system is always undergoing transformations, we might say that the only thing which is constant is change itself. Thus change or development becomes of prime scientific interest. (Similar arguments have been made in anthropology by Lévi-Strauss, 1949, and in geography by Blaut, 1961.)

II. *The Development of Fundamental Spatial Cognition*

Before we can ask the question of the development of knowledge of space in large-scale environments, we must ask the logically prior question of the genesis of the fundamental concepts of space in the child. The first question is whether space is ideal or real; that is, whether the concept of space is given innately to the child or is built up empirically from sensation. The rationalist philosopher Descartes argued that the concept of space is given immediately as an innate idea before experience, whereas the empiricist Berkeley held that reality could only be contained in sensation. Kant (1902; Hendel, 1953) argued, however, that the *matter* of all phenomena (that which corresponds to sensation) is given in experience, but that their *form* is given *a priori*. Whereas both Descartes and Berkeley assumed that one could understand the ultimate nature of reality, Kant argued that since there is no way for us to apprehend the nature of "reality" except through man, it is impossible to completely separate the process of knowing from the resultant knowledge. There can be no complete understanding of truth in either sense or reason. Therefore, Kant argued, instead of naively assuming that knowledge can ever represent exactly what is real, we are led to the conclusion that what we take to be real is a product of the act of knowing (a *construction* of thought).

CASSIRER'S THEORY OF SYMBOLIC FORMS AND SPACE

One of Kant's greatest followers was the philosopher Ernst Cassirer (1944, 1953, 1955, 1957) who attempted to understand the fundamental forms of human culture. Following Kant, Cassirer called space and time the principal modes of experience. He thus treated the problem of space extensively.

Cassirer (1944) was one of the first theorists to deal with space developmentally. He saw three fundamentally different types of spatial experience, as shown in figure 14.1. The lowest order he termed *organic* or *active space*. The second order, *perceptual space*, a farther-reaching and more complex space, is characteristic of higher animals. The integration of different kinds of sense experience—visual, auditory, tactual, vestibular, and kinaesthetic —is not possible for animals lower on the phylogenetic scale. The highest order, *symbolic* or *abstract space*, is of greatest interest to Cassirer (1944, p. 43) because it is "the borderline between the human and animal worlds." Humans alone develop the ability to comprehend and represent the idea of abstract space—the space of "pure intuition," bereft of any necessary concrete referent.

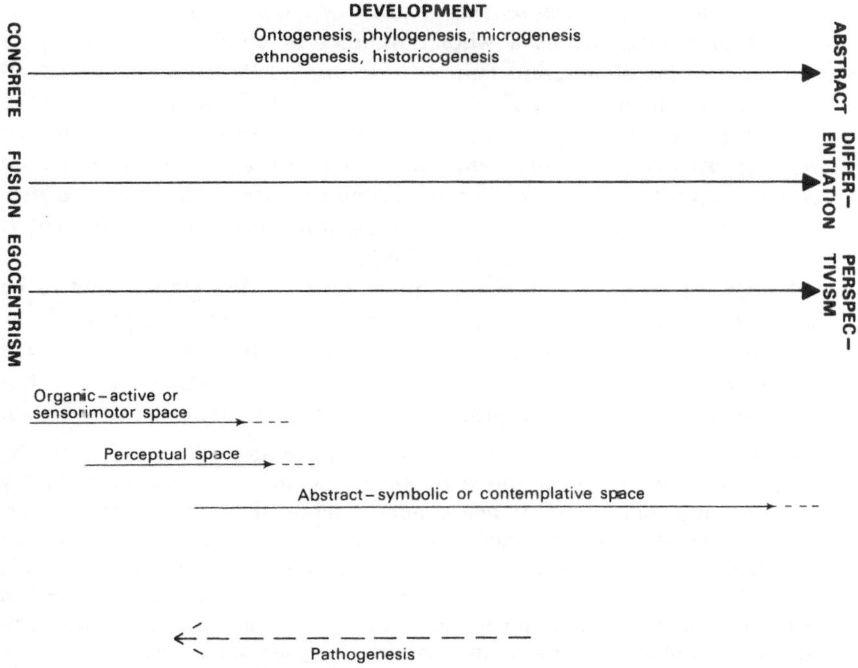

FIGURE 14.1. *Schematic representation of the levels of spatial experience recognized by Cassirer and Werner.*

Cassirer, like Werner who followed him, was a pioneer in the *comparative analysis of development*. Cassirer (1944, p. 46) suggested extensive parallels between ontogenesis, phylogenesis, ethnogenesis, historicogenesis, and pathogenesis, and on this basis made an important distinction between *concrete acquaintance* with, and *abstract knowledge* of, space and spatial relations:

> *Acquaintance means only presentation; knowledge includes and pre-supposes representation.* The representation of an object is quite a different act from the mere handling of the object. The latter demands nothing but a definite series of actions . . . it is a matter of habit acquired by a constantly repeated unvarying performance of certain acts. But the representation of space and spatial relations means . . . we must have a general conception of the object, and regard it from different angles in order to find its relations to other objects . . . and determine its position in a general system. (italics added)

WERNER'S ORGANISMIC-DEVELOPMENTAL THEORY OF SPACE

The first comprehensive theory of the development of space to be derived primarily from empirical findings was that of Werner (1948, 1957; Werner and Kaplan, 1963; see also Baldwin, 1967; Langer, 1969, 1970).

General Theory of Development. Werner wished to understand the development of mental life as a whole, whether in children, animals, psychotics, or peoples of different cultures. He argued that the essence of all forms of development is the differentiation and subordination of parts to the whole, formalized as the *orthogenetic principle*: *Insofar as development occurs in a process under consideration, there is a progression from a state of relative globality and lack of differentiation to states of increasing differentiation, articulation, and hierarchic integration* (Werner, 1948, 1957; Werner and Kaplan, 1963; Kaplan, 1966a).

Although the term "development" might seem at first glance only to be useful for the description of processes changing over time, the principle as used by Werner and by Piaget defines development in terms of the *degree of organization* and, thus, is not *limited* to processes changing over time, but also may be used for the conceptual ordering of contemporaneous systems.

> The more differentiated and hierarchically integrated a system is in the relations between its parts and between means and ends, the more highly developed it is said to be. If one system is more differentiated and hierarchically integrated than another, it is developmentally more advanced than the other. If a single system is increasing in differentiation and hierarchic integration, it is developing. . . . In this way, the orthogenetic principle is a formal definition of development and, as such, is applicable to the comparison of different systems and to the analysis of a single system in transition. (Wapner, Cirillo, and Baker, 1971)

Intimately connected to the orthogenetic principle are three developmental progressions: *progressive self-object differentiation*, *progressive constructivism* and *progressive perspectivism*. Werner and Kaplan (1963) show that development is marked by a shift in the organism-environment and means-ends relationships. In early stages of development, the infant cannot differentiate self from environment, nor means from ends. It reacts passively to stimuli. As the self becomes differentiated from the environment (usually before the age of 2), the child is capable of directing sensorimotor behavior toward goal objects, and, a little later, of differentiating means from ends so that behavior can be directed by alternate means to the same ends and new means-ends chains may be established. Nevertheless, the view of the young child is still egocentric. Subsequently (from about 2 to 8) the child becomes increasingly active in initiating and determining its own actions and perceptions, and begins construction of its own universe (the beginning of progressive constructivism). The child also becomes increasingly able to differentiate its own viewpoint from that of others (the beginnings of the shift from egocentrism to perspectivism). Finally (around 8 or 9) the child is actively able to construct knowledge about the universe, to clearly differentiate and integrate a wide range of means and ends, and to adopt the perspectives of others as well as his own. (The relative unimportance of these approximate age statements should be emphasized.

Following Werner and Piaget, we have already defined development as changes in levels of organization independent of time. Stages of development follow one from another, apparently in an invariant sequence. At what precise age a given stage occurs is seen as an uninteresting question by the major developmental theorists. It is much more important to understand *how* development takes place, than exactly *when* it can be expected to occur. These approximate age statements, and those which follow, should only serve as very rough guideposts.)

Following a line of reasoning similar to other major developmental theorists, Werner (1948; see Langer, 1969, 1970) recognized three levels of development: *sensorimotor*, *perceptual*, and *contemplative*, progressing from concrete acquaintance with, to abstract knowledge of, the world (see fig. 14.1).

Werner was less interested in charting the various sub-stages of development than he was in establishing descriptive principles across the range of developmental phenomena. Five context-free and time-free *polarities of orthogenesis* were introduced to characterize developmental shifts (Werner, 1948; Kaplan, 1966a; Baldwin, 1967, pp. 497–504). In contrast to the orthogenetic progressions above, which are actual achievements, these five polarities provide methodological means to characterize different features and levels of development. First, the dual *functions* of a behavioral act are ends and means. (1) *Ends*, (or goals) develop from being *interfused* to being *subordinated* one to another; that is, from a lack of differentiation between ends to their differentiation and hierachization, as, for example, the subordination of short-range to long-range goals, and similarly, (2) the *means* develop from being relatively fused or *syncretic* to becoming *discrete;* for example, from the fusion of perception and memory in dreams to their differentiation in waking life. Next, (3), with regard to *structure*, behavior develops from being *diffuse* to *articulate*; that is, from global acts to clearly segmented, articulated movements related to the whole of the act. Finally, with regard to the dynamics of the organism in maintaining integrity while adapting to inner and outer vicissitudes, development is a dual movement (4) from *rigidity to flexibility* and (5) from *lability* to *stability*. The more differentiated and integrated the cognitive structure of an organism, the more flexible and stable its behavior; that is, the organism is more capable of genuine modification in response to changes in the organism-environment system, yet it retains integrity in the face of rapid fluctuations in the environment.

Specific Theory of the Development of Notions of Space. The organismic-developmental theory treats the development of space from action-in-space to perception-of-space to conceptions-about-space as a function of increasing differentiation, distancing, and reintegration between the organism and its environment (Werner, 1948; Follini, 1966). Like Cassirer,

Werner (1948, pp. 167–181, 382–390, 484–486) compares the two end-points of this development—the spaces of action and contemplation (see Wapner, Cirillo, and Baker, 1971, for the development of perceptual space).

These two spaces can be compared on eight of the above progressions and polarities of development. (1) The space of the child initially is undifferentiated from his own body, but as self-object differentiation progresses, a near-space of the child's immediate reach is differentiated from proprioceptive body-space, and later, during the middle of the first year, a further differentiation leads to a far-space of perception (Stern, 1930; Werner, 1948). (2) The child's interaction with the environment progresses from the passivity of the prenatal child who reacts to stimuli from within the body but not to external stimuli (Coghill, 1929; Carmichael, 1933) to the preadolescent's active construction of "spheres of reality" through contemplation, thus indicating a shift from passive acceptance to active construction (Werner, 1948, pp. 389–390). (3) Another progression, from egocentrism to perspectivism (about 2 to 8 or 9), is from the space-of-action, bound to the child's body, to a coordinated abstract system which allows consideration of space from many perspectives. (4) The early experiments of the Muchows (1935) found that the life-space of young children is bound together by personal significance and is centered around the child rather than being determined by any abstract system, thus indicating the concrete and substantial nature of the child's space-of-action as compared to the normal adult's abstract and symbolic space-of-contemplation (Goldstein & Scheerer, 1941; Werner, 1948). Finally, Werner (1948) notes that the child's space-of-action is characterized (5) by its diffuse and undifferentiated structure (the child does not develop a space of interrelated parts), (6) by a syncresis and irreversibility of means of operating on it (the young child cannot reverse his route, and orientation must proceed as an irreversible temporal succession of movements—he gets lost if he has to start from a new point on a route), and (7 and 8) by a rigidity and instability of dynamics (the child's notion of space is rigidly tied to particular objects and relationships rather than to a stable abstract system independent of and hence flexible in relation to any particular arrangement of objects). The only dimension which does not seem to apply is that of the interfusion to subordination of ends, which concerns the function rather than the content of knowledge.

Several studies have been conducted by Werner and his students at Clark University in order to test and extend the organismic-developmental theory of space. Werner (1940) studied the development of space in mentally retarded children and found that younger children solved spatial problems in a concrete and personal manner, whereas older children solved the same problems in an abstract and impersonal manner. These findings substantiate the combined developmental shift from concrete to abstract means of representing space and from egocentrism to perspectivism. Independent experiments by Meyer (1935, 1940), showing similar shifts, lend further support for these developmental progressions.

Follini (1966) was interested in testing other predictions from the theory by studying microgenetic parallels to ontogenesis. Many studies have found that formal similarities exist between microgenesis and ontogenesis (Werner, 1957; Follini, 1966); that is, that development proceeds through the same orderly sequence of stages whether in short exposures to selected stimuli or over the long-term experiences of the life span (for microgenetic methodology, see Sander, 1930; Werner, 1956; Flavell and Draguns, 1957). In one set of studies, Follini had blindfolded subjects explore an unfamiliar room containing several large objects. After each trial, they were asked to describe the room and to draw a picture of it. Over the course of the trials, she found qualitative shifts in the organization of representations from a functional-concrete basis to a more geometric-abstract basis and from global-diffuse to articulated-integrated organization.

Following Cassirer's (1953, 1955, 1957) notion of *culture forms*, and Werner's (1957) notion of the *multidirectionality of development*, (that various forms of higher development are parallel and equal, rather than subordinated one to another), Gittins (1969) extended Werner's theory by conducting a series of studies on the relation between scientific and aesthetic styles of thought and the process of forming representations of an unfamiliar city. He found clear differences between the two styles of representation, but, as predicted, also found that both scientific and aesthetic styles led to equivalent degrees of organization. Studying the representations of Worcester taxi-drivers versus Worcester-based airplane pilots, Rand (1969) also found two styles of representation: "strip" and "relationship." However, the former were personal and uncoordinated, whereas the latter were abstract and integrated. Current work at Clark is exploring further aspects of the organismic-developmental theory of space.

PIAGET'S EQUILIBRATION THEORY OF THE DEVELOPMENT OF THE CHILD'S CONCEPTION OF SPACE

The most extensive theory of the development of the child's knowledge of space is that of the Swiss genetic epistemologist and psychologist Jean Piaget and his co-workers. Piaget's thought begins with the classical problem of epistemology—the nature of "reality" and whether there is any justification for assuming that one's impression of the world is "accurate." For Piaget (1963a, 1963b), the problem of knowledge is that of the relation between subject and object viewed biologically—how the subject comes to know (construct) the object. In this regard he has constructed a *genetic epistemology* (an epistemology studied empirically and developmentally rather than strictly philosophically). He began in the 1920's with the study of the origins of intelligence (1954a, 1962, 1963a, 1963b, 1968; Piaget and Inhelder, 1967, 1969; Piaget, Inhelder, and Szeminska, 1960) the work of developmental psychology for which he is best known.

General Theory of Development. Due to his complexity, thoroughness, and prolific output, it is almost impossible to summarize Piaget's theory and related findings. Fortunately, there are two excellent summaries by Piaget himself (1968; Piaget and Inhelder, 1969) and three other reasonably good summaries (Flavell, 1963, especially pp. 15–84; Baldwin, 1967; Langer, 1969, the last comparing Piaget's and Werner's theories).

Piaget (1963a) believes that the problem of the construction of knowledge reduces to an inherent *interaction* and series of *equilibrations* between the organism and its environment. He has empirically validated the Kantian contention that what constitutes the environment for a particular organism is an intellectual *construction* by the organism. Thus, all development is an interaction between maturation and socialization, between the organism and its environment. For Piaget, although development includes both maturation and learning, both are subordinated to the actual principle of *equilibrium*, which defines the subject-object (organism-environment) relation. The other crucial elements of his theory form a dialectic between *genesis* and *structure* (see Piaget, 1970).

Genesis: The Functional Invariants. Piaget (1963a, 1968) argues that the motivation for all biological and psychological development is *adaptation*. Adaptation is more than simply preservation and survival; it includes development from lower to higher orders of functioning. The process of development (as conceptualized by Piaget and by Werner) may be likened to a subtle mechanism which goes through gradual stages of adjustment in which the individual pieces become more *flexible* and *mobile* while the system as a whole becomes more *stable*. According to Piaget, intelligence is one form of adaptation. Although adaptation is obviously intrinsic to all living species, intelligence is not inherited; it is neither innate nor given simply through sensation; rather, it is formed through a complex interaction between the organism and its environment (Piaget, 1954a, 1963a, 1963b).

We do inherit, however, *a mode of intellectual functioning* composed of two *functional invariants*: *assimilation* (the incorporation of the external world into already-structured schemas) and *accommodation* (the readjustment of schemas to the external world). Adaptation, then, is the equilibration of assimilation and accommodation. This ever-active functioning assures the construction of knowledge and the transition from any state of temporary equilibrium to a succeeding one, or, as Piaget (1963a, pp. 6–7) says:

> Intelligence is *assimilation* to the extent that it incorporates all the given data of experience within its framework . . . that is to say, of structuring through incorporation of external reality. . . . Mental life is also *accommo-*

> *dation* to the environment. Assimilation can never be pure because by incorporating new elements into its earlier schemata, intelligence constantly modifies the [schemata] in order to adjust them to new elements. Conversely, things are never known by themselves, since this work of accommodation is only possible as a function of the inverse process of assimilation. . . . In short, intellectual adaptation, like every other kind, consists of putting an assimilatory mechanism and a complementary accommodation into a progressive *equilibrium*.

Essential to this theory is the notion of the *active organism* (Piaget, 1963a, 1963b). Piaget's findings strongly contradict the assumption of behavioristic "learning" theories that the child is a passive recipient of information from a "real" environment. On the contrary, his findings indicate that, in adapting to its environment, the organism actively initiates contacts and structures its experience. Therefore, the impetus for moving toward higher levels of equilibration comes from this *intrinsic motivation* (Hunt, 1965).

Structure: The Major Periods of Development. Both Cassirer and Werner identified three forms of development—sensorimotor (or active), perceptual, and conceptual (or abstract-contemplative). Ontogenetically, these three forms originate in an invariant sequence, yet overlap in time (see fig. 14.1). Piaget's findings are consistent in these respects. But in addition he has paid careful attention to the description of the different stages (or periods) of conceptual development and the explanation of the transitions between stages.

To define and establish a stage, Piaget utilizes two general and four specific criteria (see Flavell, 1963; Laurendeau and Pinard, 1970). Generally, the notion of stages implies: (1) *hierarchization*—all stages are in a fixed order of succession, and (2) *equilibration*—each stage is a step in a developmental progression realized in successive degrees. Specifically, to distinguish one stage from another, the following criteria are invoked: (1) *qualitative differentiation*—each stage is qualitatively different from the one preceeding, (2) *integration*—each stage integrates the acquisitions or structures of all previous stages instead of simply substituting for them or adding to them, (3) *consolidation*—each stage involves aspects of the consolidation of achievements of earlier behavior and aspects of preparation for behavior at the following level, and (4) *coordination*—each stage is a coordinated whole by virtue of ties of implication, reciprocity, and reversibility: any behavior which can be done in one direction must be able to be systematically done in the opposite order.

On the basis of numerous experiments using these criteria, Piaget has identified four major periods in the development of intelligence: *sensorimotor*, *preoperational*, *concrete operational*, and *formal operational*. Each level is composed of an organized totality of mutually dependent and reversible behavior sequences known as *schemas* (or *schemata*).

The *sensorimotor period* extends from birth to the age of 2. In this stage, the human child changes from an organism capable only of reflex activity to an individual capable of coordinated actions and internalized thoughts. Near the end of this period, the child's behavior may be considered intelligent, although this intelligence is tied to actions and the coordination of actions, and does not involve internal representation. However, in as much as intelligence is internalized action, subsequent higher order thought patterns are reflective abstractions of these sensorimotor schemas, the "action" being carried out symbolically (Piaget, 1954a, 1962, 1963a).

Once the child is able to evoke mentally what has not actually been manipulated or perceived, he is able to "think," albeit at elementary levels. This defines the early childhood period of *intuitive* or *preoperational* thought (approximately 2 to 7). The child can now represent the external world in terms of symbols and can begin to operate on them mentally, although these operations, far from being systematic, are at this level only intuitive and partially coordinated. By the end of the period, a type of loose "reversibility" develops (the child can "reverse" by starting again at the beginning), but this is a cyclical rather than a symmetrical or "true" reversibility. In addition, the child's thought is egocentric. He fails to see the necessity of justifying assumptions, has difficulty in decentering from one aspect of a situation, and focuses on particular states rather than on transformations.

The stage of *concrete operations* (roughly 7 to 12) marks a decisive turning point in intelligence. Forms of mental organization develop which are highly stabilized equilibrations of constructions begun during the preoperational period. At this stage the child is capable of logical thought. As Piaget (1968, pp. 48–49) shows, the concept and explanation of *operations* are crucial to the understanding of intelligence:

> Psychologically, an operation is, above all, some kind of action . . . rooted in the sensorimotor schemata. . . . Before becoming operational, they [actions] constitute the substance of sensorimotor intelligence, then of intuition. . . . Intuitions become transformed into operations . . . which are both *composable* and *reversible* . . . [i.e.] when two [mental] actions of the same kind can be composed into a third action of the same kind, and when these various actions can be compensated or annulled.

As a result of this formation of reversible operations, the child no longer fuses or confuses his own viewpoint with that of others; he is able to differentiate and coordinate different points of view independent of himself, such that the elements and relations are composable, associative, and reversible (Piaget and Inhelder, 1967). This progression from egocentrism to relational coordination directly parallels the progressive constructivism and progressive perspectivism of Werner's theory.

The final period of intelligence identified by Piaget is that of *formal operations* (from 11 to 14 or 15 and beyond). The logical operations of the earlier period begin to be transposed from the plane of concrete mental manipulation of "real" objects to the strictly ideational plane where they are expressed, Piaget (1968, p. 62) says, "in some kind of language (words, mathematical symbols, etc.), without the support of perception, experience or even faith." The adolescent is able to achieve a higher order of reflective abstraction—objects of thought are replaced by pure hypothetical-deductive propositions—and classifications are generalized to arrive at second and higher order classifications. This stage marks the completion of the structural aspects of intellectual development, although much content is still to be attained.

Specific Theory of the Development of the Conception of Space. Piaget and his colleagues have devoted considerable attention to the problem of space (Piaget, 1954a, 1954b, 1962, 1963b; Piaget and Inhelder, 1967; Piaget, Inhelder, and Szeminska, 1960; Inhelder, 1965); the work includes over 1200 pages of experiments, findings, and tightly reasoned explanation. There are several partial summaries and reviews (Piaget, 1953; Piaget and Inhelder 1969; Flavell, 1963, 1970; Holloway, 1967a, 1967b; Laurendeau and Pinard, 1970).

Four general conclusions arise from this research. First, Piaget has found *that the representation of space arises from the coordination and internalization of actions* (Piaget, 1954a; Piaget and Inhelder, 1967; Piaget et al., 1960). Our adult understanding and representation of space results from extensive manipulations of objects and from movement in the physical environment, rather than from any immediate perceptual "copying" of this environment. Thus, it is primarily from acting-in-space, not perception-of-space, that the child builds up his knowledge-about-space. This conclusion is consistent with the distinction, discussed in Section I, between *figurative* and *operative* aspects of knowing. (Blaut, McCleary and Blaut, 1970; Lee, chap. 5; and Stea and Blaut in chap. 12 argue the importance of active experience in understanding large-scale spaces.)

A second general finding, which comes from Piaget's (1962) earlier work on the formation of symbols, and which was corroborated by Piaget and Inhelder (1967), is that *the genesis of the image arises from the internalization of deferred imitations.* Initially, the sensorimotor child copies or imitates other people's actions. Subsequently, as this imitation schema is internalized, and hence able to be deferred, the response becomes symbolic. The role of imagery will be discussed in more detail later in relation to other modes of spatial representation.

The third finding is that there are four levels or *structures* of spatial organization: sensorimotor, preoperational, concrete operational, and formal operational space (Piaget and Inhelder, 1967; Laurendeau and Pinard, 1970).

Finally, many experiments (Piaget and Inhelder, 1967; Piaget et al., 1960) indicate that there are three classes of specific spatial relations which form the *content* of spatial cognition: topological, projective, and euclidian or metric relations. Ontogenetically, the understanding of topological relations preceeds the understanding of projective and euclidian relations. The latter two develop in parallel, although the final equilibrium of euclidian relations is achieved slightly later than projective relations. These last two general conclusions, and some of the experiments which led to them, are elaborated below. (We might note that the order of these ontogenetic developments are the exact reverse of the scientific discovery of space and of the order most often taught in elementary mathematics; Piaget, 1953.)

Structure: Levels of Spatial Organization. As stated above, there are four major stages or periods of spatial organization: Stage I, sensorimotor space; Stage II, preoperational space; Stage III, concrete operational space; and Stage IV, formal operational space, the last three each having two substages. During the first stage (from birth to about 1½ or 2), *sensorimotor space* evolves and culminates in four important simultaneous developments (Piaget, 1954a; Flavell, 1963, pp. 129–162). The first one, the genesis of the *image*, has already been alluded to. The second involves the formation of *object permanence*, also called the *object concept* (Piaget, 1954a, pp. 3–96). At birth and for the first year or so, when an object is removed from the immediate sight of an infant, he believes that it ceases to exist. Slowly, through playing with objects, the child comes to realize that an object still exists even if temporarily hidden. Thus the child develops a stable representation of objects in the world which transcends perceptual or tactual stimuli. Third, by the end of the sensorimotor period, the infant has developed from acting in a series of separate spaces centered on different personal needs and body parts (for example, postural space, buccal [mouth-related] space, tactile space, auditory space, and visual-perceptual space), to moving in a single coordinated space within which all objects are interrelated and begin to be represented (Piaget, 1954a, pp. 97–218). Because of this development, the child is able to move freely and confidently through a limited spatial terrain—for example, he can take shortcuts over routes (combination), return to a point of origin (reversibility), and detour around an object in order to get to another place (associativity) (Piaget and Inhelder, 1969, pp. 15–17). This *sensorimotor group of spatial displacements* provides the base for all subsequent developments of spatial intelligence. Fourth, and signaling the transition to preoperational space, the sensorimotor schemas of the infant become sufficiently independent of immediate action to begin to be *internalized* into thought patterns (Piaget, 1962).

From these crucial beginnings, and with the advent of the symbolic function, representational space begins to develop in the child. The child must reconquer the obstacles overcome with the equilibrium of sensorimotor space, but this time on the level of symbolic representation. This internalization occurs in three additional stages ordered by degree of organization. Stage II, the formation of *intuitive* or *preoperational* space (from about age 2 to 7), has two substages: IIa, symbolic and preconceptual thought (from 2 to about 7), and IIb, intuitive partially regulated thought (from 4 to 7 or 8). Preoperational space, although representational, is still subject to the limiting conditions of sensorimotor and perceptual activity; that is, the first representations of space merely evoke successive states that have already been carried out on manipulated or perceived objects (Laurendeau and Pinard, 1970). Although certain rudimentary and isolated transformations can be performed, the representation of space is essentially static and not immediately coordinated into a reversible structure. As with preoperational thought in general, the preschool child can only return to a point of origin in thought by tracing a cyclical rather than symmetrical or truly reversible route. Furthermore, preoperational represensations are egocentric; that is, the child's conception of space is still tied to his own point of view, although some beginning moves are made in the direction of decentering.

The child's conception of space gradually develops into the fully mobile, flexible, and reversible structures of Stage III, *concrete operational space* (from 7 to 12), also consisting of two substages: IIIa, the appearance of concrete operations (around 7 or 8); and IIIb, the organization of operations into logical structures (from about 9 to 11 or 12). During the early school years, spatial thought can finally disengage itself from images; it is transformed into operations which are, however, still concrete—that is, dependent on the presence of real or represented objects (Inhelder, 1965). The spatial structures, which by this time have achieved a considerable degree of abstraction, are formed through a logical coordination of space from multiple viewpoints (Piaget and Inhelder, 1967). Thus, the child is finally free of his egocentric orientation toward space.

Finally, in Stage IV, the *formal operational space* of adolescence and beyond, spatial operations can be completely removed from real action, objects, or space. The mathematical multiplication and coordination of space (Piaget, Inhelder, and Szeminska, 1960) allows the adolescent to survey the whole universe of spatial possibilities. In short, he has moved from the concrete spatial world to a new equilibrium, but this time in the realm of the possible, the hypothetical, and the infinite.

Content: Specific Spatial Relations to be Constructed. To know space as a geometric entity which has certain properties irrespective of transformations is the result of a series of developmental steps. On this problem alone Piaget and his colleagues have conducted over thirty experiments, summarized in two long volumes (Piaget and Inhelder, 1967; Piaget, Inhelder, and Szeminska, 1960). Following the organismic assumption introduced in Section I, the development of knowledge of spatial relations is inseparable from levels of structuring space, and from general cognitive development. Figure 14.2 illustrates some of these connections.

The three major types of relations or properties of space are defined as follows: *Topological properties* are simple qualitative relations like proximity and separation, open and closed, which remain invariant under continuous deformations excluding tears or overlaps. *Projective properties* are relations in terms of a particular perspective or point of view, such as a straight line, a triangle, or parallel lines which remain invariant under projective or perspective transformations. *Euclidian* or *metric properties* of space are relations in a system of axes or coordinates whose equivalence depends on mathematical-geometric equality; for example, an angle, an equal interval, or a distance.

Piaget and Inhelder (1967, pp. 17–79) devised a simple yet ingenious experiment to study the *sequence* of development of these fundamental spatial relations. Children ages 2 to 7 tactually explored a number of familiar objects, solids, and flat shapes without being able to see them, and tried to match them with duplicates which could be seen but not touched. Only by translating impressions from the tactile-kinaesthetic modality into an internal representation could the comparison and identification be made. Thus the experiment was an excellent test of the development of the child's ability to *represent* different spatial relations. The findings indicated that, although the infant can begin to form images near the end of the sensorimotor period, he is unable to differentiate topological shapes until he has entered the preoperational period. This is followed by a steady development in the ability to represent projective and euclidian shapes until, at the time of concrete operations, the child is able to represent complex euclidean forms easily and accurately. Thus, the construction of topological, projective, and euclidian space may be viewed as progressively including more spatial conservations. These findings were corroborated by Inhelder (1965).

But what of the development of representations of specific spatial relations? Piaget and Inhelder (1967, pp. 80–149) conducted three studies on the development of *topological space*. The findings from the three studies were similar. The child begins to gain some notion of topological properties by the preoperational Stage II: in fact, the more elementary notions of surrounding and enclosure, open and closed, are fairly well established by this time. Not until the appearance of concrete operations (IIIa) are notions of proximity, separation, between, and order clearly formed. The conception of points and continuity depends on the notion of infinity and thus is not firmly established until formal operations (IV), when thought is totally liberated from the "concrete."

These topological relations are invariant under the contraction or expansion of space. Other spatial features such as distances, straight lines, and angles are unable to be conserved under changes in shape. Hence it is impossible for topological relations alone to form comprehensive systems linking different spatial elements together by means of perspective or axial coordinates. This depends on the development of projective and euclidian space.

Piaget and Inhelder (1967, pp. 151–374) conducted seven experiments on the development of *projective space* broadly concerned with the develop-

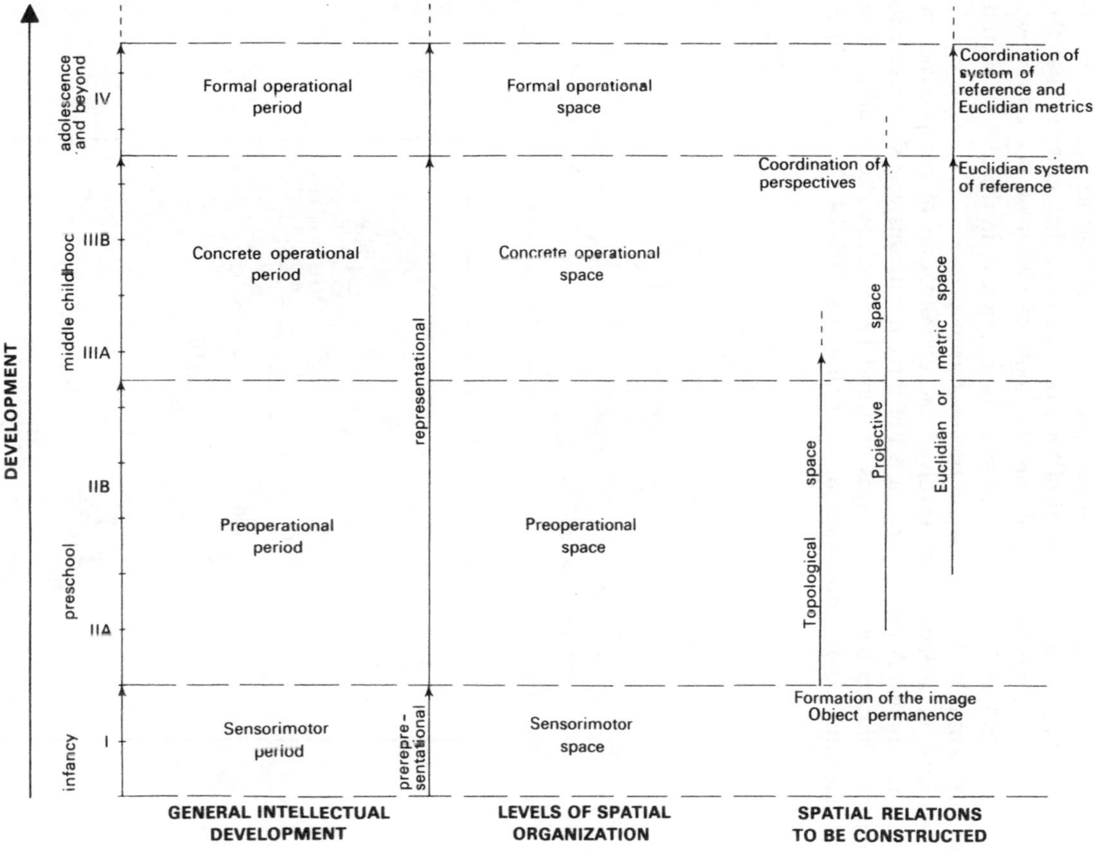

FIGURE 14.2. *Schematic representation of Piaget's theory of the development of spatial cognition in relation to overall intellectual development.*

ment of the child's awareness of perspective and the interrelations between elements seen from different points of view. One experiment involving the construction of a straight line illustrates "the essence of the projective concept: the line is still a topological line, but the child has grasped that the projective relationship depends on the angle of vision or point of view" (Piaget, 1953, p. 75). This is an important step also in the child's overall development from egocentrism to perspectivism.

An intriguing experiment involved the coordination of perspectives of three mountains. A pasteboard model of three mountains was shown to one hundred children between the ages of 4 and 12. The mountains differed in size, location, color, and object at their summits (fig. 14.3). The child is

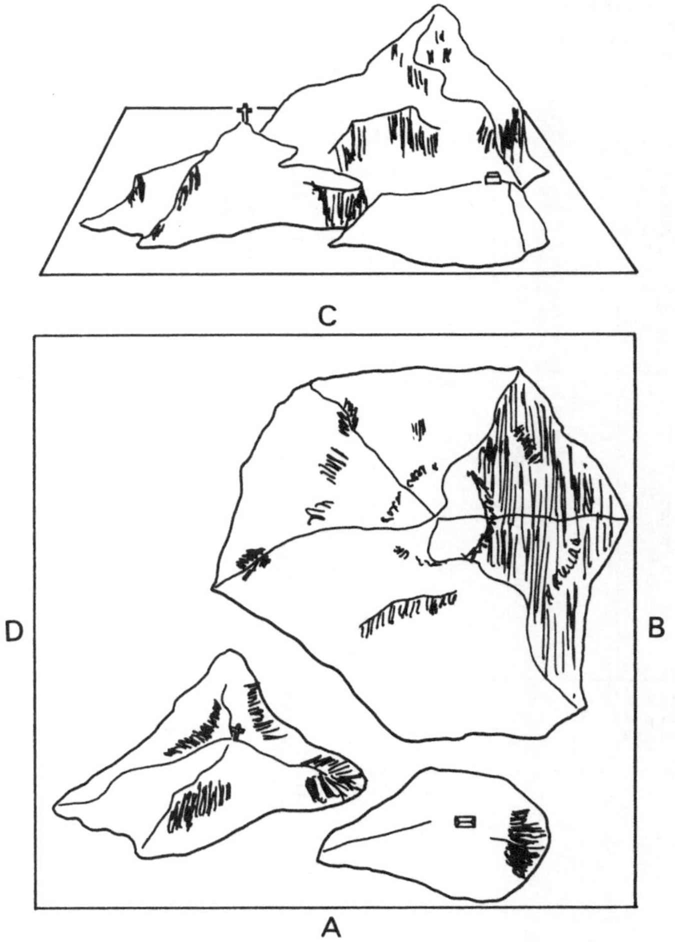

FIGURE 14.3. *Pasteboard model of three mountains used in Piaget and Inhelder's coordination of perspectives experiment (after Piaget and Inhelder, 1967).*

seated in front of the model at *A*. A small wooden doll is placed alternatively at *B*, *C*, and *D*. After each placement, the child is shown a collection of ten pictures, each representing the mountains from different viewpoints, and is asked to select the picture which would correspond to the doll's view.

Preoperational children (II: up to about 6½) select the picture corresponding to their own view. They persist even when they have the chance to check their selection by going around to the doll's position to look (Piaget, 1953). With the appearance of concrete operations (III*a*: 7 to 9), there is a progressive differentiation of points of view and certain projective relations are formed. Before and behind are correctly differentiated first, then left and right. It is not until the final equilibrium of concrete operations (III*b*: around age 9 or 10), however, that the two schemas are intercoordinated and the child masters a comprehensive coordination of viewpoints completely independent of his own point of view. These findings may be explained in terms of the developmental shift from egocentrism to relational coordination and the equilibrium of fully reversible concrete operations. Through these developments the child is able to decenter from one point of view, to differentiate between his own and other points of view, and to integrate this differentiation into a coordination or equilibration of perspectives. (Blaut, McCleary, and Blaut, 1970; Stea and Blaut's findings in chaps. 3 and 12 that first graders can interpret vertical aerial photographs, which seems to involve something akin to the coordination of perspectives, may be explained similarly.)

The final development in the child's conception of spatial relations involves euclidian or metric space. Piaget and his colleagues have focused particularly on systems of reference the child uses to organize space (Piaget and Inhelder, 1967, pp. 375–446; Piaget et al., 1960, pp. 3–26) and conservations of length, distance, surface, and volume (Piaget et al., 1960). Piaget and Inhelder (1967, 376) describe the importance of a system of reference (or a spontaneously constructed network) for the organization of space:

> However, a reference frame is not simply a network composed of relations of order between the various objects themselves. It applies equally to positions within the network as to objects occupying any of these positions and enables the relations between them to be maintained invariant, independent of potential displacement of the objects. Thus a frame of reference constitutes a euclidean space after the fashion of a *container*, relatively independent of the mobile objects *contained* within it, just as projective coordination of the totality of potential viewpoints includes each viewpoint actually envisaged.

Three interesting experiments were conducted to explore the coordination of systems of reference. In each case children were asked to make a plan or model of a large-scale environment, in a sense externalizing their "topographical schemas," or what we term "topographical representations" in Section III below.

In one experiment, eight objects are placed on a blank cardboard base to form a model village. The child is given another piece of blank cardboard and a larger set of village pieces from which to choose and arrange a duplicate village. This experiment poses problems for the child because of the lack of other objects, lines, or angles to serve as reference markers. During the early preoperational period (II*a*: until age 4), the child is unable to arrange the village pieces in any systematic way. During the middle part of the preoperational period (II*a*-II*b*: 4 to 6), the child picks out the correct number and type of pieces and tries to locate them in similar positions, although for being able only to present topological relationships, he is very unsuccessful; distances as well as perspective relationships are ignored, and there is no coordination between the arrangement of objects and the external reference system provided by the card they lie on. During the latter part of the preoperational period (II*b*: 5–7), certain sets of relations are constructed in an intuitive manner, although they are not inter-coordinated. Then, coincident with the onset of concrete operations (III*a*: 7 to 8), a system of reference is constructed, but this does not become stabilized until concrete operations reach equilibrium (III*b*: 9 to 11). Not until the development of formal operations (IV: 11 or 12) is the concept of a coordinated reference system attained which takes into account the additional projective and euclidian relations of proportional reduction to scale, accuracy of distance, and metric coordinates.

Thus, the beginnings of general projective and euclidian coordination develops co-temporally with the formation of concrete operations, whereas the final equilibrium of a truly abstract understanding of the coordination of projective and metric properties of space into a stable system of reference develops with formal thought. In another experiment, the child was asked to construct a model of the environment around his home. This experiment will be discussed in Section III.

Finally, the conservation of the metric properties of length, distance, surfaces, and volume have been considered (Piaget et al., 1960). Generally, it has been found that the conservation of length and surface are not achieved until the concrete operational stage, when the child discovers through reversibility that a quantity has remained the same although a transformation has occurred. The conservation of volume is not achieved until the formal operational stage: "the discovery of logical relationships is a prerequisite to the construction of geometrical concepts" (Piaget, 1953, p. 76). These conservations are required for final equilibration of a fully coordinated and abstracted euclidian space (Piaget et al., 1960, pp. 389–408).

In summary, we have seen from the work of Piaget and his colleagues a number of important developments in the child's conception of space: the formation of object permanence; the genesis of the image; the sensorimotor group of displacements; representation through internalization of actions; the sequence of sensorimotor, preoperational, concrete operational, and formal operational structurings of space; the sequence of topological, projective, and euclidian or metric contents of space; and the co-development of stages and specific relations of space into the coordination of perspectives and of euclidian systems of reference.

STANDARDIZATION AND REPLICATIVE STUDIES: THE WORK OF LAURENDEAU AND PINARD

One of the great merits of Piaget's theory of space is its inclusion in a complete theoretical structure which supports it and of which it is a central ingredient. Although there has been a devaluation of his experimental methodology among certain experimental psychologists, Laurendeau and Pinard (1970, p. 19) argue:

> We believe that the value of an experimental method is measured not by the abundance or subtlety of the statistical analyses it may favor but by the richness of the information it provides or the new hypotheses it suggests. In this respect, it would be futile to make any comparisons between Piaget's synthesis and the rather fragmentary studies which are currently flooding the journals.

Furthermore, there have been numerous studies (see references in Flavell, 1963 and 1970; Laurendeau and Pinard, 1970) which have replicated, with only minor discrepancies, Piaget's findings in almost every domain of intellectual development. His work on the development of space is no exception. Unfortunately, most of these studies have investigated only minor details (like age differences) unrelated to the central thrust of the theory. Only a few (Meyer, 1935, 1940; Laurendeau and Pinard, 1970) have seriously looked at Piaget's *explanation* (the functional invariants, genesis and structure, levels of spatial organization) for the phenomena uncovered.

The most comprehensive replicative work has been conducted by Laurendeau and Pinard (Pinard and Laurendeau, 1966; Laurendeau and Pinard, 1970). They have replicated five of the most central of Piaget and Inhelder's experiments on space, paying attention to standardization of the experimental materials and conditions and to the formation of subject groups representative of the general population. Two of their experiments concern the coordination of perspectives and the coordination of topographical positions. The findings on the development of specific spatial relations, on the succession of topological to projective and euclidian relations, and on the succession from preoperational to operational equilibrations of space are consistent with those of Piaget and his colleagues.

A number of less comprehensive studies have been done by other investigators. However, far from dealing directly with explanations or even structural descriptions of the developmental progressions involved in the child's understanding of space, they are partial studies, limited to very particular aspects of spatial cognition, and almost never integrated into a theoretical structure (see reviews by Thompson, 1962, pp. 326–330; J. G. Wallace, 1965, pp. 95–122; Flavell, 1963, pp. 388–392; 1970, pp. 1015–1019).

MODES OF REPRESENTATION

Before leaving the topic of fundamental spatial cognition, we will briefly review the literature on various modes of spatial representation. For convenience, we follow the three categories suggested by Bruner (1964, 1966): *enactive, iconic*, and *symbolic representation.* Bruner and his colleagues claim that these categories occur in this order in ontogenesis, each depending upon its predecessor for development. Piaget and Inhelder (1969), Bower (1970), and Rusch (1970) have also dealt (although in different terms) with these three stages of children's representations of their world.

Enactive Representation. Bruner's (1966, p. 11) idea of *enactive* representation is premised on the conviction that:

> Insofar as action is flexibly goal-directed and capable of surpassing detours, it must be based on some form of representation that transcends the mere serial linking of stimuli and responses.

Piaget recognizes the importance of action in spatial cognition but has criticized Bruner for speaking of enactive "representation"; Piaget doubts that action "stands for" or represents anything beyond itself (see discussion in Bruner, 1966, pp. 10–11). As discussed above, Piaget's theory recognizes a space which is perfectly organized and balanced at the level of action or behavior, but this sensorimotor space is prerepresentational, lacking the symbolic function which facilitates imagination and reconstruction. Coming from a different tradition of psychology, Bruner believes that action becomes representational by virtue of its "habitual pattern." Following Cassirer's distinction between knowledge by acquaintance (sensorimotor schemas) and knowledge through representation (cognition), Piaget's insistence that habitual sensorimotor schemas are not the same as cognitive representations seems more acceptable.

To explore this problem, psychologists have conducted numerous maze-learning experiments with rats. These studies have been central in the long-standing behaviorist controversy of *response learning* versus *place learning* (Tolman, chap. 2; Birch and Korn, 1958). Tolman claims that rats (and men) do not develop specific "responses" in coping with the demands of a situation; instead, something like a "field map" develops.

Brown (1932) studied maze learning with humans and reached a different conclusion. He noted that a subject begins to gain control over his movements before he is able to describe his position in the maze. After learning the maze, his subjects were asked to walk the pattern on an open floor. Brown found that although the maze consisted of right-angled turns, the reconstructed pattern had no right angles and bore only a slight relationship to the original, and concluded that the subject does not learn a pattern or "configuration" but a "succession of movements" which are executed very imperfectly when he is deprived of the guiding walls of the maze. Brown (1932, p. 134) suggested also that the maze is, in all important ways, like learning a "new place," city, or country, and that it "consists essentially in connecting together numerous elements of sense-impression and of behavior into a more or less successful pattern of action . . . [consisting] almost entirely of a series of connected movements."

Bruner (1966, p. 18) gives man more credit: for behavior to become more skillful, "it must become increasingly freer of immediate or serial regulation by environmental stimuli operative while the behavior is going on." Supporting his conviction with the experiment of Mandler (1962), Bruner suggests this freedom is achieved by a shift from response learning to place learning. In his experiments, Mandler's subjects were blindfolded and asked to master a complex maze of toggle switches. Unlike the maze experiments of Brown and most others, Mandler required his subjects to continue "running through" the maze after they had mastered it. He noted that after mastery, the subjects began to approach the task differently. What had been sequential was rendered simultaneous; the subjects reported having what they called an "image" of the true path rather than a connected series of overt movements. This discovery suggests that when motor activity becomes "regularized" or "steady," it is converted from a *serial* to a *simultaneous* form (Lashley, 1951). Bruner believes that such a shift from *response* to *place* learning could not take place without the mediation of a spatial framework, which depends, he suggests, upon vision.

In summarizing twelve reports of experiments with blind and blind-deaf-mute persons, Shemyakin (1962) made inferences about sighted persons and concluded that locomotion in space and not vision is the necessary and crucial condition for the emergence of simultaneous-type representations. In his extensive summary of evidence from the congenitally blind, von Senden (1960, pp. 285–286) suggests, however, that a spatial framework cannot be achieved without sight and that simultaneous-type representations do not exist for the blind; locomotion is but a substitute for vision:

> Nothing is given to him [a congenitally blind person] simultaneously, either by touch or the other senses; everything is resolved into successions. . . . Thus when no external happenings occur to him, he has the urge to take action himself in order to maintain connection with his environment.

Thus, before such statements as Shemyakin's can be considered as more than suggestive, systematic research is required which will compare macro-spatial cognition in congenitally blind, adventitiously blind, and sighted persons.

The notion of "representation" by action does not necessarily demand that large-scale environments be actually traversed. After analyzing children's interpretations of aerial photographs, Stea and Blaut suggest that a child's "cognitive mapping" abilities appear to antedate first-hand dynamic experiences with the larger environment (see chap. 3). They have hypothesized that this knowledge is acquired through the observation and manipulation of surrogates for direct experience, i.e., through acquiring the ability to model the real world. This hypothesis does not depart from Piaget's emphasis on manipulation in early spatial representation. Again, however, the suggestion remains to be tested.

Iconic Representation. Bruner (1966) uses the term *iconic* for representation using the medium of an image. It is commonly held that imagery is derived from visual perception (Bruner, 1966, p. 21). This view contrasts markedly with Piaget's (1962, 1969) findings that imagery is an internalization of deferred imitation (an active schematizing reproduction). (For a more general view of mental imagery, see Holt, 1964; Richardson, 1969.) Inhelder (1965, p. 4) has clearly expressed the particular value of imagery:

> It seems there is a large domain that language is unable to describe except in devious and complex ways, that is, the domain of everything that is perceived (as opposed to conceived). Sometimes it is useful to communicate things perceived, but, above all, it is necessary to retain a large part of them in the memory if future action is to be possible.

The relevance of this notion to spatial representation is probably evident to anyone who has attempted to use words to describe the spatial layout of a place.

The formation of mental images cannot, according to Piaget, precede understanding. Thus, the mental images of the child at the preoperational level are almost exclusively static; not until he reaches the concrete operational level is he capable of anticipating transformations in imagery. Piaget does not deny the importance of imagery, but challenges the over-simplified formulation of imagery as an extension of perception. The resemblance between imagery and perception is, according to Piaget and Inhelder (1969, p. 68), due to imitation: imagery "actively copies" perceptual data. Piaget, Inhelder, and Szeminska (1960, p. 12) suggest that:

> Our own comings and goings provide the framework for our memory images of districts and landscapes. Indeed, examples such as these provide the best proof that visual images are internal imitations of actions, the objects being imagined, and the actions internalized.

Symbolic Representation. We deal below only with what Bruner describes as the most specialized "natural" system of symbolic activity—language. It is widely recognized that language provides a means not only for representing experience, but also for transforming it (Vygotsky, 1962; Bruner, 1964, 1966; Piaget and Inhelder, 1969). Furthermore, language enables thought to range over vast stretches of time and space, liberating it from the necessity for immediate sensory stimuli and allowing a considerable degree of abstraction from concrete experiences (Piaget and Inhelder, 1969; Bower, 1970).

Piaget and Inhelder (1969, p. 86) noted that "thought, particularly through language, can represent simultaneously all the elements of an organized structure." Arnheim (1969, p. 247) on the other hand, in his discussion of visual thinking, states that thought as mediated by language "strings perceptual concepts in linear succession" and "dismantles the simultaneity of spatial structure." Ultimately, the development of spatial representation depends neither upon language nor imagery alone. The gradual development of the child's freedom from his own actions depends on representations that utilize a whole range of symbolic functions, including language and imagery (Piaget and Inhelder, 1969, p. 91).

Ames and Learned (1948), Lyublinskaya (1954), Olson and Baker (1969), and Kershner (1970) have studied the relationship between the development of language and spatial representational ability. A common problem in such studies is that learning the semantic markers of a word (i.e., the senses a word has or the contexts into which it fits) is a slow process (Bruner, 1966, p. 32). It is therefore difficult to determine how much verbalization alone indicates a real comprehension of spatial relations (Laurendeau and Pinard, 1970, p. 12).

Bruner and Kenny (1966) conducted a study which demonstrated the value of language in enabling a child to go beyond the *representation* of experience to the *transformation* of experience. Children between 5 and 7 were given nine glasses of three different heights and three different diameters arranged on a 3 x 3 grid. After familiarizing the child with the grid and its elements, the experimenter removed the glasses and placed one in a different corner from its previous location. When asked to reconstruct the original pattern of glasses, most of the 7-year-olds were successful, but hardly any of the younger children were. The 7-year-olds properly treated the transposition as a problem requiring flexible thought, while the 5- and 6-year-olds seemed to be "dominated by an image of the original matrix" (Bruner, 1964, p. 5). Bruner and Kenny noted that when the children's language was confounded by their mixing of a dimensional with a more global term—"that one is *tall* and that one is *little*"—they were twice as likely not to succeed. These findings illustrate the importance of language in enabling children to produce new structures based on rules. Such an ability is undoubtedly important in facilitating an individual's representation of unfamiliar environments through the application of previously learned rules or general notions of city structure (Freeman, 1916, 168–169; Griffin, chap. 15; Gittins, 1969; Appleyard, 1969a, 440; 1970, 115).

III. The Development of Spatial Cognition of Large-Scale Environments

Having reviewed research and theory on the general development of spatial cognition, we must now ask the more specific question of how representations of large-scale environments are constructed, and whether there is any difference in this construction from that of so-called "basic" spatial relations. Research on the development of representations of large-scale environments has been conducted since the turn of the century (see Gulliver, 1908; Stern, 1930; Trowbridge, 1913; Freeman, 1916). These representations have at least two functions: first, to facilitate location and movement within the larger physical environment, and second, to provide a general frame of reference for understanding and relating to this environment. Recognizing the importance of the first of these functions as part of orientation research, psychologists have implicitly been studying macro-spatial cognition for some time. The second function has been noted only recently but is equally important. For instance, Lynch (1960) suggests that "environmental images" help establish an emotionally safe relationship between men and their total environment by serving as organizers of activity and knowledge, as material for common memories which bind a group together, and as spatial referents for senses of familiarity. The importance of a spatial framework is evident from the writings of Lynch and those who have followed him (see Gulick, 1963; de Jonge, 1962; Stea and Downs, 1970; Appleyard, 1969, 1970a; D. Wood, 1969, 1971).

THE ONTOGENETIC DEVELOPMENT OF GEOGRAPHICAL ORIENTATION AND TOPOGRAPHICAL REPRESENTATIONS

Various terms have been given to an individual's cognitive representation of specifically large-scale environments: "imaginary map" (Trowbridge, 1913), "field map" or "cognitive map" (Tolman, chap. 2; Blaut, McCleary, and Blaut, 1970; Stea and Downs, 1970), "mental map" (Hallowell, 1955), "schema" (von Senden, 1960; Lee, chap. 5), "topographical schema" (Piaget, Inhelder, and Szeminska, 1960), and "topographical representation" (Shemyakin, 1962). We use the term *topographical representation* because it seems to be the most comprehensive and the least confusing: it implies neither a cartographic form (map) nor a habitual pattern (schema), but clearly refers to an internalized mental reflection of the physical environment. Shemyakin (1962, p. 193) has satisfactorily defined it as "a mental plan of some area which is a reflection in man's mind of the spatial placement of local objects in relation to each other and to himself."

Thus it is still a cognitive representation but one explicitly pertaining to large-scale environments.

An essential element of topographical representations is a reference system which spatially orients the individual in some systematic manner to the environment. We use the terms *egocentric*, *fixed*, and *coordinated* to refer to three systems of orientation suggested by the literature. As there is evidence that these three systems of reference develop sequentially in ontogenesis, they are discussed in this order below (see fig. 14.4).

Piaget, Inhelder, and Szeminska (1960, pp. 3–26) have provided the only complete account of the possible ontogenesis of topographical representations other than a nonempirical but insightful early paper by Freeman (1916) and the recent work of Stea and Blaut. Shemyakin (1962) has reviewed Russian research on geographical orientation and representation, but he offers no theoretical structure, and no detailed descriptions of research designs and statistical analyses. Nevertheless, his report is reviewed here because it summarizes a wealth of findings from largely inaccessible studies.

Egocentric Orientation. Ontogenetically, geographical orientation (like all development) is initially action-centered and egocentric. Numerous experiments have illustrated the effectiveness in man of a sensory system based primarily on proprioceptive input for the perception of body position; most do not concern us here because we are interested in the development of an individual's geographical or locational orientation rather than his "positional" orientation (see reviews of egocentric "positional" orientation by Witkin, Dyk, Faterson, Goodenough, and Karp, 1962; Howard and Templeton, 1966, pp. 272–293; Wapner, Cirillo and Baker, 1971). Some of this literature, however, demonstrates how a child first orients to the physical environment using axes or planes defined entirely with respect to his own body. This may be referred to as an *egocentric orientation system.* There is much evidence that this system is based on a sense of localization through bodily movements (see reviews by Gregg, 1939; Shemyakin, 1962; and, most comprehensive, Howard and Templeton, 1966). Freeman (1916) believed that the direction to a place (actual or imagined; Ryan and Ryan, 1940) is "represented in the mind" in terms of movement of the body through turning the head or pointing, both of which bring us into alignment with the place. Through this method, Freeman (1916, p. 164) believed, a preschool child of 4 to 5 years of age builds up a "fairly definite notion of the direction of buildings or streets in the immediate vicinity of his home." This suggestion accords with Piaget's description of preoperational visual images as "internal imitations of actions, the objects being imagined and the actions internalized" (Piaget et al., 1960, p. 12). But this system serves only for representation while the child remains stationary. What of larger areas where the landscape must be traversed by the individual? It is not enough for the child to imagine movements, for he must link his movements to reference points.

As we have seen in Section II, Piaget has systematically studied the development of systems of reference. One part of this enquiry dealt specifically with the development of the child's reference system used to facilitate the representation of changes in his position in the larger environment (Piaget et al., 1960, pp. 3–26). Children aged between 4 and 10 were asked to build in a sand box a model of their school buildings and environs, using a variety of wooden pieces and ribbons representing such things as buildings, woods, and rivers; to draw in the sand or on a piece of paper the route from school to a well-known landmark; to make a drawing of the sand model; and finally, to make necessary changes to the plan after the school building was rotated through 180° by the experimenter. They found that an action-centered egocentric reference system was characteristic of preoperational (Stage II*a*) children: landmarks were not organized in terms of a spatial whole; routes were thought of in terms of the children's own actions first, the various landmarks being fixed in terms of them, instead of vice versa; and, the plan could not be rotated through 180°, nor could the routes be reversed in thought.

From this experiment, Piaget and his associates concluded that for preoperational children, representation occurs only when they think out a route. Landmarks are mentioned by these children, but are simply "tacked on" to recollections of their own actions (the routes traversed). That is, they can anticipate the spatial relations between one landmark and the next as they are walking, but are not able to build up a representation of the environment as an ordered whole. Only topological relations can be represented at the early preoperational level. While children can represent pairs of neighboring objects topologically, they cannot arrange three or more objects successfully into a coordinate system. This latter development involves reversibility and coordination, both characteristic of the later concrete operational period.

Having been interested in the genesis of representations specifically of large-scale environments, Stea and Blaut (Blaut and Stea, 1969; Blaut, McCleary, and Blaut, 1970; chaps. 3 and 12 herein) have found what they consider a "primal" form of cognitive representation, namely the ability of first-graders and even some preschoolers to interpret and correctly identify objects from vertical aerial photographs. In addition, they have suggested that this early form of representation may in large part be due to a generalized cognitive model built up initially from post-infancy toy play as a surrogate for experience which the very young child lacks by not traversing routes in the larger environment (1969, pp. 22–26).

Based upon empirical findings, Shemyakin's (1962) account of the development of topographical representation largely accords with that of Piaget. He distinguishes two fundamentally different types of topographical representation: *route maps* and *survey maps*. Shemyakin (1962, p. 218) describes a route map as a representation constructed by mentally "tracing the route" of locomotion through an area. This seems to agree with Piaget's account of internalized actions in the form of known routes, as discussed above. By survey maps Shemyakin means representations of "the general configuration or schema of the mutual disposition of local objects." This seems equivalent to what Piaget describes as a true topographical representation utilizing a coordinate reference system (see fig. 14.4).

From observations of orientation in blind persons, and of map drawing and orientation in sighted children, Shemyakin has suggested that route-map representations are a necessary prior development in the formation of survey-map representations. Observations of children (Shemyakin, 1940; Lyublinskaya, 1948, 1956; see Shemyakin, 1962) indicated the importance of route-mapping in the early development of topographical representations. Shemyakin argued that after a child learns to walk, a new stage in the understanding of space relations opens before him, citing Lyublinskaya's description of a 1½ year old child's early mastery of the "space of the route." From observations of movement in and immediately around the house, Lyublinskaya also suggested that "the child is not only able to orient herself in space among familiar objects, but retains also a pretty accurate representation of the position of individual familiar objects in space" (Shemyakin, 1962, p. 218). Even children of 6 to 8 years of age, when asked to draw a plan of some locality familiar to them, did so mainly by drawing the route. Shemyakin (1962, p. 219) assumed a correspondence between the construction of these external representations and the initial development of internal representations, observing that:

> Children usually began their work at the lower edge of the paper, and drew "away from themselves" (so) that the right and left turns coincided with the real position of their body in space. . . . [Asking them to draw] "toward themselves . . ." led to a sharp increase in the number of errors in the reproduction of [the] turns.

Furthermore, some children chose to sit at that side of the table that faced the required locality. These results are similar to those found by Piaget and his associates in the above experiment when children were asked to draw a familiar route: both clearly indicate an egocentric preoperational level of development. Unfortunately, there are many limitations in the use of children's drawings for inferring the nature of internal representations. Drawings cannot be a simple translation of an image representation, for only two of the three spatial dimensions can be represented in the picture plane (Arnheim, 1967, pp. 165–212). Also, Piaget and Inhelder (1967, pp. 45–46) point out that drawings lag behind image representation particularly when the child is dealing with complex wholes. Nevertheless, they believe that pictorial representations are valuable methodologically if limited to the spontaneous drawing of simple everyday shapes and checked with the use of other methods.

Route mapping is not unique to children. Appleyard (1969, 1970) found this same system was commonly used by adults in drawing maps of Ciudad Guayana. Rand (1969) found evidence of *both* a prior ontogenetic development of route-mapping in young children (see also Lee, chap. 5) and a prevalence of route-mapping in adult taxi-drivers' representations of a city (in contrast to the relationship- or survey-mapping of pilots). However, following Werner's orthogenetic principle discussed in Section II, even in the latter case, route-maps may be seen to be less integrated and hence less advanced developmentally than survey-maps (see fig. 14.4).

Fixed Systems of Reference. For orientation in large-scale environments, Freeman (1916) noted the importance of the child's ability to free himself

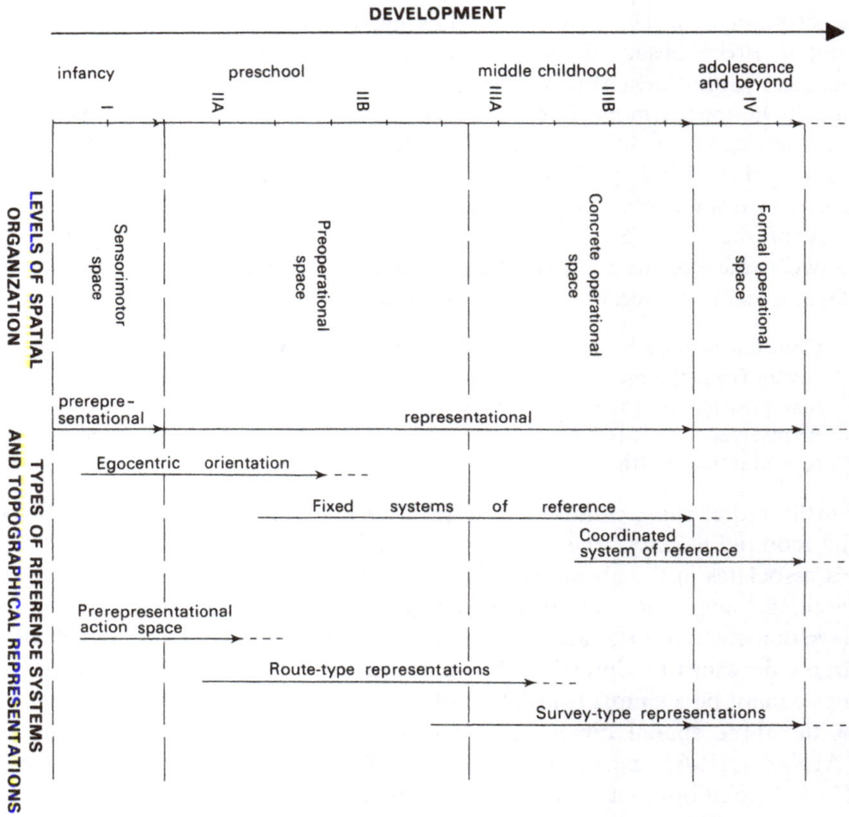

FIGURE 14.4. *Schematic representation of the ontogenetic development of topographical representations.*

from the limitations of this early egocentric orientation system. Werner (1948) also noted the transition from egocentrism to perspectivism and its importance for organizing space. This development, according to Piaget (1963a), does not begin until the onset of reversible thought in the concrete operational period. At this stage, Piaget et al. (1960) have shown that the child's position and movements begin to be oriented in terms of fixed elements in the environment rather than the elements being oriented egocentrically to the child. The next stage in geographical orientation and topographic representations appears to be a transitional stage between early preoperational egocentric orientation and concrete operational coordinated systems of reference. We refer to this stage as that of *fixed systems of reference.*

Freeman (1916, p. 165) observed that children move from orientation with reference solely to themselves to orientation with reference to the position of some fixed object or of some fixed direction: the child thereby achieves a "detached view of a region as though it were seen from a distance." This accords with the second level in the development of topographical representations found by Piaget et al. (1960, pp. 14–21) in the model-the-school experiment described above. For early concrete operational children (III*a*), the spatial representation of the landscape is partially coordinated by the use of landmarks. The children are unable to coordinate the system as a whole, either in demonstrating with the wooden models the topographical relations between landmarks, or in describing a route. They can relate objects to each other in discrete local areas, but they cannot appreciate the totality of relations between landmarks which occur in the environment because the landmarks are fixed in partially coordinated *sub*groups, each based on an independent vantage point or on a particular journey.

W. Brown (1932, p. 126) also recognized a fixed reference system as a second stage in his human maze-learning experiment, and noted the particular importance of landmarks as reference points:

> Objects which can be perceived (a rough-spot or a tilting board) are able to serve as landmarks . . . objects which are recognized in this way assume a considerable effective value. The subject often expresses great satisfaction in finding them . . . the nucleus of any *locality* is an object.

As a means of representing areas larger than those lying within an individual's visual field, the hypothesis that features are organized in relation to a fixed reference point gains support from von Senden's (1960, p. 286) summary of clinical observations of orientation by the blind:

> The subject sets out from a perfectly definite central point, resembling the center of a spider's web, whence he gains acquaintance with the routes that matter to him, up to an outer periphery that can be more or less gradually extended. Wherever he may be, he always remains mentally in conscious relation to his fixed starting point.

Trowbridge (1913, p. 890) had earlier provided a convincing account of the efficacy of this system, suggesting the home as the necessary fixed object for children; in using this system "all changes of position . . . can be referred at any moment to definite distances and angles, forming a simple trigonometric figure which gives the direction to the home." Trowbridge (1913) suggested this *domicentric* system (which may be seen as a special case of fixed reference systems) is not only found in birds, beasts, fish, and insects, but also in "uncivilized man" and young children. It was once thought to be an innate ability which remained as a vestigial sense in children (Warren, 1908; Hudson, 1922; W. F. Smith, 1933) but this has since been discounted (see Howard and Templeton, 1966).

Although domicentric orientation is not innate, children do seem to commonly orient themselves towards their home. In an aerial geography experiment, one of us (Hart, 1971) observed that children aged 8-9 years would not become involved in a geographical learning experience provided by flying and aerial photographs *until* they had first located their home or their school. Muchow and Muchow (1935), Lee (chap. 5) and Rand (1969) offered similar conclusions on domicentricity based on observations of children's drawings and their behavior with regard to home and school.

Unfortunately there is still very little known about general fixed systems of reference. Most research on the development of orientation in children has been concerned with children's understanding of the cardinal directions, even though orientation to fixed familiar features would seem from the above analysis to be more likely for preoperational and concrete operational children. In fact, Lee doubts that even adults construct a rectilinear grid of coordinates to orient themselves in space; he suggests instead that in each of their "spatial schemata" they use a *polar* system consisting of a set of radiating lines from a fixed reference point.

Reference systems are but the framework for topographical representations. These can never be fully abstracted from those objects and space relations between them which form a given locality and distinguish it from any other locality (Shemyakin, 1960, p. 191). How the topographical representations themselves develop is a question which has been given little attention in the literature. Of course, we already know that the representation of space begins with the internalization of action on space; we might anticipate that walking or cycling would be most important to the child's formation of topographical representations and that more passive modes of travel would not serve the same purpose. This point is also argued convincingly by Lee (chap. 5) and Stea and Blaut (chap. 12). As Lee says, walking is "intimate to the environment and therefore articulates the schema." Stea and Blaut suggest the importance of toy play as a surrogate for other forms of direct environmental experience. These hypotheses are supported by research findings on the role of kinaesthetic cues in geographical orientation (see Liebig, 1933; Howard and Templeton, 1966).

Given the importance of action and of domicentricity, it is not surprising that Rand (1969) found children's "trips and excursions to far-away places are remembered but not thought of as connected with, or of the same world, as their immediate habitat." Similarly, Lee found that the spatial world of primary school children in England was divided into various local "schemata" which bore a "detectable relationship to the physical world," but that "beyond this home area lay one total schema that might be called the 'elsewhere schema' in which physical dimensions were irrelevant" (chap. 5). Rand (1969) claims that a deep sense of familiarity with the home area is necessary as a basis for further exploration and discusses the notion of home as "sacred space" from which the child can make "brief excursions into the profane world" (see Eliade, 1959). This seems sufficiently obvious to be a truism, but such statements are understandable: not only has there been little research into the development of topographical representation in children but there have been very few studies of children's environmental behavior. With an increasing awareness of the importance of early environmental experience in children's development into "rich and competent human beings" (Carr and Lynch, 1968), we can anticipate a growth of research in these fields (see Lukashok and Lynch, 1956; Cobb, 1959; Parr, 1967; Shelton, 1967; Bunge, 1969; Ladd, 1970; Anderson and Tindal, 1970; Colyard, 1971).

Coordinated System of Reference. The literature on the development of reference systems of a higher order than the fixed systems discussed above (with the exception of that of Piaget and his colleagues) has been totally concerned with children's understanding of cardinal directions. This sheds little light on the "spontaneous" development of reference systems in children. While knowledge of cardinal directions has pedagogical merit for the understanding of maps and for the communication of directions, it is not necessary in the formation of topographical representations of large-scale environments. For example, none of the 4- to 10-year old children in Piaget and associates' (1960) model-the-school experiment described earlier in this section were reported to have utilized the cardinal directions in the constructing of topographical representations. Nevertheless, this experiment confirmed the findings of Piaget's model landscape and model village experiments described in Section II: children who had achieved equilibration of concrete operations (Stage III*b*) can construct a topographical representation in line with a two-dimensional coordinate system. The coordinates, when drawn, do not necessarily run parallel with the margins of a piece of paper as with a cartographic grid. Rather they are based upon physical features in the experienced environment, but the end result for children of this level is always a coordinated whole (Piaget et al., 1960). In drawing the plan of the school area, children at this level use two complementary methods: either they group the elements of the plan in terms of relations between local areas, or they select one or more common starting points and reconstruct routes which radiate from them. In this way any single part of the representation is related to all the other parts, for through associativity each point can be reached mentally by a variety of routes, and through reversibility each route can be represented in the reverse direction to that experienced. Thus, having progressed through egocentric nonrepresentational orientation and through a fixed system of reference wherein the first preoperational, partially-coordinated representations are developed, the child has advanced to the final stage of topographical representation, that of utilizing a *coordinated system of reference.*

The question now arises, how does the child coordinate the set of independent and fixed subgroups of space into a fully coordinated topographical representation? Piaget is the only researcher who has addressed this question (Piaget and Inhelder, 1967; Piaget, Inhelder, and Szeminska, 1960). What he seems to suggest is the following: the child decenters from each of the partially coordinated fixed systems of reference of the previous stage. Thereafter, through processes of reciprocal assimilation among the reference systems and reflective abstraction beyond them, he intercoordinates these structures and advances to a higher plane of thought, that of fully equilibrated concrete operations (III*b*).

Other writers have discussed what has been called *geocentric orientation* (Trowbridge, 1913; Freeman, 1916; Gregg, 1939; Lord, 1941), apparently equivalent to what we prefer to call a coordinated system of reference. A wealth of reports from anthropology reveals that neither is universally the final stage of development in ontogenesis (see reviews by Werner, 1948; Hallowell, 1955; Lynch, chap. 16). Werner (1948, p. 172) has suggested:

> There must be a complete revolution in the history of the human mind in order to uproot the idea of space from its primitive, anthropomorphic, qualitative-dynamic ground and thereby elevate it to the free sphere of physical-geometric abstraction.

From the number of studies bemoaning children's ignorance of the western system of abstract cardinal directions and suggesting ways of teaching it (Gulliver, 1908; Trowbridge, 1913; Ridgley, 1922; Howe, 1931, 1932; Gregg, 1940; Lord, 1941; Kabanova-Meller, 1956), one can reasonably assume that ontogenesis demands a similar "revolution" in geographical orientation to that suggested by Werner. As we demonstrated in Section II, in the words of Shemyakin (1962, p. 190), "the crux of the general development of the understanding of space is the transition from a [fixed] point of reading to a system with a 'freely transferable' point of reading." Trowbridge (1913, p. 894) claimed that confusion results from poor teaching of cardinal directions leading to entirely erroneous "imaginary maps," and further suggested that:

> Children should be seated in a special manner when studying geography, with the cardinal points of the compass marked in the room, and the maps in the books properly orientated . . . imaginary maps [should be] systematically corrected in childhood.

This suggestion and other similar ones (Kabanova-Meller, 1956; Ridgely, 1922; Howe, 1932) have been repeatedly reiterated by geography educators in the *Journal of Geography*. If all else fails, Gregg (1940) suggests, have the child think of places lying in specific directions and lean his trunk, incline his head, or turn his eye "vigorously" towards the place while calling out its name! None of these writers recognized the real problem: the most fundamental factor in the child's use of an abstract or coordinated system of reference is the prior development of his ability to recognize perspectives other than his own (see section II). Geographic education has only recently begun to take note of the crucial importance of considering development of the child's concept of space (e.g., Miller, 1967; Almy, 1967; Towler and Nelson, 1968; Towler, 1970; Blaut, 1969; Muir and Blaut, 1969–70; Blaut and Stea, 1969, 1971; Blaut, McCleary, and Blaut, 1970; Hart, 1971).

It remains necessary for us to describe how this system is related to the environment, for clearly, no system of reference is useful if fully abstracted from the environment in which the individual must act. Trowbridge (1913) suggests that the individual uses reference points on the horizon corresponding to the cardinal directions. As the lines from these points always intersect at the ego, he called this an "egocentric" system of orientation. We find this use of the term confusing, for the ultimate point-of-reading for all orientation is indeed and necessarily the perceiving individual. It is because of this that the egocentric system of reference is discussed as a *prior* stage which is later subsumed under more flexible, abstract systems of reference. Nevertheless, something like what Trowbridge described must exist for all those who utilize the cardinal directions.

Lord (1940) measured the orientation abilities of 317 children in the later elementary grades (10 to 14 years) by asking them to point to the cardinal directions, distant cities, and nearby places, and to draw sketch maps of the spatial arrangement of local cities, villages, and features within a familiar city. Lord found that children have two frames of reference: a "conventional map frame" for distant places (using the cardinal directions) and a "direct experience frame" for relatively near-by places (fixed system). Through the simultaneous use of these two systems a child may locate two cities, one distant and one nearby, in directions which are markedly inconsistent with each other.

In summary (see figure 14.4), there is considerable evidence that in developing topographical representations of the large-scale environment, the child utilizes a framework or system of reference for interrelating different positions, routes, patterns of movement, and himself in this environment, and that this system of reference is the most important component of spatial representation. Furthermore, the ontogenetic development of systems of reference proceeds through three stages: egocentric, fixed, and coordinated, the third stage being achieved in children with the equilibration of concrete operations. Furthermore, it appears that the topographical representations themselves, like all spatial representations, are formed initially on the basis of visual-motor connections. Though much of the evidence is anecdotal, the suggestion is that route-mapping ontogenetically precedes survey-mapping.

Much of the above work requires a stronger empirical base. Further systematic experimentation is required before we can extend with confidence the existing theories of fundamental spatial cognition to the specific development of topographical representations. Some work has begun at Clark University following Piagetian theory (e.g., Cannitello, 1970; Mark, 1970).

THE MICROGENETIC DEVELOPMENT OF TOPOGRAPHICAL REPRESENTATIONS

Following organismic-developmental theory, we would expect formal similarities between the ontogenesis and microgenesis of macro-spatial cognition (Werner, 1948; Follini, 1966). The *microgenetic* development of an individual's cognition of, and orientation in, a new and unfamiliar macro-space (e.g. when a child explores a new area or an adult moves to a new city) is an important yet relatively uncharted area of research.

Freeman (1916) claims that an individual visiting a new city constructs a system of orientation by unconsciously applying the cardinal directions to various features; only if he makes a mistake does the person become conscious of the process. While this may be true for some, it is difficult to believe that cardinal directions are used as a reference system by all adults.

In summarizing the work of Khopreninova (1956) and Sverlov (1951), Shemyakin (1962, p. 230) emphasized the prior accumulation of a certain quantity of internally represented "route-maps" as the most crucial factor in the subsequent formation of "survey-maps." With these "route-maps," he claims, "a given locality is, as it were, presented as a system of roads leading 'there' and 'back' ". This is similar to von Senden's description of the development of "schemas" in the blind. After a blind person has successfully represented a number of journeys, von Senden (1960, p. 287) says:

> He then seeks to relate a number of such routes, all starting from the same point, by means of "lateral crossties," and so to create a network of relations, to encompass and knit together the parts of the region commanded in this way.

Von Senden points out that the blind use this method even for the exploration of rooms. Others have explored the question experimentally, simulating blindness by blindfolding their subjects (Brown, 1932; Liebig, 1933; Worchel, 1951; Follini, 1966).

Some different findings have come from recent studies of urban form. Lynch (1960) observed that the sequence in which subjects' sketch maps of cities were drawn develops in either of five ways: outward from familiar lines of movement; outward from a familiar dense kernel; inward from the construction of an enclosing boundary; from a gridiron pattern; or from an initial set of adjacent regions. Lynch's (1960, p. 86) suggestion that "this might have some relation to the way in which it [the image] first develops as an individual becomes familiar with his environment" so far lacks empirical validation.

One of Lynch's colleagues, Appleyard (1969a) found that sketch maps drawn of the new city of Cuidad Guayana by 320 inhabitants were structured either *sequentially*, using roads and rivers as organizing principles, or *spatially*, using buildings and districts. His suggestion that sequential maps are of a lower developmental order than spatial maps accords with the findings of Shemyakin on route-maps and survey-maps reviewed above. Appleyard also noted that the maps ranged from primitive topological maps to positional maps. He argues that both of these are developmental distinctions, but provides no evidence of the transition between the two. It is interesting to note, however, that 75 percent of the maps were of the sequential type. If it is true that these are developmentally primitive, it could be that this large percentage resulted from the limited experience of most of the inhabitants with this new city (the survey was taken only three years after the start of construction). Therefore, it is possible that the development of a person's representations of a new environment proceeds from route-mapping (sequential) to survey-mapping (spatial).

Gittins (1969) conducted a thorough study of the development of representations of unfamiliar large-scale environments. In Section II we described his results as regards two different types of final *product* of the representation, but he was also interested in the *process* by which people form impressions. His methodology included asking his subjects (writers and scientists) at several points on a walking tour through an unfamiliar city to describe and draw the section of the city recently passed through and their current overall impression. At the end they were asked to describe in what way they approached the task of forming impressions.

He found that there were two different processes of cognitively structuring a city: *aesthetic-poetic* (characteristic of most of the writers), and *scientific* (characteristic of most of the scientists). The aesthetic process began with an initial intuition of the city which slowly but steadily became organized into a "comprehensive integrated conception" through the combining of a series of insights into a concrete, sensuous impression of the specific city traversed. The scientific process did not evidence intuition at the beginning, nor was there much organization for some time, until at mid-tour, a sudden reorganization occurred leaving a clear, articulate, abstract, "formula-like" structuring of the city, which applied to cities in general rather than to this particular city, and which was only slightly modified to incorporate aspects of the city experienced later.

IV. Conclusion

At the beginning of this review of work on spatial cognition, we saw that there were two quite separate bodies of literature, one emanating from developmental psychology, the other from behavioral geography and urban planning. Both address the same basic issue: How do we come to know the form of space? To the best of our knowledge, these two bodies of research have not previously been brought together. However, not only are the questions asked essentially the same, but, as we have tried to show, the findings are commensurate. Some of the fundamental questions are: Is the concept of space innate, learned, or constructed? What is the role of experience? How does the child come to know about the spatial form and relations of the everyday physical environment? Is there an order or sequence in this understanding?

We have seen from Kant's contention and from the research of Werner and Piaget that the concept of space is neither innate nor learned; it is constructed by the child as a series of hierarchically integrated equilibrations through his adaptive interaction with the environment and by means of the reciprocating processes of assimilation and accommodation. From the internalization and reflective abstraction of actions arises the child's first intuitive understanding of space. However, Cassirer, Werner, and Piaget have made clear that the understanding of space is more than the learning or mere accumulation of facts which have some kind of real existence. Piaget (Piaget, Inhelder and Szeminska, 1960, pp. 23–24) light-heartedly noted that if it were simply a case of accumulation one might explain the development of topographical representations in the following manner: at four years of age, a child is brought to school by his mother, and is therefore aware only of the school, his home, and the local candy store; at seven, he knows a few roads, and can therefore describe fragmentary routes; and at nine or ten, he is allowed to roam free and consequently knows the topography intimately. But as we have demonstrated and Piaget (Piaget et al., 1960, p. 24) has well expressed:

> The growth of knowledge is not a matter of mere accumulation . . . while it is true that between the ages of four and ten, children collect a good deal of information about their district, they also coordinate the picture which they have of it, which is an infinitely more complex process of development.

We have seen from Cassirer, Werner, and Piaget that there are several dimensions or polarities of the child's understanding of space. All three theorists have identified a progression from concrete to abstract space consisting of sensorimotor action in space, perception of space, and symbolic or abstract representation and thought about space (refer to fig. 14.1). Werner expressed the same development as a progression from a fusion between the child and his environment to differentiation and, ultimately, reintegration. Both Werner and Piaget have shown a progression from egocentrism to perspectivism (or relational coordination) and that this progression is a dimension which seems to underlie many of the other developments in the child's understanding of space. Piaget has identified four levels of spatial intelligence: sensorimotor, intuitive or preoperational, concrete operational, and formal operational. We have used these levels as a framework for organizing findings on the spatial cognition of large-scale environments. Furthermore, we have used the progressions of general intellectual development identified by Werner and the stages and functional invariants identified by Piaget to begin to suggest explanations for the findings on macro-spatial cognition (see figs. 14.2 and 14.4).

In the sensorimotor period of development, the infant moves in a space of action. His orientation to the larger environment is totally egocentric and he has no topographical representations.

As spatial actions become internalized through the dual process of assimilation of impressions from the environment into the sensorimotor schemas and accommodation of the schemas to the environment, the child's first images or iconic representations of space develop. This leads to a major new period, that of preoperational space, the beginning of representational space. During this period, most topological relations are formed, and projective and euclidian relations begin to be constructed. Egocentric orientation gives way to the onset of a fixed system of reference, centered first on the home (domicentricity) and later on a small number of uncoordinated routes, landmarks, and familiar places. The child no longer operates solely in a space of action for he begins to represent his routes (route-type representations). Generally, this is a period of gradual differentiation of the child from his environment, of the child's point of view from that of other people, and of elements and relationships from each other within the environment.

These differentiated but uncoordinated representations of discrete parts of a total space (e.g., routes, landmarks, barriers) begin to be coordinated with the onset of concrete operations during the early school years. Many other specific developments occur, such as the concept of the straight line, paralleleity, proportional intervals, and angles. As we have shown, the child's understanding of these relations is a result of the grouping of partial structures into a coordinated whole through the equilibration of assimilation and accommodation. The child is now able to coordinate perspectives and construct a euclidian system of reference. Both of these are most important for his understanding of the large-scale environment. Furthermore, the child's individual route-type topographical representations become coordinated into a comprehensive survey-type representation.

Finally, with the onset during early adolescence of formal operations, which are a reflective abstraction from concrete operations, the individual is not only able to act in space and mentally coordinate his thoughts about concrete objects and spatial relations, but he is also able to reflect on these accomplishments and consider a theoretical space totally abstracted from any concrete particulars. The concepts of length, distance, area, and volume, all of which depend on the formal concept of infinity, are also able to be constructed and conserved. Thus, both a true metric space and a totally abstract space are possible for the adolescent.

It seems from the research literature on the ontogenetic development of spatial cognition that there are five domains of parallel development: levels of organization of spatial cognition (sensorimotor, preoperational, concrete operational, and formal operational); types of spatial relations (topological, projective, and euclidian); modes of representation (enactive, iconic, and symbolic); systems of reference (egocentric, fixed, and coordinated); and types of topographical representations (route and survey). Each of these developments in turn parallels the four periods of general intellectual development discovered by Piaget and the orthogenetic principle and developmental progressions elucidated by Werner (all as illustrated in figs. 14.1, 14.2, and 14.4). Furthermore, as we have tried to show, there seem to be certain direct correspondences and functional and explanatory relations between the different domains of the development of spatial cognition.

This review has focused on a structural description of the child's understanding of space and has suggested some of the dynamics involved in the process of development between structures. Drawing on the theories of Piaget and Werner, the next step should be the construction of complete explanations of the process involved both across the domains of the development of spatial cognition and between the stages within each domain. Piaget and Werner have established the groundwork for this effort through a systematic description of the development of the fundamental concepts of space and by suggesting explanatory concepts of a comprehensive nature. However, within the context of the cognition of space in large-scale environments, these explanatory relationships have yet to be systematically investigated (although some hypotheses have been advanced by Stea and Blaut) and the application of these theories is yet to be tested, remaining an issue of crucial importance. In addition to reviewing the literature on the development of spatial cognition, we have tried to suggest a direction consistent with the major theories of cognitive development in which these investigations might further proceed.

V

Geographical and Spatial Orientation

Getting lost was a frustrating experience because I couldn't find any way of becoming unlost, of attaching myself to the environment again. I would either have to ask somebody, which wasn't very successful because they couldn't communicate turns, etc., to me, or watch for signs, which were completely ambiguous. Then there was the pressure of traffic lights, and horns blowing. The consequences of a wrong direction are severe and cumulative. Trying to find things on a map where the grid changes direction is frustrating as Hell.

Do I dislike being late more, or is it annoyance about not being able to do such a simple thing? Why can't Americans have decent road signs? Why are they trying to keep the locations of towns hidden? If you want to get to Winchester in England, you find periodic signs showing just how far away Winchester is as you travel the road. That's much more interesting; I mean, it's totally impossible to use a map while driving, unless you stop, so the map is of use only when you *can* stop. The whole trip through New Jersey to New York was 90 percent chance—there are no standard road number signs, no standard colors, no fixed intervals, no mile posts, no things to *predict* and *rely* upon. Even junctions are badly signed. Given the speed of traffic approaching junctions, they're not signed far enough in advance. I consciously planned the route, talked to somebody who lives in New York and got totally wrong directions. There are no good road maps, since they show roads which are proposed but not yet built. I'm going to join AAA. Yesterday was as annoying as Hell. Street signs in New York are small and never lighted, and one can't read street numbers.

—*Anonymous Sufferer from Chronic Disorientation*

Introduction

This commentary on the experience of being lost partly answers the question: "Why a section on orientation?" Orientation, like the weather, is a problem that everyone talks about, but to which no one has a practical solution. As Lynch (1960, pp. 3–4) states:

> Structuring and identifying the environment is a vital ability among all mobile animals. Many kinds of cues are used: the visual sensations of color, shape, motion, or polarization of light, as well as other senses such as smell, sound, touch, kinesthesia, sense of gravity, and perhaps of electric or magnetic fields. These techniques of orientation, from the polar flight of a tern to the path-finding of a limpet over the micro-topography of a rock, are described and their importance underscored in an extensive literature. . . . Despite a few remaining puzzles, it now seems unlikely that there is any mystic "instinct" of way-finding. Rather there is a consistent use and organization of definite sensory cues from the external environment. This organization is fundamental to the efficiency and to the very survival of free-moving life.
>
> To become completely lost is perhaps a rather rare experience for most people in the modern city. We are supported by the presence of others and by special wayfinding devices: maps, street numbers, route signs, bus placards. But let the mishap of disorientation occur, and the sense of anxiety and even terror that accompanies it reveals to us how closely it is linked to our sense of balance and well-being. The very word "lost" in our language means much more than simple geographical uncertainty; it carries a tone of utter disaster.

In order to understand the widely scattered research literature on this topic, we must first distinguish the cognitive use of orientation—which coincides with Griffin's (chap. 15)—from perceptual orientation, or awareness of bodily position in space: the latter includes the sense of the vertical and associated physiological aspects and mechanisms of balance. Midway between the two types of orientation study has been research on animal spatial behavior dealing with such problems as "homing" (Barlow, 1964). But the human problem illustrated by the introductory quotation is still an open issue, despite its importance to everyday spatial behavior.

One obstacle is the lack of a useful definition of orientation; we will employ a definition from English and English (1958, pp. 363–64):

> The discovery or knowledge of where one is and where one is going, either literally in space and time, or figuratively in relation to a confusing situation or a puzzling problem. The orientation is cognitive when it consists chiefly in knowing the situation.

The ability to orient is a process which ties cognitive maps to the spatial environments which generate and reinforce them. Specifically, we are concerned with the ways in which cognitive maps get "keyed" or related to actual environments. The city of Boston, for example, has water boundaries giving it the shape of a "funnel" seen in cross-section with the Central Business District and Government Center at the funnel's apex; cognitive maps are verbally *keyed* to the environment by such terms as "inbound" (toward the apex) or "outbound" (away from the apex and toward the wide mouth). Manhattanites refer to locations on their island, whose major axis is close to North-South in orientation (or "up-down," as maps of the island are conventionally viewed) as "uptown" or "downtown." A more exact keying device is utilized in Puebla, Mexico (Stea and Wood, forthcoming): the streets are labeled "Norte," "Sur," "Oriente," and "Poniente" (North, South, East, and West). Inhabitants refer to points *beyond* the city in terms of this local coordinate system even though the system is rotated 45 degrees from conventional map or compass coordinates.

We shall employ the term *frames of reference* for methods of keying cognitive maps to environments such as the Puebla example. The components of such frames—the represented environmental features upon which the frame is built—may be points, lines, or areas. We classify frames of reference according to three types: (1) based in the representing individual; (2) based in the environment represented; (3) external to both the environment and the representing person. Hart and Moore (chap. 14) argue that these form a developmental sequence in the growing child as a result of environmental learning and psychological maturation.

An obvious example of a *frame of reference based in the representing individual* is a three-dimensional coordinate system based upon the differentiation of the human body wherein objects in the environment are described as being to the right or left, front or back, above or below. To utilize this system, the representing individual (or an individual to whom the environment is being described) must keep in mind his position and initial "heading" (which way he is facing) with reference to environmental cues and/or the heading of the individual giving the description.

Frames of reference based in the environment depend on "natural" features and man-made features such as urban landmarks (Lynch, 1960). Thus, "landscape" frameworks may key a representation to a given locale by means of the mountains, bays, lakes, and rivers which characterize that locale. As Lynch (1960, p. 29) indicates, fishermen in parts of the South Seas find "islandward" (toward the middle of the island, frequently the loftiest and most visible point of land) and "seaward" a useful, simple system. Environments with such a clearly definable center and periphery are amenable to frames of reference based on *polar coordinates* (established upon a "center," an imaginary line of fixed direction originating at that center, and angular deviations of other directional lines from the fixed line —see chap. 5). The simplest polar coordinate systems incorporate only the notion of directionality with regard to a central point. Thus, what matters is not the *angular deviation* of removal, but rather "going towards" versus "going away."

However, angular direction takes on importance to the island-dweller when attractions or dangers surrounding the island lie only in certain sectors; he must go "this way" to find fishing grounds or "that way" to avoid being wrecked on a reef. This landscape framework requires at least one other fixed point in addition to the center of the island since no reference line of fixed direction exists from which angular deviation may be established. Such a reference system is *specific to a particular locale*, and, while extendable via features in the line of sight, is not generalizable to another area.

Of the *frames of reference external to both the environment and the representing individual*, the most familiar are those based upon compass coordinates. They may be derived from "true" North, South, East, and West; magnetic compass coordinates; or a "local" coordinate system. The first two are independent of the specific environment represented, the last, partially so. In Puebla, Mexico, local North is at 45° to true North, but the former serves perfectly well for orientation *within* any part of the city.

This three-fold division of frames of reference is less common than another, in two categories, suggested by several investigators (Angyal, 1965; Shemyakin, 1962). Some have used the terms "egocentric" and "objective" to label these two categories. Trowbridge (1913) differs in using the term "egocentric" to refer to the "earth's ego," i.e., to a compass frame external to both the environment and the representing individual, and "domicentric" to refer to an environmental frame centering about a starting point, or home. However, few frameworks are purely egocentric or objective in all environments and circumstances. In choosing to use the three-fold division, we have extended the meaning of "egocentric" to include systems centered on transportation aids, man-occupied objects in motion. Thus, a frame of reference is egocentric if its relation to a specifiable environment changes with the position of the observer or his vehicle. It is objective to the extent that such a relation does not change.

The single most comprehensive review, to date, of the scattered psychological work on geographical orientation has been provided by Howard & Templeton (1966), who distinguish between two classes of geographic orientation:

> On the one hand are those tasks which involve the ability of a person to maintain a sense of direction when moving about in strange surroundings. Such tasks do not require any prior intellectual knowledge of the spatial position of particular objects. On the other hand, there are those tasks, such as drawing a map or pointing to the North, which do require intellectual knowledge (Howard & Templeton, 1966, p. 257.)

The most common "malfunction" of the first ability is popularly called "walking in circles," supposedly common to humans and animals lost in unfamiliar or featureless surroundings. Results of studies performed during the first half of the twentieth century were taken to indicate a tendency, greater among women than men, to veer to the right, but provided no basis for determining which of three alternative explanations accounted for the behavior: (1) that some "circling mechanism" exists in all animals; (2) that the body's physical asymmetries cause circling; (3) that veering may be caused by asymmetries in the vestibular apparatus of the ear. Another question asked about geographical orientation concerns the role of visual experience. In seeking an answer, attention has been focused on two issues: (1) how do blind and blind-folded people compare on orientation tasks, and (2) what is the role of the *age* at which blindness occurs? No unequivocal statement on either issue has yet been supplied:

> There is a need for a series of studies comparing blind subjects, vestibular-deficient subjects, and normal subjects, under conditions of active and passive movement, on a variety of very simple orientation tasks, such as estimating a distance moved or an angle turned, etc. Only when such analytic, detailed, and comprehensive studies have been done will we be able to assess the role of various modalities in different types of orientation behavior. (Howard & Templeton, 1966, p. 262)

The second ability to which Howard and Templeton call attention—involving such tasks as pointing in a given direction, drawing maps, etc.—was studied late in the 19th century by Binet (1894) and early in the present century by Claparede (1903). Gulliver (1908) discussed problems and methods of teaching compass orientation to children, but it is the ground-breaking work of Trowbridge (1913) which has served as the major source of inspiration for later research efforts. The most significant result of the studies is that consistent group and individual distinctions can be identified in modes of representing geographical directions. In this regard, Howard and Templeton (1966, p. 263) suggest that two fundamental questions remain to be studied:

> Are those people who choose one mode of map drawing incapable of using other modes if asked, or does their consistency in performance merely reflect their preference to use one of several methods of which they are equally capable? If people do differ in their ability to use one or other mode, do these differences reflect fundamental differences in spatial intelligence, or do they merely reflect differences in the way the people have been taught?

Inherent in the above questions are issues of learning and development, which, in the more general context of cognitive mapping, are treated elsewhere in this book. It may be useful to divide development into "normal," leading to more ordered geographical orientation, and "abnormal" (usually the result of neurophysiological dysfunction) leading to "disordered" orientation. How, we then ask does "normal" geographical orientation develop? No evidence has thus far been presented that any organism possesses a "special sense" for compass direction:

> There have been those who were convinced that man has a separate modality for compass direction, which gives direct awareness of the earth's magnetism . . . and a popular belief persists that primitive people and children possess an innate sense of direction. Darwin (1859) speculated on the possibility of

> an instinct of orientation, but omitted such speculation from later editions of his book. No good evidence has been produced in support of this hypothesis . . . (nor) to suggest that primitive people have an orientation ability which a civilized man could not match, given training in the use of natural landmarks, dead reckoning, and other skills . . . of course there may still be inherited differences in orientation aptitude. (Howard and Templeton, 1966, pp. 265–266)

Disorders of geographical orientation have rarely been studied in their own right; rather, we know about them because they have been used to diagnose certain forms of brain damage, particularly of the parietal lobes. Patients suffering from unilateral visual inattention, for example, who ignore objects in one half of their visual fields, draw maps of familiar places with landmarks on one side shifted toward the other, or altogether eliminated. Other patients suffer loss of topographical memory, failing to recognize familiar landmarks; still others demonstrate parietal disorders by losing themselves. "Losing oneself" in this sense takes two forms, the first, often called "Topographical Agnosia", involves loss of memory for the *positions* of objects; the second involves the inability to *relate* objects which are individually familiar—the less the familiarity, the greater the degree of the disorder. Still another spatial disorder, believed to be associated with aphasia, results in a loss of the ability to *describe* location; the patient may not be able to *say* where he is located or where other places are located, but is often quite capable of finding his way around.

The role played by vision in these disorders has often been greatly overvalued. On this subject, Semmes, Weinstein, Ghent, and Teuber (1955) concluded

> that vision as such has no role in long-term spatial disorientation and, on the other hand, that disorientation rarely occurs without somaesthetic defect, although the converse is common. . . . They suggested that disorientation may involve a pathological inattention to backgrounds, since performance on [a] route-finding task was correlated with performance on a conditional-reaction task in which the background of the stimuli had to be taken into account. (Howard & Templeton, 1966, 271).

In summary, Howard and Templeton (1966, 271) conclude that:

> The clinical literature as a whole is characterized by a tendency to "symptom naming" which produces a welter of named disorders with little attempt at a theoretical or experimental analysis. The answers in this field are not simple; man's skills of geographical orientation are very complex and very idiosyncratic.

Of the contributions in this section, Griffin, in one of the few chapters on spatial orientation ever included in an American introductory psychology text, provides an excellent coverage of information on both human and animal orientation, although only the portions dealing with human orientation are included here. Lynch's discussion of orientation suffers from being an appendix to the popular *Image of the City*, and is overlooked by most readers. He draws heavily upon anthropological evidence pointing to the existence of a multiplicity of specific systems of orientation. In part IV, The Development of Spatial Cognition, Hart and Moore review the development of orientation and systems of reference in the child.

These articles are heuristic as well as informative and they raise more questions than they provide answers. For example, why is one frame of reference chosen in preference to another, or why is one chosen at one time (place) and another at a different time (place)? Can morphological characteristics of "natural" and "designed" environments be defined such that one frame of reference is appropriate to one kind of morphology and another frame to another morphology? More generally, what is the influence of a cognitive map established in a familiar environment upon the establishment of frames of reference in a new environment (or vice-versa)?

15

Topographical Orientation

DONALD R. GRIFFIN

To be successful in your daily business you must know where you are going, and your brain has few more important tasks than keeping you oriented in space. To know where you are going, you must know where you are now, and also know something of your immediate surroundings. This knowledge of special relationship with objects in your environment is called *topographical orientation.* A minimum of it is required when you move about by merely following a road, a trail or a leader. This type of locomotion is often called "blind"; its opposite is the well-oriented behavior of a person who knows where he is and where he is heading. So basically true is this statement in its literal sense that the term *well oriented* has taken on the familiar connotation of competence to deal with social surroundings.

The Topographical Schema

Let us consider first the simplest type of topographical orientation, finding the way by following a path. It may be a footpath on a college campus, a city street, a numbered route on a highway or a blazed trail in a forest. At all times you need clues which tell you how to stay on the path; and usually they are obvious—the edge of the path or street, the undergrowth on both sides of the forest trail. If the path is well-defined, it may be sufficient to follow your nose, provided it was pointed in the right direction to begin with; yet most paths have branches and crossings where some other procedure is required. Suppose that you are driving an automobile and seeking a certain

This article is part of Chapter 17, pages 380–386, entitled 'Topographical Orientation' in Boring, G. G., Longfeld, H. S., and Weld, H. P., *Foundations of Psychology,* (New York: John Wiley and Sons, 1948). Reprinted by permission of the author and publisher.

numbered highway which you know leads to your destination. You may reach a crossroad which bears the correct route number; yet this clue is insufficient by itself. You must also know which way to turn. For example, U.S. Route 1 runs from Maine to Florida along the Atlantic coast; if you are driving east in Virginia and come to a highway marked "Route 1," you must decide whether to turn right (towards Miami) or left (towards Boston). Once this decision has been made correctly you need merely follow the signs.

Thus it is clear that even a simple path cannot be followed blindly without some idea of its spatial relationship to other objects. In practice the following of a marked path is supplemented by some sort of *mental map* or *schema* in which this path is perceived in spatial relationship to other paths and places. In familiar surroundings people are usually quite unaware that they are using such a schema, even though they can orient themselves correctly when placed in an unexpected part of this familiar territory. They can recall the landmarks and other topographical clues in such a well-organized relationship to one another that each landmark can be related to any one of several others and serve for guidance along a variety of routes of travel.

Components of the Schema

Our mental map may include all sorts of memories of objects, once perceived as we moved about—visual memories of the appearance of buildings, trees or hills, bird's-eye views of buildings, pathways or streets and, in some cases, kinesthetic or auditory clues. The schema may also involve the realization that certain places or objects not immediately perceived lie in the same direction as other objects close at hand which we can perceive directly. We may think of the post office as beyond a seen window, or to our right, or over that hill yonder. Thus familiar places or parts of our own bodies become points of reference for other more distant objects or places. London lies in the same direction as the post office but beyond it; China is beyond the railroad station. If our schema has been much influenced by the use of maps, it will include the cardinal points of the compass.

Graphic Representation of the Schema

A map is, in fact, a graphic representation of such a schema, rendered more accurate by the techniques of surveying. The better oriented a person is, the more closely his schema is likely to resemble a map. For many people, however, geographical accuracy is not necessary or even important. A person who does all his traveling by street car may have a schema built around the street car lines. If asked to sketch the schema, he will simplify the street car routes, presumably smoothing curves into straight lines, and tending to show most turns as right angles. His customary stops are likely to be prominent, and other sections of the line may be telescoped or omitted altogether. Thus his schema will be quite distorted when compared with an accurate map, yet schemata of this kind fit the psychological needs of the person who travels over fixed routes. They are exemplified by the "maps" inserted in railroad timetables, where the railroad route is usually made to appear straight with geography forced to conform to it. The motivation behind the drawing of railroad maps is partly a desire to make the route of the particular railroad appear short and direct, but the gross distortion of geographical fact also serves a real purpose for the traveler, for the railroad map conforms more closely to the topographical schema of the average traveler than the geographer's accurate maps. The traveler can turn this sort of schematic map into a mental schema more readily because it is already in the right form. The traveler is interested primarily in his starting point and destination. He has a secondary interest in the stations along the way and cares little or nothing about the twists and turns of the route. The railroad map and the mental map do not confuse him with geographical information which is of no immediate concern to him.

Not only are directions distorted in the topographical schemata of most people, but distances and the relative areas of various regions also reflect their importance to the individual rather than geographical reality. This is best illustrated by asking an adult person who has not for some years studied geography to draw a map of the United States. Almost invariably the area in which he lives will be drawn larger and in more detail than the rest of the country.

Extension of the Schema to New Territory

When a person goes to a strange place he takes with him parts of his old schema, which may, depending upon the circumstances, aid or hinder him "in learning his way about"—in developing a new schema for the new locality or in extending the old one to include it. The old schema is an aid when it includes the points of the compass or other geographical relationships such as the bearings of shorelines, rivers and mountain ranges which remain pertinent when the old schema is extended to include the new area.

Often, however, the old schema does not fit because there were unnoticed turns in the road, and hence the extension is in error. Then part at least of the person's schema conflicts with the realities of his new environment. "When I was in Wilkinsburg, Pa.," a psychologist writes, "the trolleys for the eastern suburbs ran north through the town, but the east-bound trains ran south through the town. I arrived on the train and, of course, took the trolley going the wrong way to get to the west. My frame of reference was in conflict all that summer."

In learning a new schema, one may be led astray by the assumed directions of streets or paths. If the assumption is wrong and the road, instead of being straight, runs at an angle or bends in its course, and if these deviations are not noticed, the topographical orientation becomes utterly confused. The Boston Common is surrounded by five streets which make approximately right angles with each other. Some of the streets are bowed inward enough to make this circumstance possible, but the stranger rarely notices the bends. Hence those who take the Common as a point of reference may often be ninety degrees out of reckoning.

What we call a "sense of direction" is a skill at retaining orientation or expanding the topographical schema sufficiently to keep it up to date as we travel about. People vary greatly in this ability, which depends upon the rapid and accurate assimilation of new scenes into the schema. Like any type of learning, its perfection depends partly upon practice and partly upon motivation and attention. It is common experience that, in a strange city, we remain much better oriented if we find our own way about than if we cover the same ground in the company of a local inhabitant who leads the way. Active participation demands attention and facilitates learning.

Nonvisual Clues

We may use a topographical schema to find our way about even when the landmarks must be perceived through a different sensory modality from the one when they were first encountered. Often the landmarks are first learned as visual clues and later felt by means of other senses. A familiar instance of this is our ability to walk about in our own homes in the dark. We are likely to know, without conscious effort or counting, such topographical relationships as three steps from the bed and it is time to reach for the knob on the bedroom door, six more steps to the left brings us to the stairs, after four steps down we reach a landing and must run right before continuing down ten more steps to the downstairs hall. (Perhaps the ten steps have to be counted, but the others usually can be perceived correctly without enumeration.) All this information gets organized into so coherent a topographical schema that a man can walk safely anywhere in his own house in pitch darkness even though he is nearly or quite asleep. In such cases the mental schema was probably first erected on the basis of visual clues, but in the dark it gets transferred to substitute senses. We can recognize the round door knob, the hard hall floor and the mounting stairs by the somesthesis of hands or feet, the somesthesis of what it is like to take six steps. Hearing may also provide clues. The sounds of walking on a carpet and on a bare floor are different. We never have an auditory map, but we can have a kinesthetic one, the feel of how it is to cover the whole route; and we may have, not a verbal map, but a verbal guide by which we instruct ourselves how to go. In other words, the equipment of imagery that we use in other kinds of thinking is available for use in mental map making.

16

Some References to Orientation

KEVIN LYNCH

We can look for references to the environmental image in many places; in literature ancient and modern, in books of travel or exploration, in newspaper accounts, or in psychological and anthropological studies. Such references are generally scattered, but are frequent and revealing. While skimming through them, we will learn something about how such images are formed, what some of their characteristics are, and how they seem to play a social, psychological, and esthetic, as well as a practical, part in our lives.

From the accounts of anthropologists, for example, we infer that primitive man is normally deeply attached to the landscape he lives in; he distinguishes and names its minor parts. Observers refer to the multitude of place names, even in uninhabited country, and to the extraordinary interest in geography. The environment is an integral part of primitive cultures; the people work, create, and play in harmony with their landscape. Most often, they feel completely identified with it, are loath to leave it; it stands for continuity and stability in an uncertain world (Best, 1924; Jackson, 1956–57; Porteus, 1931; Reichard, 1950). The people of Tikopia (Santa Cruz Islands) say, "The land stands, but man dies; he weakens and is buried down below. We dwell for but a little while, but the lands stands in its abiding-place" (Firth, 1936). These environments are not only highly meaningful, but their image is a vivid one.

Certain holy areas may become very highly charged, so that there is a strong focusing of attention, a fine differentiation of parts, a high density of names. The Athenian Acropolis, saturated with a long cultural and religious history, was evidently named and parceled to the gods small area by small

From Kevin Lynch, *The Image of the City* (Cambridge, Mass., 1960) Appendix A, pp. 123–139, by permission of The M.I.T. Press.

area, almost stone by stone, making renovations extremely difficult. The Emily Gap, a small gorge 100 yards long by 30 yards wide in the MacDonnell ranges of central Australia, is to the native people a veritable gallery of legendary locations (Spencer and Gillen, 1899). In Tikopia, the Marae, a sacred cleared space in the forest, was used ritually only once a year. It was a small rectangle, yet contained over twenty locations with regular fixed names (Firth, 1936). Among more advanced cultures an entire city may be holy, such as Meshed in Iran, or Lhasa in Tibet (Donaldson, 1938; Shen and Lu Shen Ci, 1953). These cities are full of names and memories, distinctive forms, and holy places.

Our environmental image is still a fundamental part of our equipment for living, but for most people it is probably much less vivid and particular today. In a recent story of fantasy, C. S. Lewis (1957) imagines that he has entered someone else's mind, and is moving about in her image of the outside world. There is a gray light, but nothing that could be called a sky. There are vague, dingy green shapes, blob-like, without anatomy, that he peers at and finally identifies as Shoddy Trees. There is soft stuff underneath, of a dull grassy color but without separate blades. The closer he looks, the more vague and smudged it all becomes.

The environmental image has its original function in permitting purposeful mobility. A correct map might mean life or death to a primitive tribe, as when the Luritcha of central Australia, driven from their territory by four years of drought, survive by the precise topographic memory of the oldest men (Porteus, 1931). These elders, from experience gained years before, and from the instructions of their grandfathers, knew the chain of tiny water holes that led them out across the desert to safety. The value of being able to distinguish stars or currents or sea-colors is obvious to the South Sea navigator, for when he sets out to strike his tiny goal he engages in a gamble with death. Knowledge of this kind allows mobility, which may make possible a better standard of living. On Puluwat (Caroline Islands), for example, there was a famous native school of navigation. Because of this skill the people of Puluwat were pirates, able to raid the islands within a wide circle.

Although such skills might seem unimportant today, we see things in a different light if we consider the cases of men who, through brain injury, have lost the ability to organize their surroundings (Colucci, 1902; Marie and Behague, 1919; Paterson and Zangwill, 1945). They may be able to speak and think rationally, even to recognize objects without difficulty, but they cannot structure their images into any connected system. These men cannot find their own rooms again after leaving them, and must wander helplessly until conducted home, or until by chance they stumble upon some familiar detail. Purposeful movement is accomplished only by an elaborate memorization of sequences of distinctive detail, so closely spaced that the next detail is always within close range of the previous landmark. Locations normally identified by many objects in context may be recognizable only by virtue of some distinctive, separate symbol. One man recognizes a room by a small sign, another knows a street by the tram car numbers. If the symbols are tampered with, the man is lost. The whole situation parallels, in a curious fashion, the way in which we proceed in an unfamiliar city. In the case of brain injury, however, the situation is inescapable, and its practical and emotional significance is manifest.

The terror of being lost comes from the necessity that a mobile organism be oriented in its surroundings. Jaccard (1932) quotes an incident of native Africans who became disoriented. They were stricken with panic, and plunged wildly into the brush. Witkin (1949) tells of an experienced pilot who lost his orientation to the vertical, and who described it as the most terrifying experience of his life. Many other writers (Binet, 1894; Peterson, 1916; Trowbridge, 1913), in describing the phenomenon of temporary disorientation in the modern city, speak of the accompanying emotions of distress. Binet (1894) mentions a man who took care to arrive at one particular railroad depot in Lyons when coming from Paris, because, although it was less convenient, it concurred with his (mistaken) image of the side of Lyons which lay toward Paris. Another subject felt a slight dizziness throughout his stay in a small town, because of the persistence of a mistaken orientation. The uncomfortable tenacity of an original and incorrect organization of the environment is attested in many sources (Gatty, 1958). On the other hand, in the highly artificial and seemingly neutral situation of a laboratory maze, Brown (1932) reports that subjects developed affection for such simple landmarks as a rough board, which they recognized as familiar.

Way-finding is the original function of the environmental image, and the basis on which its emotional associations may have been founded. But the image is valuable not only in this immediate sense in which it acts as a map for the direction of movement; in a broader sense it can serve as a general frame of reference within which the individual can act, or to which he can attach his knowledge. In this way it is like a body of belief, or a set of social customs: it is an organizer of facts and possibilities.

The differentiated landscape may simply exhibit the presence of other groups or symbolic places. Malinowski (1922), in discussing agriculture in the Trobriand Islands off the New Guinea coast, describes the tall groves which rise above the jungle brush and clearings, and which indicate villages or tabooed tree clumps. In a similar way, tall campaniles mark the locations of towns throughout the flat Venetian plain, or grain elevators the settlements of the American Midwest.

The environmental image may go further, and act as an organizer of activity. Thus, on the island of Tikopia, there were several traditional resting-places on a trail that the people used to and from their daily work (Firth, 1936). Such locations gave form to the daily "commute." In the sacred Marae on this island, a small clearing packed with place names, the minuteness of distinction of locale was an essential feature of the complexly organized rituals. In central Australia since the legendary heroes of the natives moved along certain "dreamtime" roads, these channels are strong parts of the landscape image, and the natives feel safe in traveling them (Pink, 1936). In Pratolini's (1947) autobiographical novel, he gives a striking example of people who in their daily walks continued to follow streets that no longer existed but were only imaginary tracks through a razed and empty section of Florence.

At other times, distinguishing and patterning the environment may be a basis for the ordering of knowledge. Rattray (1927) speaks with great admiration of the Ashanti medicine men who strove to know every plant, animal, and insect in their forests by name, and to understand the spiritual properties of each. They are able to "read" their forests as a complex and ever-unfolding document.

The landscape plays a social role as well. The named environment, familiar to all, furnishes material for common memories and symbols which bind the group together and allow them to communicate with one another. The landscape serves as a vast mnemonic system for the retention of group history and ideals. Porteus (1931) denies that the Arunta tribes of Australia have any special memory ability, although they can repeat extremely long traditional tales. Every detail of the countryside is a cue for some myth, and each scene prompts the recollection of their common culture. Maurice Halbwachs (1950) makes the same point in reference to modern Paris when he remarks that the stable physical scene, the common memory of Parisians, is a potent force in binding them together and allowing them to communicate with each other.

The symbolic organization of the landscape may help to assuage fear, to establish an emotionally safe relationship between men and their total environment. A quotation regarding the Luritcha people of central Australia illustrates the point.

> To each Luritcha child born in the shadow of these huge strange rocks, vast enough to awe the imagination of the white man whose eyes have beheld many wonders, the legends which identify them with the history of his own people must seem a source of great comfort. If these great rocks arose merely to mark the wanderings of his own spirit ancestors, it puts them in familiar relation with him. Legends and myths are more than tales told to while away the hours of darkness, they are part of the means by which the savage fortifies himself against the fear of the awesome and the unknown. Naturally tormented as the primitive man's mind is by the fears that are the product of loneliness, it is little wonder that he eagerly seizes on the idea that this vast, indifferent, if not inimical, Nature commemorates in many of its most striking features his tribal history, and is by the practice of magic subject to his control. (Porteus, 1931)

Even in situations less lonely or frightening, there is a pleasant sense of familiarity or rightness in a recognized landscape. The Netsilik Eskimo put this well-worn idea in their own way: "to be surrounded by the smell of one's own things."

Indeed, the very naming and distinguishing of the environment vivifies it, and thereby adds to the depth and poetry of human experience. Passes in Tibet may have such titles as "The Vulture's Difficulty," or the "Pass of the Iron Dagger," which are not only highly descriptive but are poetic evocations of parts of Tibetan culture (Bell, 1928). An anthropologist makes this comment about the Arunta landscape:

> No one who has not experienced it can appreciate the vivid reality of the myths. The whole country through which we passed was apparently only mulga scrub, a few gum creeks, a low or high range here or there, or some open plains, yet it was made a scene of much activity by aboriginal history . . . So vivid are the tales that the investigator has the feeling of an inhabited area with much activity around: people hurrying hither and thither. (Pink, 1933)

While today we may have more organized ways of referring to our environment—by coordinates, numbering systems, or abstract names—we often miss this quality of vivid concreteness, of unmistakable form (Kepes, 1956). Wohl and Strauss (1958) give many examples of the efforts of people to find a shorthand physical symbol for the city they live in, both to organize their impressions of it and in order to carry on their daily activity.

The feeling and value of an imageable environment are well summed up in Proust's (1954) moving description (in *Du Coté de chez Swann*) of the church steeple in Combray, where he spent many childhood summers. Not only does this piece of landscape symbolize and locate the town, but it enters deeply into every daily activity, and remains in his mind as an apparition for which he still searches in later life:

> It was always to the steeple that one must return, always it which dominated everything else, summing up the houses with an unexpected pinnacle.

Types of Reference Systems

These images may be organized in different ways. There may be an abstract and generalized system of reference, at times explicit, at times rather a habitual manner of referring to the locations or relations of features. The Chukchee of Siberia distinguish 22 compass directions, three-dimensional and tied to the sun. They include the zenith and the nadir, midnight (north) and midday (south), all of which are fixed, plus 18 others which are defined by the sun positions at various times of the day or night, and therefore change with the seasons. This system is of sufficient importance to control the orientation of all sleeping rooms (Bogoraz-Tan, 1904). The Micronesian voyagers of the Pacific used a precise directional system, which was not, however, symmetrical but was tied to constellations and to island directions. The number of directions varied up to 28 or 30 (Finsch, 1888; 1891; 1893).

The system used on the North China plain is a strictly regular one. It has deep magical connotations: north being equated with black and evil, south with red, joy, life, and the sun. It controls very strictly the placing of all religious objects and permanent structures. Indeed, the chief use of the "south-pointing needle," a Chinese invention, was not for navigation at sea, but for the orientation of buildings. So pervasive is this system that the country people on this flat land give directions by compass points, and not by right or left, as would be natural to us. The organizing system does not center on the individual, moving and turning with him, but is fixed, universal, and outside the person (Winfield, 1948).

The Arunta of Australia, in referring to an object, habitually give its proximity, orientation, and visibility with reference to the speaker. An American geographer, on the other hand, once presented a paper on the necessity for orientation to our own four cardinal points, and was surprised to find from his audience that for many city people, accustomed to orient to conspicuous urban features, this is no necessity at all. He himself was brought up in open country, in the sight of mountains (Peterson, 1916). For an Eskimo or an inhabitant of the Sahara, constant directions may be recognized, not by heavenly objects, but by prevailing winds, or by sand or snow formations which are the products of such winds (Jaccard, 1932).

In parts of Africa the key direction may not be an abstract, constant one, but rather the direction toward the home territory. Thus Jaccard (1932) cites a joint encampment of several tribes, who spontaneously grouped themselves into sectors which pointed toward their own territories. Later he mentions the case of French commission merchants who have business in a succession of cities strange to them. They assert that they pay little attention to names or landmarks, but simply keep a continuous mental record of the direction back to the railroad station, and strike out for it directly when their work is done. Australian gravemounds, as another example, are shaped with reference to the direction toward the individual's totem center, or spiritual home (Spencer and Gillen, 1899).

The island of Tikopia is an example of another sort of system, which is neither universal, egocentric, nor directed toward a base point, but is tied to a particular edge in the landscape. The island is small enough so that one is rarely out of sight or sound of the sea, and the islanders use the expressions inland or seaward for all kinds of spatial reference. An axe lying on a house-floor is localized in this way, and Firth reports overhearing one man say to another: "There is a spot of mud on your seaward cheek." This reference pattern is so strong that they have difficulty in conceiving of any really large land mass. The villages are strung along the edge of the beach, and the traditional terms of guidance refer only to "the next village" or the one beyond the next, and so on. This is an easily referenced, one-dimensional series (Firth, 1936).

Sometimes the environment is organized, not by a general directional system, but by one or more intensive foci, toward which other things seem to "point." In Meshed (Iran) extreme sacredness attaches to every object near the central shrine, including the dust which falls on the precinct. The high point on the approach to the city, from which the traveler first sees this mosque, is in itself important, and within the city it is proper to bow when crossing every street that leads to the shrine. This sacred focus polarizes and organizes the entire surrounding area (Donaldson, 1938). This is comparable to the custom of genuflecting in a Roman Catholic Church when crossing the axis of the altar, which orients the church interior.

The city of Florence was organized this way in its centuries of greatness. At that time, description and locational references were made in terms of the "canti," that is, the focal points, which were such things as loggias, lights, coats of arms, tabernacles, important family houses, and key stores, especially pharmacies. Only later, the names of the canti attached themselves to the streets, which were subsequently regularized and signposted in 1785. Progressive house numbering was introduced in 1808, and the city shifted over to reference by paths (Casamorata, 1944).

Imaging and referencing by districts was very common in older cities, where quarters and their populations were relatively stable, isolated and distinctive. In Imperial Rome, addresses were given solely by small defined districts. Presumably, arrival at such a district allowed one to proceed to one's final destination by personal inquiry (Homo, 1951).

The landscape may be patterned by the lines of movement. In the case of the Arunta in Australia the entire territory is magically organized by a network of mythical paths linking together a series of isolated totemic "countries" or clan estates, and leaving waste areas between. There is normally only one correct trail to the sacred storehouse containing totemic objects, and Pink (1933) tells of the long detour made by one of his guides to approach a sacred place properly.

Jaccard (1932) speaks of a famous Arab guide in the Sahara, who could follow the faintest trail, and for whom the entire desert was a network of paths. In one instance he followed painstakingly the continuous twists of the scarcely-marked way, even while his destination was clearly visible to him across the open desert. This reliance was habitual, since storms and mirages often made distant landmarks unreliable. Another author writes of the Saharan *Medjbed*, the transcontinental path worn by camels that goes for hundreds of kilometers over the empty land from water hole to water hole, marked by piles of stones at crossing points. It may mean death to lose it. He speaks of the strong personality, the almost sacred character, that this trace acquires (Gautier, 1908). In quite another landscape, the seemingly impenetrable African forest, the tangle is intersected by elephant paths, which natives learn and traverse as we might learn and traverse city streets (Jaccard, 1932).

Proust (1925) gives a vivid example of the sensation of a path reference system in his description of Venice:

> My gondola followed the course of the small canals; like the mysterious hand of a Genie leading me through the maze of this oriental city, they seemed, as I advanced, to be carving a road for me through the heart of a crowded quarter which they clove asunder, barely dividing with a slender fissure, arbitrarily carved, the tall houses with their tiny Moorish windows; and, as though the magic guide had been holding a candle in his hand and were lighting the way for me, they kept casting ahead of them a ray of sunlight for which they cleared a path.

Brown (1932), in his experiments in putting subjects blindfolded through a maze for the feet, found that even in this very restricted situation subjects seemed to use at least three different kinds of orientation: a memorization of the sequence of movements, usually difficult to reconstruct except in correct sequence; a set of landmarks (rough boards, sound sources, rays of sunlight that gave warmth) which identified localities; and a general sense of orientation in the room space (for example, the solution might be imaged as a general movement around the four sides of the room, with two excursions into the interior).

Formation of the Image

The creation of the environmental image is a two-way process between observer and observed. What he sees is based on exterior form, but how he interprets and organizes this, and how he directs his attention, in its turn affects what he sees. The human organism is highly adaptable and flexible, and different groups may have widely different images of the same outer reality.

Sapir (1912) gives an interesting example of this differential focus of attention, in the language of the southern Paiute. They have single terms in their vocabulary for such precise topographical features as a "spot of level ground in mountains surrounded by ridges" or "canyon wall receiving sunlight" or "rolling country intersected by several small hillridges." Such accurate reference to topography is necessary for definite locations in a semi-arid region. He goes on to note that the characteristic Indian vocabulary does not contain the English lumping-word, "weeds," but has separate terms for these sources of food and medicine, terms which for each species distinguish whether the specimen is raw or cooked, as well as its color and stage of growth: as in the English calf, cow, bull, veal, and beef. He mentions, on the other hand, one Indian tribe whose vocabulary does not distinguish between the sun and the moon!

The Aleuts have no native names for the great vertical features of their landscape: the ranges, peaks, volcanoes, and the like. Yet the tiniest horizontal aqueous feature—rill, streamlet, or pond—had its own name. Presumably this is because the tiny waterways are the environmental features which are vital for travel (Geoghegan, 1944). The attention of the Netsilik Eskimo seems also to be riveted on the aqueous features. In a group of twelve sketch maps done by natives for Rasmussen (1931), there are 532 place names indicated by the draftsmen. Of these, 498 designate islands, coasts, bays, peninsulas, lakes, streams, or fords. Sixteen refer to hills or mountains, and only eighteen make scattered reference to rocks, ravines, swamps, or settlement sites. Yung (1918) makes an interesting reference to a trained geologist who was able to march unhesitatingly through foggy Alpine country, simply by his recognition of the patterning of the geologic type of the exposed rocks.

Still another, and rather unusual, area of attention is sky reflection. Stefánsson (1914) says that in the Arctic low-hanging clouds of uniform color reflect the map of the earth below: those above open water being black, above sea-ice white, above soiled land-ice somewhat darker, and so on. This is of great value in crossing wide bays where the landmarks are below the horizon. These sky reflections are commonly used in the South Seas, not only to locate an island below the horizon but even to identify it by the color and shape of the reflection. Some idea of the great range of forms which are available for orientation may be gained by a reading of Gatty's (1958) latest book on navigation.

These cultural differences may extend not only to the features receiving attention but also to the way in which they are organized. The Aleutian Islands have no generic name in the native tongue, since the Aleuts do not recognize what to us appears to be the obvious unity of the chain (Elliott, 1886). The Arunta group the stars quite differently than we do, frequently putting bright, close stars in different groups, while linking faint and distant ones (Maegraith, 1932).

So adaptable is our perceptual mechanism, moreover, that every human group can distinguish the parts of its landscape, can perceive and give meaning to significant detail. This occurs no matter how undifferentiated that world may seem to an outside observer. This is true of the endless grey mulga thicket which is part of the Australian landscape; the flat snow-covered land of the Eskimo, where even the distinction between land and sea is lost; the foggy, shifting Aleutians; or the open "trackless" ocean of the Polynesian navigator.

Two primitive groups developed a science of direction-finding and geography which was only recently surpassed by western map-making. These are the Eskimo and the navigators of the South Seas. The Eskimo are able to construct usable maps, freehand, covering territories sometimes 400 or 500 miles in one dimension. This is a feat of which few people anywhere are capable, without prior reference to constructed maps.

Similarly, the trained navigators of the Caroline Islands in the Pacific had an elaborate system of sailing directions which were carefully related to constellations, island locations, winds, currents, sun positions, and wave directions (Finsch, 1888, 1891, 1893; Lyons, 1928). Arago stated that a celebrated pilot once represented all the islands in the archipelago for him by grains of maize, marked their relative position, named each one, and stated the accessibility and products of each. This archipelago is some 1500 miles from east to west! Furthermore, he made a compass from bamboo, and indicated the prevailing winds, constellations, and currents by which he guided himself.

Both cultures producing these triumphs of abstract ability and perceptual attention had two things in common: first, their environments of snow or water were essentially featureless or differentiated only subtly, and second, both groups were forced to be mobile. The Eskimo must travel seasonally from one type of hunting to another, if he is to survive. The best seafarers in the South Seas did not come from the fertile high islands but from the tiny low islands, where natural resources were scanty and famine was always close. The nomad Touareg, in the empty Sahara, are a similar group, and have an almost equal ability. On the other hand, Jaccard (1932) notes that native Africans, of sedentary agricultural habit, easily become lost in their own forests.

The Role of Form

Nevertheless, having said so much about the flexibility and adaptability of human perception, we must now add that the shape of the physical world plays its part, as well. The very fact that skilled navigation arose in what would seem to be perceptually difficult environments indicates the influence of these outward shapes.

The ability to distinguish and orient in these resistant environments is not achieved without cost. The knowledge was usually limited to specialists. Rasmussen's informants who drew his maps were chiefs—many other Eskimos could not do it. Cornetz (1913) remarks that there were only a dozen good guides in all south Tunisia. The navigators in Polynesia were the ruling caste. Knowledge was transmitted from father to son, and there was, as mentioned above, a formal school in the subject on the island of Puluwat. It was customary for the navigators to eat at a separate mess, where the talk was constantly of directions and currents. This is reminiscent of Mark Twain's (1917) Mississippi River pilots, who were constantly discussing and riding up and down the river, and thus keeping abreast of its treacherous shifting landmarks. Admirable as this skill is, it is some distance from an easy and familiar relation with the environment. Polynesian sea voyages were evidently accompanied by real anxiety, a usual sailing formation being a long row of canoes abreast to aid in finding land. Among the Arunta of Australia, as another example, it is only the old men who can lead from water-hole to water-hole or who can correctly locate the proper sacred path in the mulga thicket. On the well-differentiated island of Tikopia, the problem could hardly come up.

We have frequent accounts of native guides who lose orientation in featureless surroundings. Strehlow (1907–1920) describes floundering for hours in the Australian mulga thicket with an experienced native, who repeatedly climbed trees in the effort to get his bearings from distant landmarks. Jaccard (1932) recites the tragic cases of lost Touareg.

At the other end of the scale, there are visual qualities in some landscape features which make them the inevitable subjects of attention, despite the selective power of the eye. Most often, sacredness is concentrated in the more striking natural features, such as the connection of the Ashanti gods with the great lakes and rivers, or the common reverence attached to great mountains. So in Assam there is a famous hill which is the legendary site of Buddha's death. It is described as bold and picturesque by Waddell (1895), rising directly from a plain to which it is in sharp contrast. He then notes that it was worshipped by the aborigines long before, and has become holy for Brahmans and Mohammedans as well.

For physical reasons, the great mountain on Tikopia Island is the central organizing feature. It is the crowning point of the island, both sociologically and topographically, the place of descent of the gods. It marks the location of home from far out to sea, and has an aura of the supernatural. Since the crest is rarely cleared and planted to taro, there is a peculiar flora here which is lacking down below, and this reinforces the special interest of the place (Firth, 1936).

Occasionally, a landscape may be so fantastically differentiated as to compel attention. Kawaguchi (1909) describes the banks of a river near Lake Kholgyal in Tibet:

> Rocks piled up here and there, some yellow, some crimson, others blue, still others green, and some others purple . . . the rocks were highly fantastic, some sharp and angular, others protruded over the river. The nearer bank . . . was full of queerly shaped rocks, and each of those rocks bore a name . . . all these were objects of veneration to the common people.

To take a humbler example, the territories defended by nesting birds in a meadow have been mapped over a succession of years. These territories show wide fluctuations and reorganizations, as might be expected from their occupation by different individuals. But certain perceptually strong boundaries of fence or brush remain stable throughout the shifts (Nice, 1941). Migrating birds, advancing in a general direction over a broad front, are known to direct their flight and follow along major "leading lines," or edges formed by topographic features such as seacoasts. Even swarms of locusts, who maintain cohesion and direction with reference to the wind, become disorganized and scattered when they move out over featureless water surfaces.

Other features may not only be noticeable and distinguishable but even have a "presence," a sort of animation or peculiarly vivid reality, that is felt by peoples of utterly different cultures. Kawaguchi (1909) speaks of a holy mountain in Tibet, seen for the first time as "sitting with an air of great solemnity," and likens it to his own Buddha Vairochana, flanked by Bodhisattvas.

A similar experience, closer to home, was the original impact of a particular escarpment along the Oregon Trail:

> As the west-bound party drew abreast of the bluffs a wave of astonishment swept through it. . . . Numerous observers discovered lighthouses, brick kilns, the capitol at Washington, Beacon Hill, shot towers, churches, spires, cupolas, streets, workshops, stores, warehouses, parks, squares, pyramids, castles, forts, pillars, domes, minarets, temples, Gothic castles, "modern" fortifications, French cathedrals, Rhineland castles, towers, tunnels, hallways, mausoleums, a Temple of Belus, and hanging gardens . . . Taken at a glance the rocks had the appearance of Cities, Temples, Castles, Towers, Palaces, and every variety of great and magnificent structures . . . splendid edifices, like beautiful white marble, fashioned in the style of every age and country. (Shepard, 1956–57)

Many observers are quoted, to indicate the common and overwhelming impact of these special geological shapes.

Therefore, while noting the flexibility of human perception, it must be added that outer physical shape has an equally important role. There are environments which invite or reject attention, which facilitate or resist organization or differentiation. This is analogous to the ease or difficulty with which the adaptable human brain can memorize associated or unassociated material.

Jaccard (1926) mentions several "classical locations" in Switzerland where people are consistently unable to maintain direction. Peterson (1916) notes that the organization of his image of Minneapolis typically breaks down each time the street gridiron changes its orientation. Trowbridge (1913) finds that most people are unable to point to distant cities from New York without gross errors, but that Albany is an exception, since it is visibly linked by the Hudson River.

In London, a small development called Seven Dials was built about 1695, consisting of seven streets which converged on a circular junction containing a Doric pillar bearing seven sun dials, each facing one of the radiating streets. Gay (1922) refers to the confusing shape of this area in his *Trivia*, although he implies that it is only the peasant, the stupid outsider, who could be befuddled by it.

Malinowski draws a sharp distinction between the differentiated volcanic landscape of Dobu and the Amphletts in the D'Entrecasteaux Islands near New Guinea, versus the monotonous coral islands of the Trobriands. These island groups are connected by regular trading expeditions, and the concentration of mythical meaning in the Dobu area, as well as the reactions of the Trobrianders to this imageable volcanic landscape, is described in his pages. Speaking of the trip from the Trobriands to Dobu, Malinowski (1922) says:

> The low strip of land, which surrounds the Trobriand lagoon in a wide sweep, thins away and dissolves in the haze, and before them the southern mountains rise higher and higher. . . . The nearest of them, Koyatabu, a slim, somewhat tilted pyramid, forms a most alluring beacon, guiding the mariners due south. . . . Within a day or two these disembodied misty forms are to assume what for the Trobrianders seems marvellous shape and enormous bulk. They are to surround the Kula traders with their solid walls of precipitous rock and green jungle. . . . The Trobrianders will sail deep, shaded bays . . . beneath the transparent waters a marvellous world of multi-colored coral, fish and seaweed will unfold itself . . . they will also find wonderful heavy, compact stones of various shapes and colors, whereas at home the only stone is the insipid, white dead coral . . . besides many types of granite and basalt and volcanic tuff, specimens of black obsidian, with its sharp edge and metallic ring, and sites full of red and yellow ochre. . . . Thus the landscape now before them is a sort of promised land, a country spoken of in almost legendary tone.

In a similar way, although the "dream-time" roads of Australia pass in every direction over a land which is largely level mulga plain, yet the legendary camp sites, the nodes of sacred history and attention, seem to be heavily concentrated in the two regions of differentiated landscape: the MacDonnell and the Stuart's Bluff Ranges.

Parallel to these comparisons of primitive landscapes we may put Eric Gill's (1941) comparison of Brighton, England where he was born, to Chichester, to which he moved in his adolescence:

> It had simply never occurred to me before that day that towns could have a shape and be, like my beloved locomotives, things with character and meaning. . . . [Chichester] was a town, a city, a thing planned and ordered—no mere congeries of more or less sordid streets, growing, like a fungus, wherever the network of railways and sidings and railway sheds would allow. . . . I only knew that Chichester was what Brighton was not: an end, a thing, a

place. . . . The plan of Chichester is clear and clean. . . . Over the Roman wall you could look straight out into the green fields. . . . Four straight wide main streets dividing the city into nearly equal quarters and the residential quarter similarly divided by four small streets and these almost completely filled with seventeenth- and eighteenth-century houses. . . . But Brighton, as we knew it . . . well, there is simply nothing to be said about it. When we thought of Brighton, it was of a place of which the center was our home . . . there was no other. But when we lived in Chichester the center was not No. 2 North Walls, but the Market Cross. We gained not only a civic sense but a sense of ordered relations generally. . . . Brighton wasn't a place at all. It had never occurred to me that any other sort of town could exist.

The perceptual clarity of the island of Tikopia, due to the presence of Mt. Reani, has already been mentioned. How a differentiated shape can be used in detail, is illustrated by this quotation:

When a Tikopia sets out from his native land his first estimates of distance he has travelled are based on the portions of the island still showing above the horizon. There are five principal points on the scale. The first is the rauraro, the lowland near the shore. When this disappears, the voyager knows he is some distance out. When the cliffs (mato) arising 200 to 300 feet in various spots round the coast become lost, another point is reached; then the uru mauna, the crests of the chain of hills ringing the lake, perhaps 500 to 800 feet in height, sink below the waves. When the uru asia (the last break in the contour of Mt. Reani, about 1000 feet) goes down, then the voyager realizes he is far out to sea; and when at last he sees the uru ronorono, the tip of the mountain itself, vanish from sight, he greets the moment with sorrow. (Firth, 1936)

With the aid of a favorably differentiated landscape profile, this familiar phenomenon of parting has been regularized into accepted intervals, each with both practical and emotional meaning.

When a character in a novel by Forster (1949) returns from India, he senses with sudden shock, on entering the Mediterranean, the sheer form-quality of his surroundings, their imageability:

The buildings of Venice, like the mountains of Crete and the fields of Egypt, stood in the right place whereas in poor India everything was placed wrong. He had forgotten the beauty of form among idol temples and lumpy hills; indeed, without form, how can there be beauty? . . . In the old undergraduate days he had wrapped himself up in the many-colored blanket of St. Mark's, but something more precious than mosaics and marbles was offered to him now: the harmony between the works of man and the earth that upholds them, the civilization that has escaped muddle, the spirit in a reasonable form, with flesh and blood subsisting. Writing picture post-cards to his Indian friends, he felt that all of them would miss the joys he experienced now, the joys of form, and that this constituted a serious barrier. They would see the sumptuousness of Venice, not its shape.

Disadvantages of Imageability

A highly visible environment may have its disadvantages, as well. A landscape loaded with magical meanings may inhibit practical activities. The Arunta face death rather than move to a more favorable area. The ancestral grave-mounds in China occupy desperately needed arable land, and among the Maori some of the best landing-places are forbidden because of their mythical import. Exploitation is more easily accomplished where there is no sentiment about the land. Even conservative use of resources may be impaired where habitual orientation does not allow easy adaptation to new techniques and needs.

Geoghegan (1944) refers to the richness of place names in Aleut but follows this with the interesting comment that there are so many particular names for each tiny feature that very often the Aleuts of one island have scarcely heard of the place names on another. A highly differentiated system, lacking abstractness and generality, may actually reduce communication.

It may have consequences of another sort. Strehlow (1907–20) says of the Arunta:

> Since every feature of the landscape, prominent or otherwise, is already associated with one or the other of these myths, we can understand the utter apathy of literary efforts . . . their forefathers have left them not a single unoccupied scene which they could fill with creatures of their own imagination . . . tradition has effectually stifled creative impulse . . . native myths ceased to be invented many centuries ago . . . they are on the whole uninspired preservers . . . not so much a primitive as a decadent race.

If it is desirable that an environment evoke rich, vivid images, it is also desirable that these images be communicable and adaptable to changing practical needs, and that there can develop new groupings, new meanings, new poetry. The objective might be an imageable environment which is at the same time open-ended.

As a peculiar example of how this dilemma can be resolved, even in an irrational way, we may take the Chinese pseudoscience of geomantics (de Groot, 1912). This is a complicated lore of landscape influence, systematized and interpreted by professors. It deals with winds of evil that can be controlled by hills, rocks, or trees that visually seem to block dangerous gaps, and with good water spirits that are to be attracted by ponds, courses, and drains. The shapes of surrounding features are interpreted as symbolizing various spirits contained therein. This spirit may be accounted useful, or it may be inactive and useless. It can be concentrated or dispersed, deep or at the surface, pure or mixed, weak or strong, and must be used, controlled, or enhanced by planting, siting, towers, stones, and so forth.

Possible interpretations are many and complex; it is an endlessly expanding field which experts are exploring in every direction. Divorced from reality as this pseudoscience may be, yet it has for our purposes two interesting features: first, that it is an open-ended analysis of the environment: new meanings, new poetry, further developments are always possible; second, it leads to the use and control of outside forms and their influences: it emphasizes that man's foresight and energy rule the universe and can change it. Perhaps there are hints here as to ways of constructing an imageable environment that is not at the same time stifling and oppressive.

VI

Cognitive Distance

In its foreign-policy cover story, *Newsweek* (Dec. 14, 1970) reports that a "Marxist has taken over on the U.S.A.'s doorstep in Chile."
Have any of your comrades looked at an air-distance chart lately? Santiago is over 5,100 miles from New York. Or, if you prefer, Moscow is over 400 miles *closer* to New York than Santiago.
Ultimately, we may have reasons to be bugged by developments in Chile, but proximity surely won't be one of them.

DONALD KLEIN

Introduction

Mr. Klein's letter (*Newsweek*, January 18, 1971), a wry observation on the difference between the world as we *see* it and as it really *is*, illustrates how we cognitively transform distance. This letter also suggests that we know relatively little about cognitive distance since it is so unconscious, so automatic that it is relegated to "second nature" status. It is safe to assume that people think about cognitive representations not only in a qualitative way, but also *quantitatively*. Such quantitative information, part of the geometry of cognitive maps (or locational information), is necessary for adaptive spatial behavior (see chap. 1). In this introduction, we will link cognitive distance and spatial behavior and outline the provocative, though fragmentary, attempts to understand cognitive distance functions.

Initially, we must distinguish between two approaches to space and distance existing side by side in psychology. The first considers *perceived* distance, focusing on the metrics and geometry of visual space (Forgus, 1966; Ittelson, 1951; Luneburg, 1947). Since the emphasis is on purely visual, perceived cues indicating the relative distance of objects from an observer, this approach is of little interest here. Of more significance is the approach to *cognitive* distance which considers distance estimates and beliefs made or held in the *absence* of the objects and which relies upon memory, stored impressions, judgments, and beliefs.

Given psychology's long interest in perception, physiology, and psychophysics, it is not surprising that perceived distance has been rigorously and completely analyzed. However, we know little about cognitive distance, an omission perhaps excusable from a psychological viewpoint but incomprehensible in a geographic context. Distance is a basic component of models of human spatial behavior (Olsson, 1965). People, as effort minimizers, are assumed to be aware of distance effects in terms of effort expenditures and to base their decision making upon distance-minimizing principles. Thus, the gravity model of human spatial interaction relates interaction directly to the mass or population of two locations and inversely to the distance separating them. But what is an operational measure of distance? Miles, miles squared, miles to other exponents, time, dollar costs, and many other measures have been advanced as indicators of what Deutsch and Isard (1961) call "effective" or functional distance. It is surprising, however, that there have been no attempts to use cognitive transformation of distance in models of spatial behavior.

The urban sociology and planning literature (Webber, 1964b; Chapin, 1968) has described the differences in spatial behavior patterns (in terms of trip length, frequency, and type) between diverse socio-economic groups. There have been attempts to relate such behavior differences to particular spatial arrangements of environmental objects, such as the pioneering studies by Festinger, Schachter, and Back (1950) and Whyte (1956), and the study of site planning by Gutman (1966). Obviously, there is a cognitive component to this behavior-environment relationship, and planners have sought to isolate and manipulate it:

> The effect of distance can also be created. People can be made to act on the basis of separation from other people and activities by certain manipulations of the environment which create the illusion of distance. Dense plantings, impenetrable railroad tracks, and paths that lead people in divergent directions are all manipulations of physical environment that lead to social phenomena premised on the assumption of separation in space. (Michelson, 1970, 48)

The suggested manipulations mix perceived with cognitive distance—more importantly, Michelson emphasizes the role of distance *as people see it* in determining spatial behavior.

Although we are only beginning to appreciate the role of cognitive distance at the macro-environmental scale, on a micro-scale significant inroads have been made (Esser, 1971). Because of the links between personality, small group behavior, and space, psychologists have explored such questions as the optimal spacing for conversation or the effect of room arrangements on group interaction, using a variety of conceptual labels: "proxemics" (Hall, 1966), "personal space" (Sommer, 1969), and "body-buffer zone" (Horowitz, Duff, and Stratton, 1964). It is at this micro-scale that the most searching and provocative studies have been made, although the cognitive component is not stressed. On the macro-environmental scale, two lines of research are represented in this book: the first is the work of Lee (1962, 1970, chap. 5) and the second the psychophysics approach of Lundberg, Lowrey, and Briggs (chap. 17–19).

The Brennan-Lee Approach

This work was prompted by Brennan's (1948) observation that women shoppers in an English city were violating the principle of least effort in that, given two similar facilities, they would patronize the facility on the downtown side of their residence even though that on the out-of-town side was closer to them. Lee (1962) made a preliminary supportive test of this empirical regularity and a fuller, more rigorous test in 1970 indicating that people over-estimate distances to out-of-town facilities. The differential cognitive transformation is firmly established, and leads to some obvious predictions relating cognition and behavior.

Lee also suggests that the estimated length of a trip will vary with the number of corners in the route, such that the degree of over-estimation will increase with an increasing number of corners. Lee's finding *conflicts* with that of Stea (1969), who argues that the degree of overestimation decreases with increasing corners, but *agrees* with psychological studies involving direct perceptual distance judgments. Although we cannot resolve this contradiction, it does raise the question of the effect of route segmentation on distance estimation. The impact of route segments can be intuitively appreciated when one considers the "true" time saving associated with air travel. If one overlooks the travel time to and from airports and the waiting time, then air travel *is* quick compared with road or rail. However, should we "cognitively overlook" such time components? Davis (1967) found an interesting effect of segmentation on cognitive distance in a hospital. Office-to-office trips which involved a part of the trip outdoors were considered longer than trips of similar *physical* length which were *all* indoors. Both of these observations pose the question of the continuity of the cognitive distance metric—do we consistently over- (or under-) estimate distances by a constant proportional factor? Life for the researchers would be relatively simple if this were true but the evidence seems to point away from the simple answer. For example, the maps of Lynch (1960) contain a significant number of edges, which represent evidence of cognitive discontinuity, and the Brennan-Lee argument (see Stea, 1969) indicates that, as far as urban distance is concerned, it does matter which way one is facing!

The Psychophysics Approach

The work of Lundberg, Lowrey, and Briggs is grouped together because of a shared methodological approach, although the degree of contact and borrowing is low. Lundberg summarizes an extensive research effort undertaken by the Psychological Laboratory at the University of Stockholm.

Using sophisticated psychophysical and multidimensional scaling procedures, the Stockholm studies have related the degree of emotional involvement with places to the degree of distance underestimation—the higher the emotional involvement, the more accurate the distance estimates. The environmental scale is macro, involving international distances. However, although the research has delineated an interesting set of empirical regularities, there are no satisfying theoretical frameworks for understanding the regularities.

The Stockholm research is independent of the work on urban distance done by Briggs and Lowrey, but there are many parallels between the latter studies; one major operational difference is that Briggs uses road distances and Lowrey, "as-the-crow-flies" distances. A methodological question in both approaches concerns the attempt to turn a problem in cognition into a psychophysical one. The task assumes that the subject can take a cognized standard distance (say between home and downtown) and scale it down to a *perceived* scale (or line), where the reduction is in the order of 1:1000 or more. He is then required to scale down all cognitive distances to the same perceived scale. The assumption that the subject can perform such a task remains untested.

Although the above comment indicates the numbing operational difficulties faced by researchers in this area, the papers by Briggs and Lowrey represent the first significant answers to the questions on cognitive or effective distance raised earlier. It is important to recognize the degree of compatibility between the answers—for once, geographers are beginning to cumulate knowledge by consciously replicative and overlapping studies.

Concluding Remarks

Although the papers in this section are a significant step forward, it is important to isolate those questions as yet unanswered. We have little knowledge of the metrics of cognized space. Lowrey suggests that people have an interval-scale based metric available *if* they want to use it, whereas Lynch indicates primitive distance metrics based upon the degree of removal from a *fixed* point which suggest an ordinal scale. Not only are we uncertain of the mathematical properties of our distance metrics, but we do not have any firm indication as to the type of units employed: the question of miles or cost or time is unresolved.

We are uncertain as to whether people construct a cognitive representation which has the mathematical properties of a metric space. As Stea (1969) indicates, a crucial characteristic of a metric space is that $d(a, b) = d(b, a)$, or that the distance from point a to point b is the same as that from b to a, and hence that distances based upon a supposedly metric world should be commutative. However, there are several sources of violation of this characteristic: for example, end points of routes may be differentially attractive—consider the difference between going home and going away, and the findings of Lundberg on emotional involvement and of Lee (1970) on downtown orientation.

Briggs suggests that studies of cognitive distance have two themes, one in terms of the phenomena per se and the other in terms of modeling spatial behavior. We are a long way from the latter, especially given the lack of information about metric properties and questions of group differences in distance estimation. However, the latter is significant from an environmental design and planning viewpoint since the way in which people cognize and use the environment is at least as significant as the planner's intentions (see Carr, 1970; Michelson, 1970; and Harvey, 1970 a,b).

17

Emotional and Geographical Phenomena in Psychophysical Research

ULF LUNDBERG

Faced with problems regarding variation and inconsistency in psychological scaling, many researchers have attempted to avoid quantification in favor of more qualitative approaches to psychological problems. Nevertheless, much work has been devoted to improving the existing scaling techniques in, for example, problems involving perception, where the relation between stimulus intensity and sensory magnitude is studied (for a review see Ekman & Sjöberg, 1965; Zinnes, 1969). During the last few years these techniques have also been used to quantify more complex variables (e.g., Ekman, 1962; Stevens, 1966). The aim of the present paper is to emphasize the necessity for the quantification of psychological variables, for although the methods used may be relatively primitive and the variables themselves very complex, many interesting phenomena may otherwise remain undiscovered.

The work described in this paper was carried out recently at our laboratories; the methods used were the ones just mentioned and the results support the view just given. Geographical and temporal effects on the intensity of emotional reaction are dealt with and some other psychological relations in a geographical context are presented. The results suggest that there are some interesting and simple mechanisms at work. These mechanisms give rise to certain implications which may be of practical importance and these are briefly discussed.

I am greatly indebted to Professor Gösta Ekman, who has provided me with most of the ideas described in the text. However, the responsibility for the interpretation made here rests on the present author.

Emotional Involvement as a Function of Distance in Time and Space

These investigations were initiated from the well-known fact that we are much less affected emotionally by events taking place at a great distance, than by what happens close by. Although people in the developed countries are nowadays more aware of and more dependent on events in other countries than they were in the past, it will be shown that this statement still seems to remain valid.

The first study was carried out by Ekman and Bratfisch (1965). A group of students estimated the subjective distance to a number of cities using the complete method of ratio estimation (e.g., Ekman, 1958). The students were also instructed to imagine important events taking place in these cities and to estimate their degree of emotional involvement in these events by using the same method. The scales obtained were related to each other and it was found that the intensity of emotional involvement decreased according to the square root of the increasing subjective distance (see fig. 17.1). The "inverse square root law" has been supported by a number of subsequent studies using different groups of subjects, different methods, and

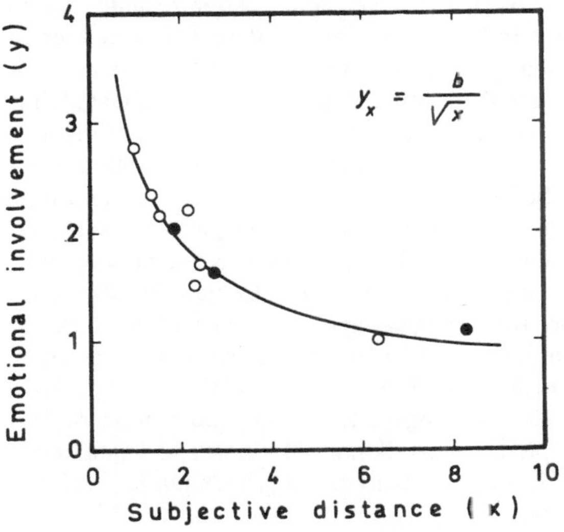

FIGURE 17.1. *Emotional involvement plotted as a function of subjective geographic distance. The trend of the data is described by a power function. (Reproduced from Ekman & Bratfisch, 1965.)*

different countries (Dornič, 1967; Bratfisch, 1969). This invariance is a very unusual finding in psychophysical research and has been interpreted as being the result of having investigated the phenomena at the proper level; the purely psychological level suggested by Ekman (1961, 1970a). On the other hand the relation between emotional involvement and objective distance was not invariant between the different studies.

When subjective distance was plotted as a function of real geographic distance, the relation was describable by a simple power function.

$$d = a\, D^n, \tag{1}$$

where d is subjective distance, D is physical distance, n is the empirically found exponent of the function determining the curvature, and a is introduced to account for the arbitrary unit of measurement. The exponent was found to vary between the different studies, but it was most often less than one, indicating a negatively accelerated trend. Three examples of this relation were obtained by Bratfisch (1969, Experiment III) and are shown in figure 17.2, where the left diagram (*A*) is given in linear coordinates and the right diagram (*B*) in logarithmic coordinates. If the fit to the power function is satisfactory, the logarithmic plot will show a linear trend. The same type of power function has been used to describe psychophysical relations for a large number of physical continua (Stevens, 1957; Stevens and Galanter, 1957; Stevens, 1960; Ekman, Hosman, Lindman, Ljungberg and Åkesson, 1968). However, the scatter around the fitted curves is relatively large with the present kind of data as can be seen in figure 17.2*A*.

The instructions given to the subjects were formulated so they would base their estimates on their immediate and unsophisticated impressions. The importance of this instruction is illustrated by a study in which Stanley (1968) failed to verify the relation between emotional involvement and distance when he used Armidale, Australia, as the central point for his investigation. The instructions that Stanley gave his subjects differed from those that Ekman, Bratfisch, and Dornič used; Stanley asked his subjects to estimate the actual distances, while Ekman, Bratfisch, and Dornič emphasized a more psychological approach. The latter approach has also produced interesting results in a number of studies regarding the mechanism of similarity (Eisler and Ekman, 1959; Eisler, 1960; Ekman, Goude and Waern, 1961; Ekman, Engen, Künnapas and Lindman, 1964) and in investigations of preference (Ekman, Hosman and Lindström, 1965; Ekman and Åkesson, 1965). This kind of "psychodynamic" research was first suggested by Guilford (1939).

Subjective geographic distance has also been related to other variables in a study by Stapf (1968). The certainty in the estimates as expressed by the subjects was found to be inversely related to the distance when the names of German cities were used as stimuli. He also found a linear relation between subjective and objective geographic distance.

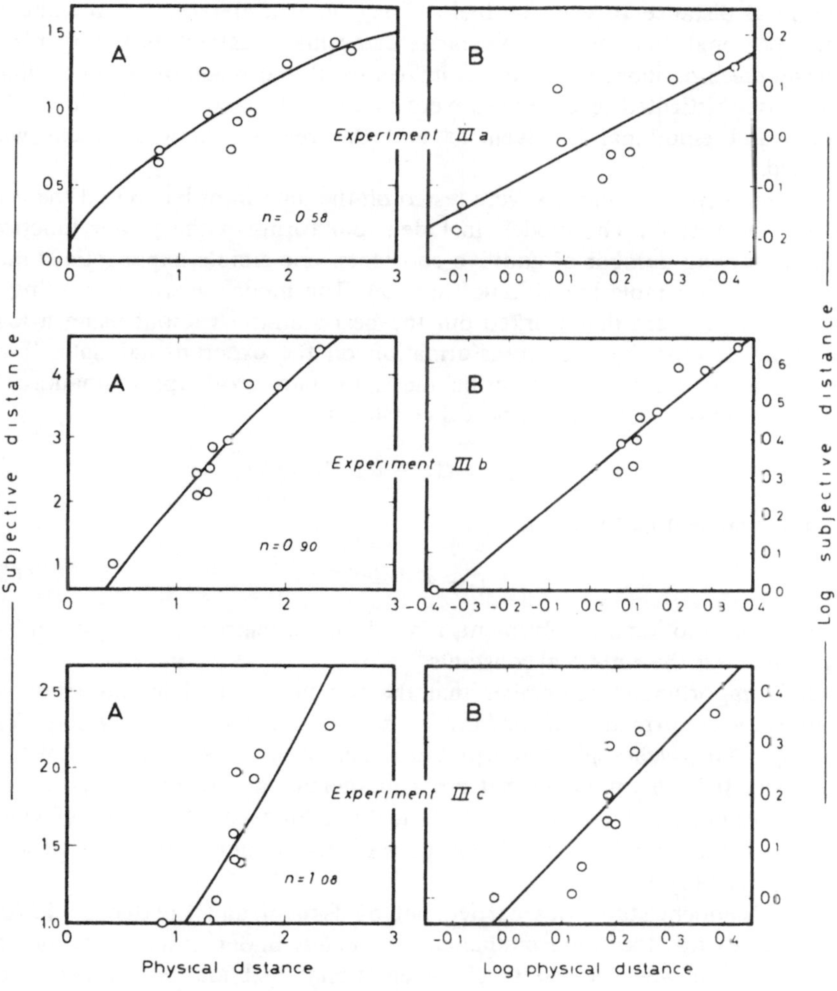

FIGURE 17.2. *Subjective distance as a function of geographic great circle distance (the unit being 1000 kilometers). The coordinates are linear in all A-diagrams and logarithmic in all B-diagrams. The trend of the different data has been approximated by power functions. (Reproduced from Bratfisch, 1969.)*

Emotional involvement has also been studied as a function of temporal distance in order to see if a simple relation also exists between these variables. The first study of this kind was carried out by Bratfisch, Ekman, Lundberg, and Krüger (1970) and the stimulus periods were defined as "Bismark's time", "The time of Charlemagne", etc. The subjects were instructed to think of these historical periods and to estimate the subjective temporal distance to each of them. They were also required to indicate their personal emotional involvement in events imagined as taking place during these periods. The latter estimates were also made on an immediate and unsophisticated level. Scales were then constructed for subjective distance and emotional involvement and the relation between them was studied.

Six mathematical models were tested on the data in order to find the one giving the best fit. The models included four forms of the power function with a varying number of additive constants, one simple exponential function, and one simple logarithmic function. The models were fitted using a computer program that worked out the best solution without using a logarithmic or other type of transformation on the experimental data. Two functions were found to describe the data with good approximation: a power function with three empirical constants,

$$y = a(t + b)^{-n}, \tag{2}$$

and an exponential function,

$$y = a + bc^{t}, \tag{3}$$

where y is emotional involvement, t is subjective temporal distance, and a, b, c, and n are the empirical constants.

It is important to remember that the functions found in this study as well as the confirmation of the "inverse square root law" are only empirical findings. No psychological interpretation has been given of the parameters involved. It is also possible that other mathematical models may be found which are preferable from a psychological and/or theoretical point of view. However, the models we used may be regarded as quite useful for present descriptive purposes.

A subsequent study was carried out by Ekman and Lundberg (1970) in order to test the same mathematical models under other experimental conditions. It was decided to study emotional reaction as a function of temporal distance and to include both future and past events. Different sets of stimuli were chosen in order to reduce the influence of "secondary variables" such as interest and representativeness connected to specific historical periods and names. Secondary variables were found to be of some importance in the previous study. The points in time used specific years, e.g., "1892", "2041".

In the study by Bratfisch et al. (1970) only one of the tested mathematical models was found to be markedly inferior to the others. Therefore the new data was tested with all the remaining five models. The main interest was centered on the relation between emotional involvement and subjective temporal distance as gained according to the psychological approach mentioned above. The same two models which had fitted satisfactorily in the previous study accounted for the largest portion of the variation found regarding future time. When past time was considered, the exponential function was found to be the most appropriate one for describing the data. In the study now under consideration, the scatter around the fitted curves was much less than in the previous study (Bratfisch et al., 1970). Figure 17.3 shows emotional involvement as a function of subjective temporal distance with all the five models fitted to the data. Models *A* – *D* represent different forms of the power function (2), and Model *E* the exponential function (3). The point at present time was excluded for various reasons (see Ekman and Lundberg, 1970). The scatter was also very moderate when subjective distance was plotted as a function of objective temporal distance and equation (1) is shown fitted to the data in figure 17.4. The exponents for past and future time are 0.89 and 0.72, respectively.

The scatter around the fitted curves being much smaller with specific years than with historical periods indicates that, as intended, the influence of "secondary variables" was reduced.

The exponential function suggested has also been investigated in two

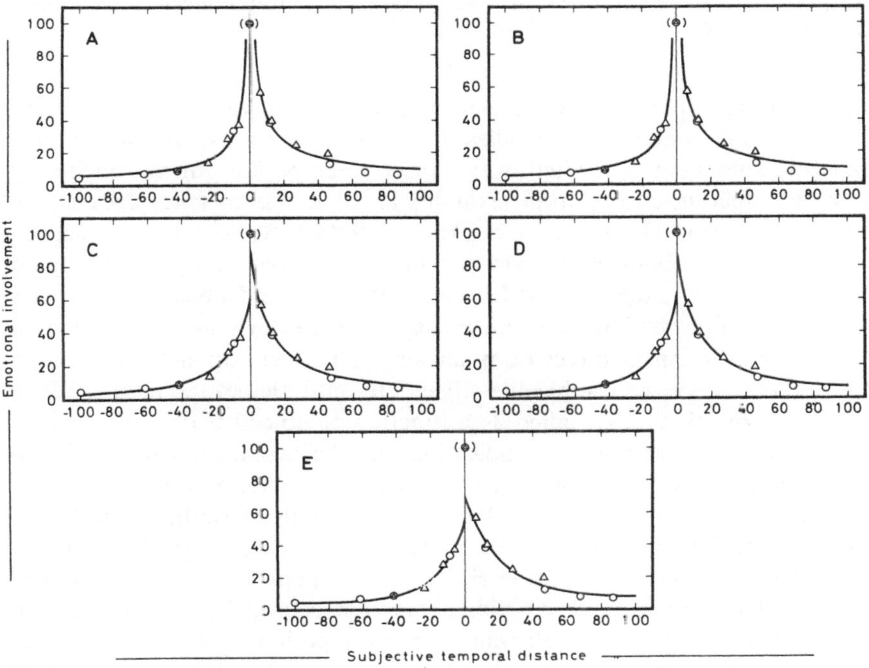

FIGURE 17.3. *Emotional involvement plotted as a function of subjective temporal distance. The curves A – D represent different forms of the power function and E an exponential function fitted to the data. (Reproduced from Ekman & Lundberg, 1970.)*

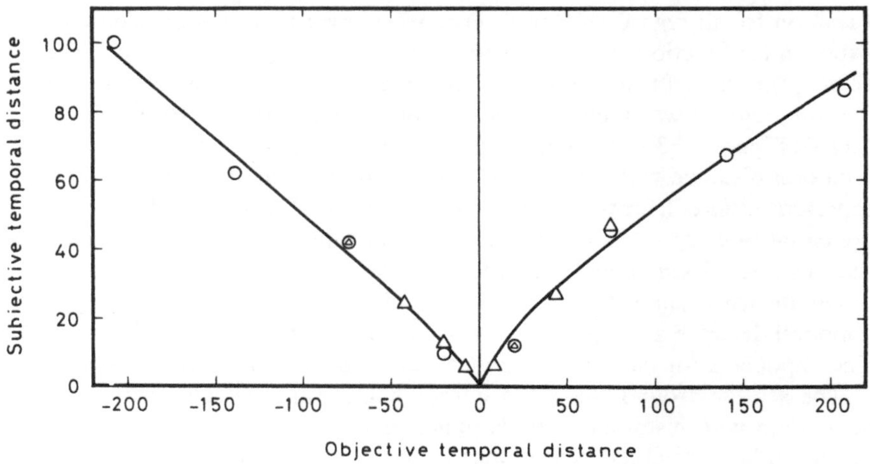

FIGURE 17.4. *The relation between chronological and subjective temporal distance. The curves represent the simplest form of the psychophysical power function fitted to the data for past and future time. (Reproduced from Ekman & Lundberg, 1970.)*

studies using stimuli that were expected to produce real emotions in the subjects. In the first of these studies the intensity of emotional reaction of a group of students was studied over a three week period during which they were anticipating an important examination that was to take place at the end of the period (Lundberg and Ekman, 1970a). Scale values of elapsed time were also obtained. In contrast to the previous study (Ekman and Lundberg, 1970), the relation between subjective and objective time was found to be a power function with an exponent greater than one, emotional reaction increasing in intensity as an almost perfect exponential function (3) of the decreasing subjective time left until the examination. When emotional reaction was studied as a function of objective time, the scatter around the curve was greater, indicating that the exponential function describes an intra-psychological rather than a stimulus-response relation.

The second study was carried out in a laboratory setting (Lundberg, Ekman, and Frankenhaeuser, 1970). On arrival the subjects learned that they were to receive an electric shock after a period of 45 minutes. Estimates of the intensity of emotional reaction caused by the anticipation were obtained at irregular intervals during the period. Half of the subjects had to estimate the time remaining before the shock, while the other half could follow the actual time elapsing on a clock. The emotional reaction was found to increase in intensity as the period before shock decreased, both for the group ignorant of real time and for those who were informed by the clock. When emotional reaction was plotted against physical time, no trend was visible; however, when emotional reaction was plotted against subjective time, a systematic trend appeared in both groups. The trend followed the same exponential function (3) obtained in the previous studies.

The studies mentioned above show that emotional involvement can usually be described as a simple inverse exponential function of subjective time. This was shown to be the case for intervals of time varying from less than one hour to more than two thousand years, for different groups of subjects, for different kinds of stimulation, and for different methods. This suggests the existence of a mechanism of considerable generality. Furthermore, the relation appeared to be less complex at the psychological level, i.e. between emotional involvement and *subjective* temporal distance, indicating a pure psychological, rather than a direct stimulus-response mechanism. The results from these latter studies support the previous findings with geographic distance concerning the advantage of a psychological approach to this kind of problem.

Besides the basic importance of scaling variables of the present kind, a quantitative approach to social and emotional phenomena has many important political and humanitarian implications. National reforms must often be worked out in anticipation of future needs, and as has been shown, present engagement in problems that won't be acute for fifty years is very poor. Serious international problems such as air pollution, lack of sufficient food and fresh water, and the rapidly expanding population of the earth, have only recently [and possibly too late] come close enough in "psychological time" to raise the amount of public engagement to a noticeable level. A similar reasoning applies to problems which are geographically distant. Our attitudes towards serious problems in other parts of the world still tend to be characterized by personal detachment.

One important field of research in the present context is that aimed at finding factors which affect the relations appearing in these experiments. Some factors have already been discovered: if an event takes place in a well-known and important city, the involvement variable is higher than would be expected from the distance alone. This is also the case for events taking place in one's own country. Preliminary results from a study of political attitudes indicate that, as one might expect, conservative people are more concerned about past events than others. The effect of the amount of information a person receives about an event has not yet been investigated in any of these contexts, but it is probably of great importance. It is possible that the great flow of information about the war in Vietnam has had the effect of raising the intensity of emotional reaction and decreasing the psychological distance to this country. Questions remain as to when changes are considered to be desirable and when they are not. Too much concern with stressful events possibly interferes with rational planning, threatening our mental and physical health, so we take advantage of the barriers of time and space to shut out as much as possible from our immediate attention.

A Multidimensional Study of Subjective Geographic Distances

Cities, countries, and continents are often represented on flat maps or on globes, and most of our mental impressions are probably formed by these models rather than from experience. The next experiment was carried out in order to investigate the dimensionality and structure of our subjective representation in this field. The names of various cities and an island were used as stimuli, and two different methods of analysis were applied: conventional factor analysis and Kruskal's multidimensional technique (Kruskal, 1964a, 1964b).

In the experiment, sixty students estimated the subjective interdistances of thirteen places situated in different parts of the world. The distance between the North and the South Pole was called 100 and used as the standard in relation to which all seventy-eight interdistances were estimated. Thus, the responses were of the form "the distance between Calcutta and London appears to be about 75% of the distance between the North and the South Pole". No physical distance was, of course, longer than 100; however, if a subject's mental representation of the earth is not a perfect sphere, higher estimates might occur (and this actually happened in a few cases). The analysis was applied to the mean values among subjects for the interdistances.

For the factor analysis, the interdistances were assumed to be scattered out on the surface of a sphere and the angles that projected from the center of the sphere to the pairs of assigned places were calculated. The cosines corresponding to these angles were then factorized according to the principal component solution (see Horst, 1965; Harman, 1960). Three factors were rotated and the final matrix produced is represented in the left half of table 17.1.

If the spherical model fits satisfactorily, the communalities should be fairly constant, as they should approximate the square of the radius (see table 17.1). The communalities vary from 0.58 to 1.12, corresponding to a range of square roots from about 0.76 to 1.06. The communalities higher than one indicate that the data are slightly inconsistent, but this inconsistency is considered to be of very little importance in the present context and will not be investigated any further as the criteria of invariance for communalities was not violated to any serious extent (the standard deviation is 0.17).

The Kruskal technique, which is an improvement of a similar technique constructed by Shepard (1962a, 1962b), only requires an assumption of the data being a monotonic function of the actual distances, so that it is sufficient for the dissimilarities used in the analysis to be rank ordered (the

TABLE 17.1. *Dimensional analysis of subjective interdistance data*

Variable	Factor Matrix				Kruskal Configuration			
	I	*II*	*III*	h^2	*I*	*II*	*III*	h^2
Athens	−0.4475	−0.3053	0.5615	0.61	−0.5129	−0.3694	0.5717	0.73
Calcutta	0.2564	−0.7655	0.1046	0.66	0.3490	−0.9158	0.2493	1.02
Capetown	0.4802	−0.1384	0.8378	0.95	0.5005	−0.2027	1.1013	1.51
Kiruna	−1.0491	0.1398	−0.0685	1.12	−1.2105	0.0745	0.0818	1.48
London	−0.7527	0.2445	0.4271	0.79	−0.7949	0.2312	0.5028	0.94
Montreal	0.0904	0.7992	−0.2541	0.71	−0.0396	0.9655	−0.3400	1.05
Moscow	−0.7550	−0.4827	−0.0930	0.81	−0.7722	−0.5078	−0.0260	0.85
Panama	0.4290	0.6308	0.0335	0.58	0.5853	0.6986	0.2560	0.90
Peking	0.2407	−0.8175	−0.5015	0.98	0.1848	−0.9164	−0.4959	1.12
Saigon	0.3647	−0.8115	−0.2982	0.88	0.3421	−0.9543	−0.2733	1.10
San Francisco	0.4872	0.7631	0.3223	0.92	0.5629	0.8385	−0.3067	1.12
Stockholm	−1.0290	0.1197	0.1436	1.09	−1.0707	0.1074	0.2386	1.21
Teneriffe	−0.1785	0.0933	0.7884	0.66	−0.2534	0.1196	0.8706	0.84

The left half of the table shows the loadings obtained by a factor analysis, while the right half shows the Kruskal configuration after transformation and rotation to a similar structure.

distance estimates were treated as dissimilarities in the present case). The final configuration of the data is assumed to have metric properties based on the restrictions embedded in the rank order. Other methods of this kind are now available (McGee, 1966, 1968; Young and Torgerson, 1967; Guttman, 1968; Roskam and Lingoes, 1970).

For the experiment being described different numbers of dimensions and different kinds of distance transformations were tried with the data, Minkowski *r*-metrics being used for the transformations. The *r*-value is described by the formula

$$d(x, y) = \left[\sum_{i=1}^{n} |x_i - y_i|^r \right]^{1/r}, \qquad (4)$$

where n is the number of dimensions, d is the distance between stimulus x and stimulus y, and r ($r \geq 1$) is determined by the kind of distance used. Ordinary Euclidean distances are obtained when $r = 2$, and "city-block" when $r = 1$ (see Attneave, 1950; Torgerson, 1958). Higher values of r indicate that curved distances are used by the subjects in their ratings. Non-Euclidean distances have not been used very often in psychological research, with the exception perhaps of "city-block" distances. However, curved distances are not out of the question with the present data, as the geographic great circle distances may have had a considerable influence on the estimates given by the subjects. That the distances actually lie along geodetical lines was one of the assumptions used in applying the factor analysis.

Using Euclidean distances, a three-dimensional configuration was obtained by Kruskal's method. Kruskal presents a measure of the goodness of fit called "stress" (1964a, p. 3), which in the present configuration is about 3.0%—not far from an "excellent" fit according to his verbal evaluation. The goodness of fit is slightly better when $r = 3$ (stress $= 2.2\%$). However, this small difference is of no great importance for practical purposes, and for reasons of simplicity in calculating, the three-dimensional Euclidean solution was used in the subsequent analysis.

The Kruskal configuration was obtained using a computer program that gives the coordinates of the points within a particular coordinate system. In order to fit a sphere to the data, one makes the sum of the squares of the coordinates for each point as constant as possible, for in the perfect case each of them is equal to the radius of the sphere.

Let h'_i be the sum of squares of the coordinates for point i. The problem with the data is to minimize the variance of $(h')^2$, where

$$\begin{aligned} (h'_1)^2 &= (x_1 + a)^2 + (y_1 + b)^2 + (z_1 + c)^2 \\ (h'_2)^2 &= (x_2 + a)^2 + (y_2 + b)^2 + (z_2 + c)^2 \\ &\text{etc.,} \end{aligned} \qquad (5)$$

and i ranges from 1 to 13. The values a, b, and c are the constants to be added, when fitting the sphere, to the corresponding coordinates x, y, and z, obtained in the analysis. The variance of $(h')^2$ was formed and derived for a, b, and c, respectively, and the derivatives put equal to zero. The expressions are of the fourth degree and quite complicated, but they can be reduced into the following system of equations:

$$\begin{pmatrix} \sum (x_i - \bar{x})^2 & \sum (x_i - \bar{x})(y_i - \bar{y}) & \sum (x_i - \bar{x})(z_i - \bar{z}) \\ \sum (y_i - \bar{y})(x_i - \bar{x}) & \sum (y_i - \bar{y})^2 & \sum (y_i - \bar{y})(z_i - \bar{z}) \\ \sum (z_i - \bar{z})(x_i - \bar{x}) & \sum (z_i - \bar{z})(y_i - \bar{y}) & \sum (z_i - \bar{z})^2 \end{pmatrix} \begin{pmatrix} 2a \\ 2b \\ 2c \end{pmatrix}$$

$$= \begin{pmatrix} \sum (x_i - \bar{x})(S_i - \bar{S}) \\ \sum (y_i - \bar{y})(S_i - \bar{S}) \\ \sum (z_i - \bar{z})(S_i - \bar{S}) \end{pmatrix}, \quad (6)$$

where $S_i = (x_i)^2 + (y_i)^2 + (z_i)^2$ and $\bar{x}$, $\bar{y}$, and $\bar{z}$ are the arithmetic means of the original values for each dimension.

The only unknowns, a, b, and c were easily calculated and have been added to the present data, thus placing the origin of the coordinate system in the center of the sphere. This configuration was then rotated to obtain maximum similarity with the factor solution (see Veldman, 1967) and is presented in the right half of table 17.1. A comparison was made between the two figurations using Tucker's coefficient of congruence (see Harman, 1960, p. 270). The values of these coefficients are 0.994 (Factor I), 0.998 (Factor II), and 0.978 (Factor III), showing that the two methods of analysis yield almost identical results.

The comparison described here is not identical with that made between "content" and "distance" models as presented by Ekman (Ekman & Sjöberg, 1965; Ekman, 1970b). When using the "content" model, the subject is requested to estimate the proportion of one percept contained in another, and when using the "distance" model, he has to indicate the perceived degree of similarity between two stimuli (see also Ekman, 1963; Ekman, Engen, Künnapas & Lindman, 1964; Ekman, 1965). The "distance" model is, of course, most likely the appropriate one in the experiment described here.

According to the Kruskal technique, the analysis also showed that a two dimensional Euclidean space could describe the data in a way which he would consider to be fair, the "stress" being about 10%. This suggests that the subjective representation of the assigned places is not very far from plane mapping. The two-dimensional configuration is shown in figure 17.5. Inspection by eye indicates that the distances are scattered out in a way similar to their actual spacing on an ordinary flat map; however, distances far from Stockholm (where the investigation took place) seem to be underestimated relative to distances near to Stockholm. Consequently, it may be tentatively suggested that when thinking of stimuli of the present kind, individuals are influenced by two-dimensional mapping. However, a better fit is obtained when part of a spherical surface is used, where the curvature is not too pronounced.

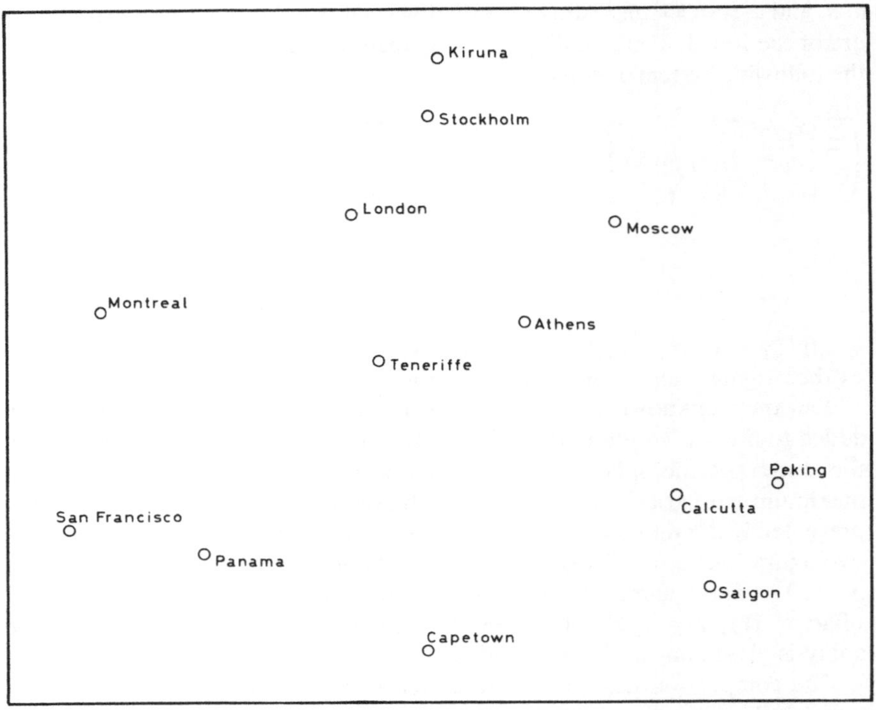

FIGURE 17.5. *Subjective mapping of twelve cities and one island. The configuration was obtained in two dimensions using Kruskal's multidimensional technique.*

Subjectively Perceived Area, Number of People, and Population Density

In another study 33 students were asked to estimate the subjective area, the number of people, and the population density of forty-four well-known countries (Lundberg and Ekman, 1970b). One country was chosen as the standard of comparison for each variable and given the value of 10. The countries used as standards were: Sweden (area), Spain (number of people), and France (density of people). All estimates were made during one session and in a random order. The median value was calculated for each country and variable.

When the stimulus names were treated separately for European and non-European countries, the subjective scales were found to be roughly describable as power functions of the corresponding physical scales. The data were split into two different stimulus groups as a trend appeared indicating that the subjects had treated these groups in different ways. For example, figure 17.6 shows the subjective population density plotted as a function of the actual number of people per unit area for European (A) and non-European countries (B). The fitted power functions are of the form shown above (1). The exponent is 0.31 for European (A) and 0.32 for non-European countries (B). Both functions have about the same degree of curvature, while the unit of measurement turned out to be different for European and non-European countries. Two points in figure 17.6B show an interesting deviation from the general trend. These points, both of which show that the population density was greatly overestimated relative to the other countries, represent China and India. The reason for these high values is probably associated with the fact that the very large populations of these countries have been the focus of international attention. Another reason could be that these countries are very unevenly populated; they have large relatively empty areas and some very densely populated ones. The latter point is presumably the better known of the two and may have had a considerable influence on the estimates of the average population density.

The data from this experiment made it possible to calculate a measure of expected subjective density through dividing the subjectively estimated number of people by the subjective area of each country. Empirical estimates of subjective population density are plotted against the "expected subjective density of people" in figure 17.7. A power function (1) has been fitted to the data and the exponent is 0.63. However, a linear function with a positive intercept might also fit the data to about the same degree of correspondence. The two points representing China and India again deviate from the general trend, which shows that the high estimates of density are

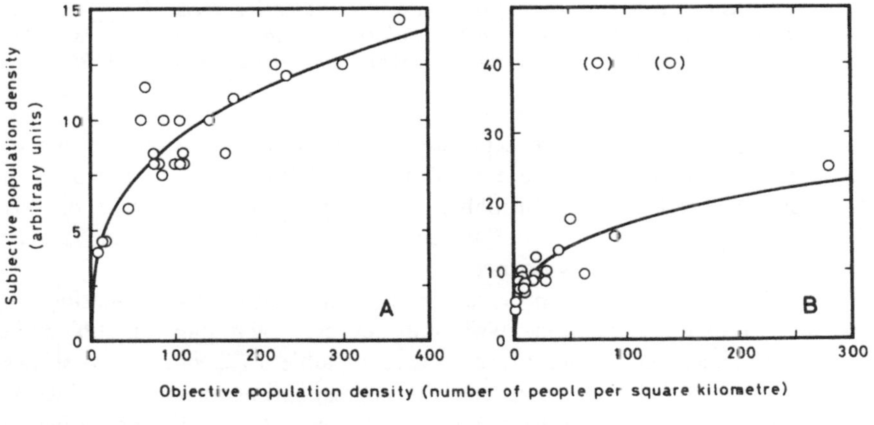

FIGURE 17.6. *Subjectively estimated population density plotted against the actual number of people per unit area for European (A) and non-European countries (B). A power function has been fitted to the data in both diagrams. The two points given in brackets were excluded from the fitting procedures. (Reproduced from Lundberg & Ekman, 1970b.)*

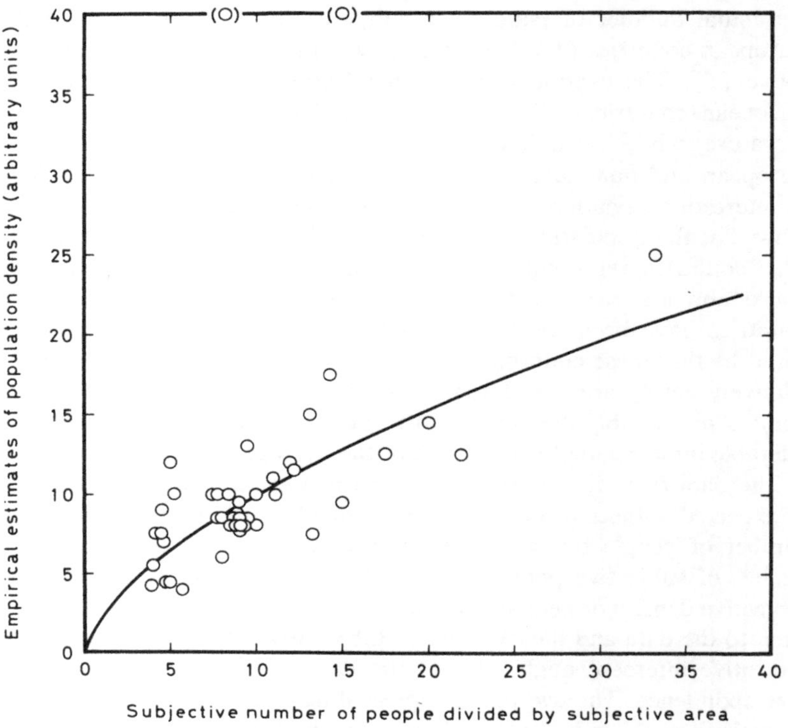

FIGURE 17.7. *Empirical estimates of subjective population density plotted against subjective number of people divided by subjective area. A power function has been fitted to the data. The two points given in brackets were excluded from the fitting procedures. (Reproduced from Lundberg & Ekman, 1970b.)*

not compatible with either the separate estimation of area or the number of people. It seems as if the cause of these high values is only connected to the density variable. Most probably, the subjects are not aware of this inconsistency, even though the difference is large enough for it to be very unlikely that it occurred by chance.

The data in this experiment have been further analysed by calculating the product moment correlation coefficient between each pair of subjective variables. These coefficients are presented in table 17.2 which also shows the correlation coefficients between the corresponding physical variables. There are two cases where the correlations between the subjective variables greatly exceed those for the corresponding physical variables and one case where the correlations between the subjective variables is positive to about the same degree as the correlation between the physical variables is negative. Both of these latter outcomes indicate that there could be an under-

TABLE 17.2. *Product moment correlations between the variables at the subjective and physical levels*

Variables	*Subjective level*	*Physical level*
Number of people/area of country	0.80	0.49
Population density/area of country	0.26	−0.33
Population density/number of people	0.75	0.03

lying factor influencing all the subjective variables. This factor may be connected to certain social and political phenomena associated with the stimulus countries, e.g., lack of food is a possible indication of too many people within one country, or political importance may suggest a country with a large population, with a large total area, and with many people concentrated within limited areas (=large cities).

The studies that have been presented here demonstrate the application of certain quantitative psychological methods and at the same time give the results of these methods when they are applied to some emotional and geographical contexts.

The mathematical functions presented above and the specific parameters obtained when they were fitted to the empirical data, should not be taken too seriously. The scaling methods used are still characterized by deficiencies such as response bias and error variance, and it is probable that other models will be found which fit the data just as well or better. However, in spite of these deficiencies many interesting and meaningful findings may be obtained by these methods. Particularly promising are the nonmetric techniques developed by Kruskal and others for multidimensional scaling (Shepard, 1962a, 1962b; Kruskal, 1964a, 1964b; McGee, 1966, 1968; Young and Torgerson, 1967; Guttman, 1968; Roskam and Lingoes, 1970). The remarkable similarity between the final structure obtained with one of these techniques and that obtained by a conventional factor analysis shows the strength of the method, particularly when one considers that only weak assumptions are made concerning the data involved. The nonmetric methods will probably be very useful in future research with psychological scaling, partly due to the fact that many psychological variables are of a multidimensional character and may be most appropriately scaled in this way, and partly to the relatively easy task they permit for the subjects, this probably helping them to be able to give more accurate estimations of their experiences. One thing is clear; these and other studies show that it is quite possible to directly and quantitatively investigate psychological problems at the psychological level without being forced to include their physical or physiological correlates.

18

A Method for Analyzing Distance Concepts of Urban Residents

ROBERT ALLEN LOWREY

I. Background

Investigators of social phenomena have given more and more attention over the past ten years to individual and group concepts of the urban environment, sometimes called urban space. The impetus for this trend has been a series of varied and complex events, including not only changes in cities themselves, but also changes in various research disciplines. Two kinds of such events have been: (1) increased pressure on public decision-makers to provide structures and facilities which are useful and acceptable to their intended users; and (2) an increased recognition by officials and researchers of the importance of small residential areas within cities, sometimes called neighborhoods, and of the design problems associated with these areas (Montgomery, 1965; Katz, 1967; Lynch, 1962). The neighborhood concept has been a mainstay in planning for many years (Dahir, 1947; Keller, 1965). Problems encountered in the processes of urban renewal and large-scale changes in land-use patterns (Fried and Gleicher, 1961; Hartman, 1964; Montgomery, 1965; Thompson, 1965, 362–368) have emphasized the notion that identifiable physical features are important in certain kinds of resident behavior.

The purpose of this report is to present a method for measuring some physical features of urban space and peoples' behavior related to them. Measurement of the way in which people compose and construct their thoughts about their surroundings has intrigued investigators for several years (Hudson, 1960; Kates, 1962; Saarinen, 1966), and recently some of these investigators have centered their efforts around people's concepts of scenes in towns and cities (Appleyard, Lynch, and Myer, 1964; de Jonge, 1962; Lee, chap. 5; Lynch, 1960). These urban settings are generally defined as arbitrary collections of buildings, streets, and other structures. Such configurations are difficult to define in a concise way, and the measurement and analysis of people's thoughts about them, or "images" of them, is equally difficult.

Several kinds of decisions and actions by urban residents have been attributed to these images of urban areas. The composition and meaning of these images are thought to be factors influencing residential, recreational, shopping, and other choices having economic and planning implications (Foley, 1950; Chapin and Hightower, 1966). There are various theories as to how these images have evolved and become meaningful (Lynch, 1960; Lowenthal, 1961; Webber, 1964a). Meaning is usually connected to experience and past behavior. In other words, the images are shaped over time, and groups of urban features become meaningful to the individual person as well as in a group or cultural sense.

The addition of the notion of group meaning complicates the theory of urban images, as well as the economic implications of peoples' decisions based on them. Expansion of the theory to a sociological level connects it with concepts such as "class" and "neighborhood" that have abundant backgrounds of theory of their own (Bracey, 1964; Bell, 1953). Social planning continues to become entangled in the difficulties of trying to comprehend large-scale group phenomena as composites of small-scale familial or individual actions.

Psychological techniques for analyzing behavior relative to urban scenes promise improvements in the understanding of commonly held group or cultural concepts (Stevens, 1966). Thus, suppose an urban scene is analyzed according to its distances, forms, colors, and other attributes. Next, judgments by local residents are obtained on several of these attributes. It is reasonable to designate the attributes on which the residents agree as elements in a commonly held cultural concept.

II. General Concepts

Distances to locations in an urban setting are one type of element or attribute of urban space that is measured in two experiments reported in this paper. These locations were various widely-used urban facilities, consisting of ten types or classes: bus stops, shopping centers, libraries, transportation terminals, schools, parking lots or garages, parks, expressway interchanges, post offices, and hospitals.

In an interview situation, residents named examples of each type of facility and judged distances from their homes. Later, the location of each person's residence and the facilities he named were located by latitude and longitude coordinates and direct surface distances were calculated.

A wide range of size, use, and frequency of occurrence is represented in the ten classes named. Simplifying assumptions were used to make people's distance judgments independent of these three dimensions, which are more precisely conditions of the facilities themselves than of the distances.

The size of each facility was eliminated from consideration by defining the facility as a point located within its area approximately at the center. These points were specified by the assignment of latitude and longitude to the point location. Assigned values were usually rounded to the nearest five seconds. A sample of distances calculated using latitude-longitude coordinates was compared with corresponding distances measured directly on a map. Errors were found to be less than six percent, a level of accuracy considered adequate for the present type of study.

Some sort of assumption was necessary with regard to the various uses of facilities that were named by subjects. Not only did the uses, or functions, of the facilities vary greatly, but the uses which people made of a particular type of facility, such as a park or shopping center, were diverse. The assumption made was that whatever diversity existed in people's uses of facilities would not produce a systematic bias in distance judgments across facility types, when data were grouped by type.

The frequency with which facilities occurred in an urbanized area was expected to vary widely; transportation terminals would be far fewer than bus stops. The assumption was that these differences in frequency would not have a large effect on people's distance judgments.

The concepts relevant to the analysis of distance judgments were more difficult to define than the concepts of facility classes. Here the essential assumption was that the urban dweller could visualize or think about distances to facilities in the abstract; for example, while sitting in his living room. This notion includes the idea that no matter how he has learned about the spatial features of his environment, he enjoys a flexibility which enables him to combine information from different trips, different time periods, and sources other than his own movements, and to respond consistently in an abstract situation.

To summarize, the assumptions in the study are:

1. Different sizes of facilities do not bias distance judgments.
2. Variation among people in their use of facilities does not bias their judgment of one facility class in relation to any other facility class.
3. Different frequencies of facility examples for different classes do not bias the judged distance to the example named.
4. Urban dwellers visualize or think about distances in abstract settings.

III. Experiment I

A. PROCEDURE

An interview technique was devised to obtain data on two questions: (1) With what facilities does the respondent consider himself familiar? (2) How are his distance concepts related to the physical distances of these facilities from his home? The interview technique was a vocal, person-to-person procedure in which the interviewer read the questions to the respondent and recorded facility names and locations. Figure 18.1 shows the form used for this part of the interview. Thus, the respondents (subjects) were allowed to select their own specific facilities (stimuli) within broad limits for each facility class.

Questions:	11.1.	Have you recently visited or used one of the (parks, etc.) in the city?
If "Yes"	11.2.	Which one was it? 11.3. Which phrase would you use to describe its distance from your home?
If "No"	11.4.	Are you familiar with any (parks, etc.) in the city?
	11.5.	Do you happen to know the name of one?
	11.6.	Which statement would you use to describe its location from your home?

Type of Facility	Name or Location of Example
1. Shopping center	
2. Bus stop	
3. Library	
4. Bus station or other terminal	
5. School	
6. Park	
7. Public parking lot or garage	
8. Expressway interchange	
9. Post office	
10. Hospital	

FIGURE 18.1. *Page 2 of interview.*

Two diagrammatic forms were developed for recording a respondent's distance judgments. These forms were designated *G* (fifteen pages) and *T* (ten pages). Sample pages from each form are shown in figures 18.2 and 18.3. Form *G* is based on the psychophysical technique known as ratio estimation (Torgerson, 1958, pp. 104 ff.). Form *T* is based on the same technique but uses a more compact experimental presentation. The forms were presented to the respondent who marked them himself after the interviewer had reminded him of the specific facilities he was comparing. Admittedly, the forms incorporate some arbitrary decisions about grouping of items, lengths of lines, and location of labeling. About six months was spent testing about four different form designs in which no particular design ever gained a clear advantage.

Since Form *T* contained all pairs possible from ten items, that is, N (N-1) pairs, the data from it produced a ten-by-ten matrix except for the principal diagonal. A value of 1.0 was assigned to each cell in this diagonal on the assumption that a distance paired with itself would result in a ratio judgment of 1.0. The data from Form *G* do not fill a ten-by-ten matrix, since Form *G* consists of a halved paired-comparisons experiment, that is, N (N-1)/2 items. The empty cells were filled with the reciprocal of the value in the cell diagonally opposite, an acceptable procedure in the present study since scales from Form *T* would serve as a check.

YOUR HOME	Suppose this is how far it is from your home to School Mark on this line how far it would be from your home to Hospital
YOUR HOME	Suppose this is how far it is from your home to Park Mark on this line how far it would be from your home to Post Office
	Suppose this is how far it is from your home to Expressway Interchange Mark on this line how far it would be from your home to Public Parking Lot or Garage

FIGURE 18.2. *This is a page from a "G" judgments form, except that the pages in the form were 8.5 by 14 inches. The length of the "standard" or "fixed" line is 100 mm.*

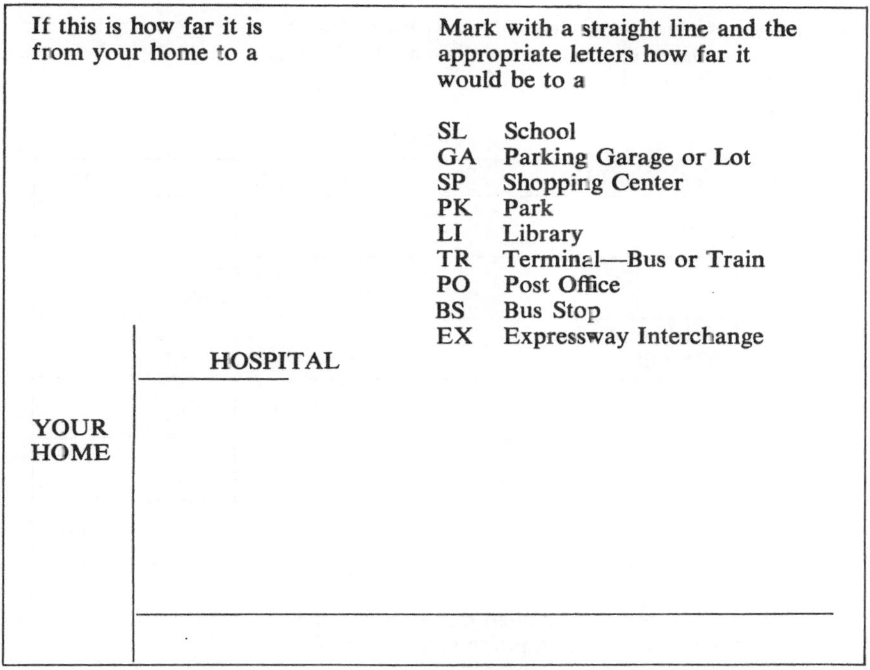

If this is how far it is from your home to a

Mark with a straight line and the appropriate letters how far it would be to a

SL	School
GA	Parking Garage or Lot
SP	Shopping Center
PK	Park
LI	Library
TR	Terminal—Bus or Train
PO	Post Office
BS	Bus Stop
EX	Expressway Interchange

HOSPITAL

YOUR HOME

FIGURE 18.3. *This is a page from a "T" judgment form, except that the pages in the form were 8.5 by 14 inches. The length of the "standard" or "fixed" line is 95 mm.*

Scaling was accomplished under the rule that the geometric mean of a row of a subject's ratio judgment matrix is the best estimate of his subjective value for the stimulus represented by that row (Torgerson, 1958, pp. 108–111). That is,

$$\log S_j = \frac{1}{n} \sum_{k=1}^{n} \log w_{jk} \qquad k = 1, 2, \ldots, 10$$

where w_{jk} equals the observed response for facility k paired with facility j, entered in row j, column k, and S_j is the best estimate of the subjective scale value or dimensional process for the distance ratios relative to facility k. Scale values were left in log form for convenience.

The design used for this experiment involved two judgments interview forms, each of which was filled in by thirty-two subjects. As indicated in table 18.1, twenty-four subjects filled in a total of forty-eight forms. Forms *GT* and *TG* were merely combinations of Form *T* and Form *G* in reverse order.

The subjects were not all college students; they were selected in two

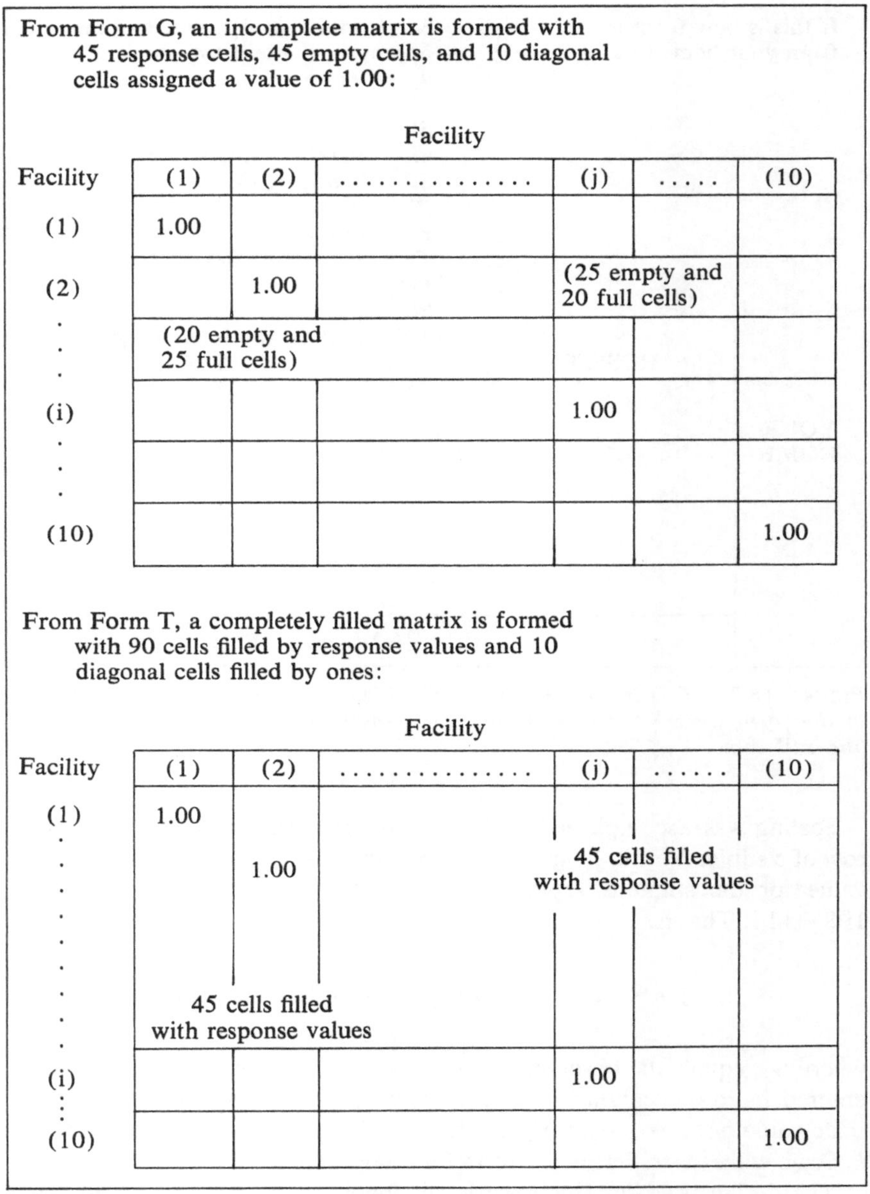

FIGURE 18.4. *Matrix formats for Forms G and T.*

different ways. Groups *T* and *G* were subsampled from a larger sample of respondents in a study already underway. The larger sample had been randomly selected from geographic strata, and the sixteen subjects in groups *G* and *T* were also sampled according to this principle. The sixteen subjects

TABLE 18.1. *Experimental groups*

Forms Filled In	*Group Designation* (N *for each group* = 8) *T*	*G*	*GT*	*TG*
T	yes	no	filled in Form T second	filled in Form T first
G	no	yes	filled in Form G first	filled in Form G second

Each group filled in the forms indicated by its designation. No member of group G filled in Form T and no member of group T filled in Form G.

in groups *GT* and *TG* were arbitrarily chosen, by four separate interviewers, from different parts of the city of Baltimore, Maryland, usually from among acquaintances (the combined Forms *G* and *T* required up to 45 minutes of judgment making).

B. METHODOLOGICAL CONSIDERATIONS

In mixing interview procedures with psychophysical techniques, it is appropriate to consider some methodological difficulties before looking at the degree of success of any scaling procedure. One of the conventional concepts from psychophysics is the notion of variable errors; in the present study, it is possible to consider these variable errors as indicators of response stability, or adaptation to the interview situation over time.

In this study, the ten pages of Form *T* provide a way to look at such effects, since each page consists of nine judgments of the same ten facilities (see fig. 18.4). Figure 18.5 shows a plot of data in table 18.2, indicating the page-by-page change in standard deviations of judgment values for three groups of eight subjects each. The abscissa on this graph is labeled with facility type abbreviations, but is also a time axis for each subjects' progress through the judgments response presentation. Note that no strong trend is apparent in the size of the standard deviations.

If the distance concepts of people are based not only on the physical distances but also on various portions of a person's experience, the abstract presentation of the distance in the form of a diagram may constitute a complex stimulus. Presentations of such stimuli in a paired-comparisons

TABLE 18.2. *Standard deviations of columns of Form T matrices. Means are calculated for groups. Data are means of logs of subjective ratios.*

	BS	*PO*	*SP*	*SL*	*PK*	*EX*	*TR*	*HO*	*GA*	*LI*
Group T	.0927	.1178	.1630	.1483	.1231	.2049	.1268	.1102	.1536	.2847
Group TG Form T	.1042	.1344	.1385	.1153	.1364	.1536	.1606	.1305	.1639	.1574
Group GT Form T	.1003	.1392	.1197	.1064	.1278	.1494	.1718	.1175	.1881	.1078

experiment may result in various kinds of inconsistent or contradictory judgments (David, 1963; Sjoberg, 1968). One such contradiction is the reversal of the second judgment in two judgments of the same pair of stimuli.

A simple way of looking at these binary contradictions is to count the number of pairs of judgment values which have resulted from two contradictory judgments (which can only occur in a complete paired-comparisons experiment, and which results in an asymmetrical table). If stimulus *A* is judged farther away than (greater than, better than, and so on) stimulus *B* on one presentation, and closer than stimulus *B* on another presentation, the subject is said to have made a contradiction. Since he was allowed only two possible responses, he is said to have made two contradictory binary judgments. If he makes no contradictions in a complete paired-comparisons experiment, his responses will produce a symmetrical matrix of plus and minus signs, ones and zeros, or whatever symbols are used. If ratios also result from the judgments, the signs of their logarithms will show whether the binary judgments were contradictory. In a matrix in which ratios have been transformed to natural logarithms, the contradictory values may or may not be equal in value, but they will be diagonally opposite to each other and will have the same sign. Recall that in both Forms *T* and *G* (figures 18.2 and 18.3), the respondent is required to make both a binary "farther than, closer to" judgment and a ratio judgment.

Conceptually, these two judgments represent responses of: "*A* is farther from home than *B*" and "*A* is closer to home than *B*," for the same *A* and *B*. Table 18.3 shows the results of counting the occurrence of diagonally opposite values of like sign for the twenty-four subjects who received Form *T*.

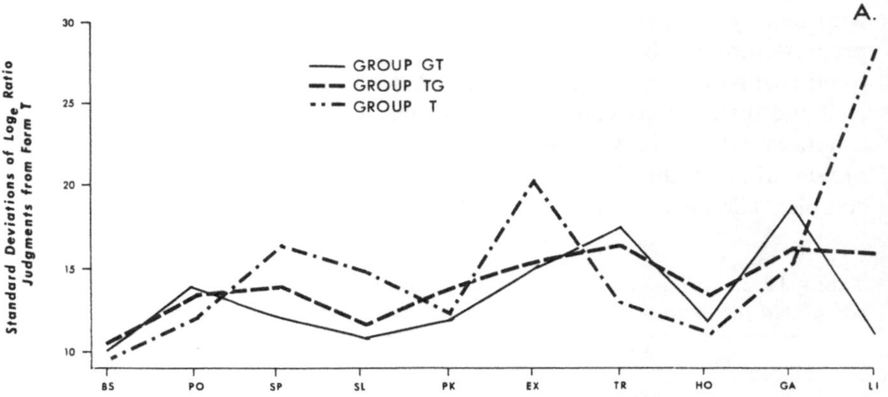

FIGURE 18.5. *Variable errors from Form T judgments. Each point represents a mean of 8 standard deviations for each facility type. The order of the facility types for plotting is the order in which the page occurred for a facility, i.e., the page on which it was the standard stimulus for comparison with 9 other facilities.*

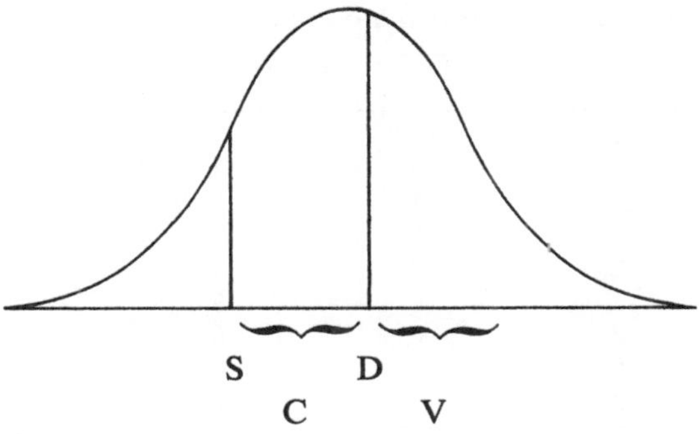

For one subject:

$$V_j = \left[\frac{1}{n}\sum_{k=1}^{n}(\log d_{jk} - D_j)^2\right]^{1/2}$$

where:

$$D_j = \frac{1}{n}\sum_{k=1}^{n}\log d_{jk}$$

and for more than one subject:

$$V_j = \frac{1}{N}\sum_{g=1}^{N} V_{jg}$$

j, k = stimuli

g = subjects

$n = 10, N = 8$

FIGURE 18.5*A*. *Diagram for table 18.2.*

Diagram of hypothetical distribution of values on subjective continuum, with value of the physical distance, or stimulus, located at S. D is the mean of judgments and V is the standard deviation or variable error. C is the constant error (see Guilford, 1954: 88). A *t*-test on the difference between the means of cells 1–5 and cells 6–10 was not significant, $p < .35$, $df = 14$.

In looking at table 18.3, a slight difference appears between group *GT* and group *TG*: the group which received Form *G* first was slightly more confused by the time they got to Form *T* than the group which completed Form *T* first. Furthermore, for groups *GT* and *TG*, the number of pairs in which both values had positive signs was larger than the number with negative signs. Since the logarithms of ratios were being counted, this means that there were more "greater than one" ratios than there should have been for a perfectly symmetrical set of responses. In other words, more facilities were judged "closer to home than" reliably than were judged "farther from home than" reliably. Another way of stating this interpretation is to think of two facilities of which *A* is closer to the subject's home than *B*. Then the pair presented *BA* is more likely to be ordered correctly than the pair presented *AB*.

TABLE 18.3. *Number of contradictions in responses to the same pair of stimuli*

T		*GT*		*TG*	
+, +	−, −	+, +	−, −	+, +	−, −
7	1	0	1	1	0
6	20	3	2	0	5
8	10	3	1	0	0
3	2	0	1	5	1
2	3	3	3	1	0
0	2	6	0	0	0
6	0	0	0	3	0
6	2	6	0	3	0
38	40	21	8	13	6
Total 78		29		19	

Contradictions are counted from judgments matrices by counting the number of pairs in which both values have the same sign. For this table, judgments resulting in zero values are not counted as part of a pair of contradictory judgments.

Since each respondent made 90 judgments in Form *T*, the total number of paired responses which could be contradictory was 45, and for eight subjects, 360. The percentages of contradictions for the three groups *T*, *GT*, and *TG* were 21.6, 8.1, and 5.3, respectively. Although the group *T* rate is high, the rates for the other two groups are relatively low and indicate a consistent ordering of urban distances in spite of possible confusion from route and direction factors.

Although both Form *T* and Form *G* present about the same line lengths, they are different both in format and in number of responses required. Correlations between sets of corresponding values (actual responses for Form *G*) are high, ranging from 0.777 to 0.967 for 16 subjects. These values suggest that the complete paired-comparisons format has no advantage over the halved format; similarly, the actual responses from Form *G* correlate well with the values in the diagonally opposite positions of Form *T* ($-0.727 > r > -0.947$, $N = 16$).

To summarize the methodological considerations involved in the present distance judgments experiment:

1. The paired-comparisons diagrams allowed subjects to make binary contradictions which introduced a small amount of bias into scale values.
2. Variable errors, while not showing any trend during the interview, are slightly larger for group *T*.
3. Comparison of the half-form (*G*) with the complete form (*T*) shows no difference between forms.

C. RESULTS: DISTANCE JUDGMENTS

In analyzing the physical distance data, an unconventional procedure involving a unit transformation was used. The physical distances were first calculated using a spherical trigonometry solution after converting the latitude and longitude data to azimuths. Then a physical distance ratio matrix was constructed by dividing each value by every other value. The row means of this derived matrix were unit transformations of the original physical measurements (ten for each person). However, the derivation of these physical distance matrices for each subject provided more values for determining the characteristics of the regression equation than if the ten-point scales had been used.

In addition to the judgment ratios and physical distances, an error matrix was calculated from Form *T* data by subtracting corresponding ratio values in the two matrices. Constant errors were obtained by calculating the column means of this error matrix, giving a constant error for each facility and for each page of the form.

Constant errors decreased significantly from the first part of the interview to the last, suggesting an improvement in "skill" of the respondents. The appearance of this effect in the present type of experiment is discussed further elsewhere (Lowrey, 1970a) but more investigation is required to determine its importance.

Figures 18.6, 18.7, and 18.8 contain plots of row means of logarithms of subjective ratios against row means of logarithms of real distance ratios. In performing regression analysis, 100 pairs of values for each subject were used. (Recall that each subject's responses to Form *T* generated a complete ten-by-ten matrix except for the main diagonal; the matrices for Form *G* were completed by filling in empty cells with the inverse of the ratio in cells located opposite them, and the physical distance matrix was completed by division.) The regression parameters for each group are shown in table 18.4. Note that the subject in group *T* somewhat misinterpreted the task. The means for this group, less that one subject, are also shown.

The learning effect occurring during the experiment appears in differences between forms for groups *GT* and *TG*. Each group had a higher regression coefficient and a lower error score on the form they received

TABLE 18.4. *Means for regression coefficients, intercepts, and deviations from regression lines for Form G*

				Means for Form T Data		
	b	a	σ	b	a	σ
Group G	0.602	.0	0.723			
Group TG	0.602	.0	0.723	0.632	−0.176	0.710
Group GT	0.579	.0	0.678	0.676	−0.169	0.677
Group T*				−0.041	−0.263	0.790

* For $N = 7$; deleting one subject having extreme values.

Note:

$$Y_j = a + bX_j, \quad \sigma = \left\{\frac{1}{n}\sum_{j=1}^{n}[Y_j - (a + bX_j)]^2\right\}^{1/2} \qquad n = 100$$

second. In spite of these differences, figures 18.9 and 18.10 show close agreement between groups (and forms) on scale position. (Although the bus stop looks out of place, its distance is only slightly underestimated.)

D. DISCUSSION AND CONCLUSIONS

Subjects in the present study have readily produced judgment "ratios" which yield good estimates of kilometer distances from home to facilities when scaling procedures are applied to the judgment values. Furthermore, these judgment-real world connections are relatively invariant over characteristics of interview or "response diagram" forms, and to a lesser extent, over groups of people.

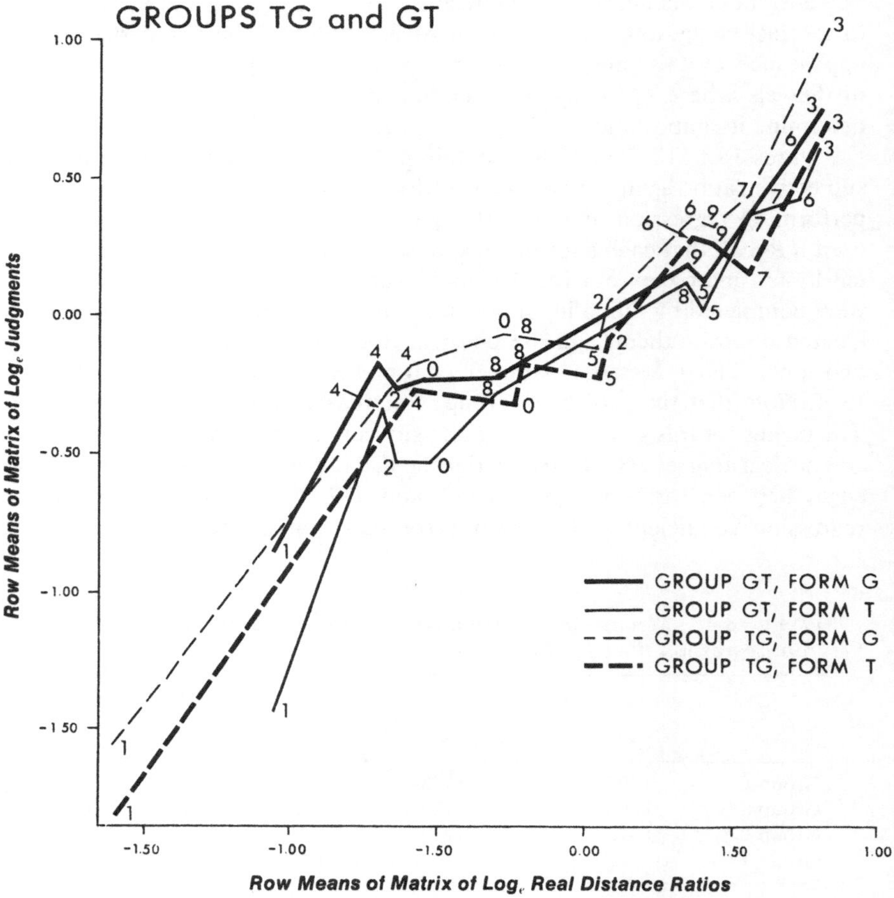

FIGURE 18.6. *Each point represents 8 subjects.*

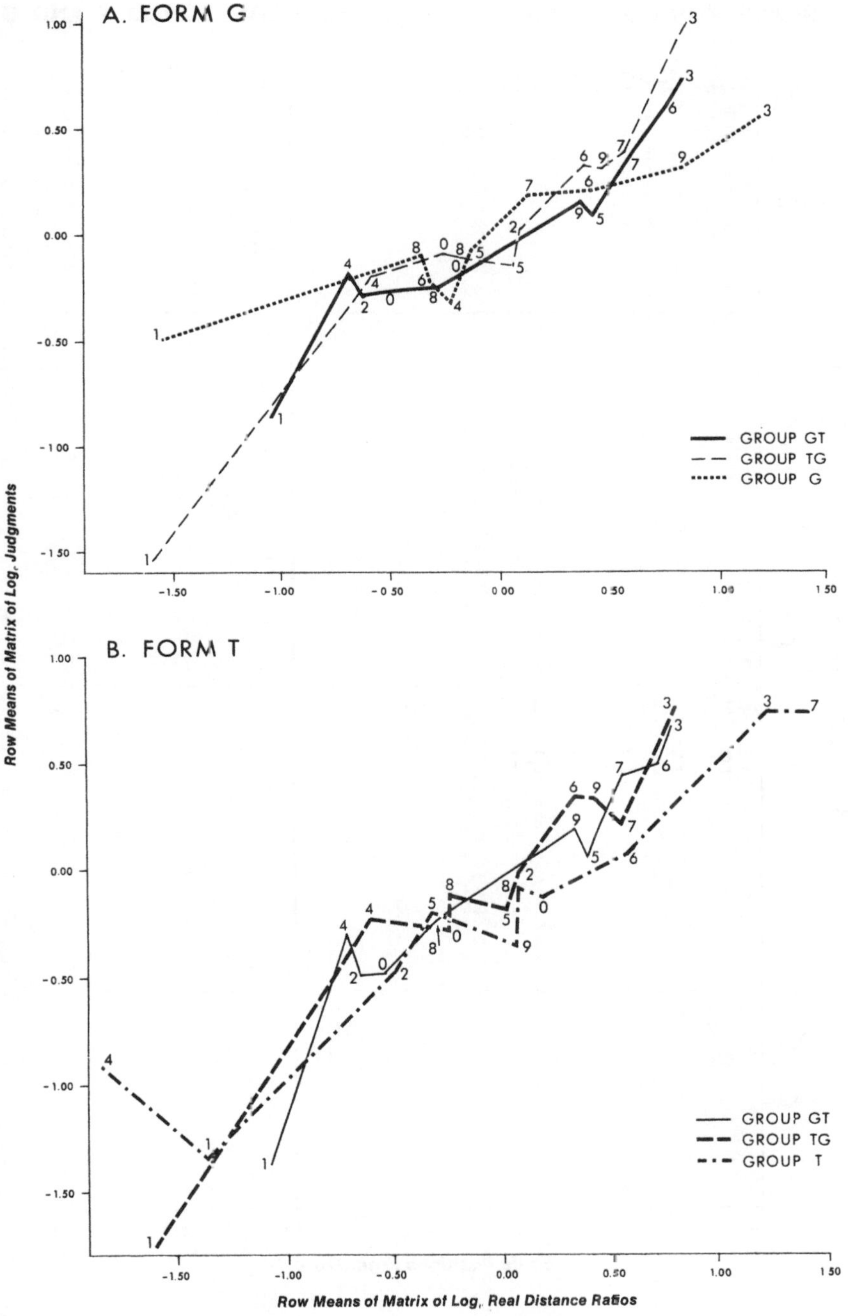

FIGURES 18.7 and 18.8. *Groups TG and GT. Each point represents 8 subjects.*

SCALE POSITIONS FOR TEN FACILITIES FROM FORMS T AND G

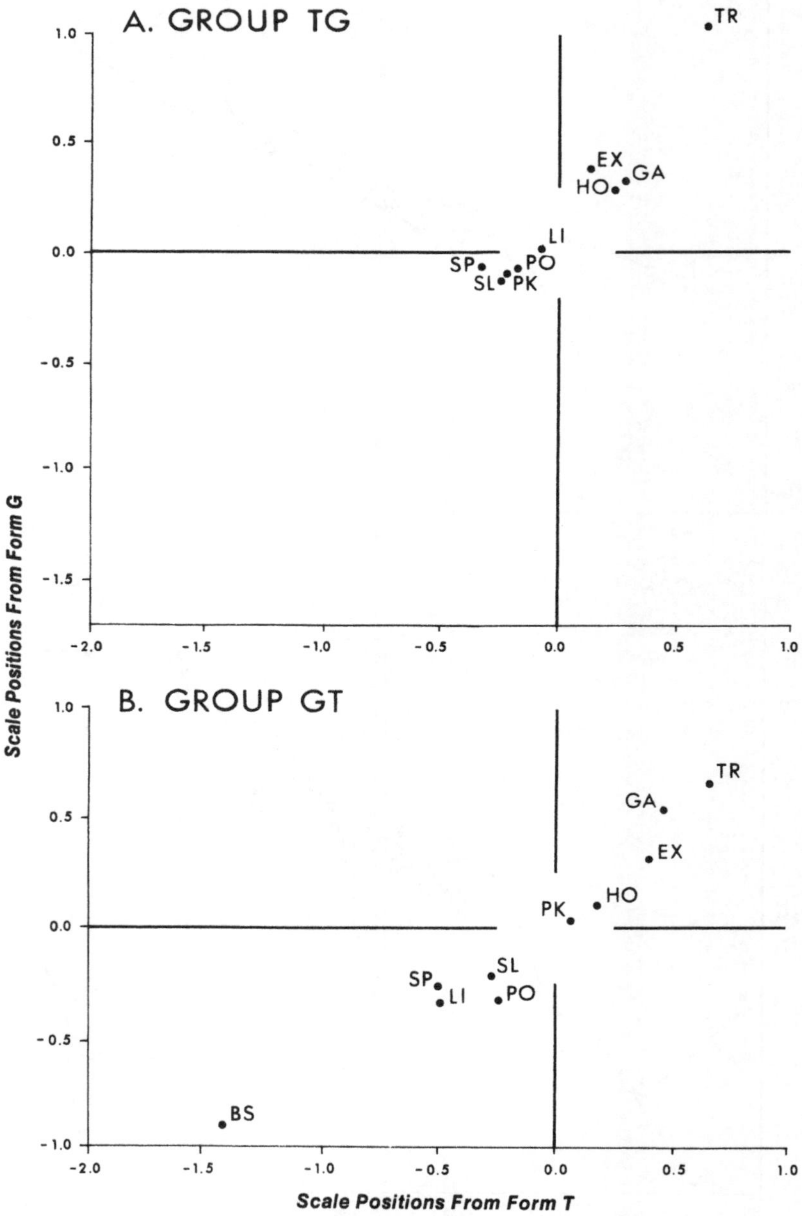

FIGURES 18.9 and 18.10. *Data are $\log_e$ subjective ratios.*

The regression coefficients for logs of subjective ratios fitted to logs of real distance ratios ranged from 0.112 to 2.065, not counting the extreme value of –3.149. One possible interpretation of this wide range is that many different personality characteristics were operating in the way subjects made judgments. The means of the coefficients suggest that in spite of possible environmental influences, such as urban topography, barriers, neighborhood effects, and travel time, the judgments were closely related to geographic distances. In other words, whatever internal procedure a subject followed in making his judgments, either there was no universal rule, or if there was, it seems to have been "use geographic distances."

The small number of "reversed pair" unreliable judgments (table 18.3) shows that people have extremely good knowledge of ordinal location relations. Perhaps ordinal judgments are what all people tend to make, although ratio judgments are available if a person is called upon to make them.

IV. Experiment II

A. PROCEDURE

As indicated in the discussion of general concepts above, the basic procedures for the two experiments were the same. In the first experiment, attention was concentrated on developing reliable connections between distances in urban space and residents' behavior. These connections were stated in the form of regression equations, and some attention was given to the properties of the judgment data and the derived regression parameters.

In the second experiment, more attention was given to differences among facility types which appear in people's judgments. In addition, some preliminary tests were made of the relation between distance judgments and people's socioeconomic characteristics. In other words, attention was directed toward the problem of separating the effects of the facilities themselves from the effects of socioeconomic characteristics. The specific questions asked were: (1) Does a person's judgment behavior differ for different types of facilities? (2) Does a person's judgment behavior differ according to socioeconomic characteristics?

One hundred thirty-eight subjects were included in Experiment II. These were selected as a subsample in a larger ongoing survey of the Baltimore, Maryland, area. The sample was geographically designed to cover most of the Baltimore area. The 138 subjects from whom complete distance judgment interviews were obtained were dispersed over 42 census tracts. The median family income of the sample was about $6,300.

As in Experiment I, data were obtained in a personal interview. The interviewer asked the subject to name, for each type of facility, an example which he had recently visited or with which he was familiar. He then asked the subject to locate the facility by street or block.

All subjects together named 511 different facilities and defined their locations for later measurement by the latitude-longitude scale. The smallest physical distance obtained, 0.01 kilometers, was between a subject's home and a bus stop, but various other facilities named, including post offices and an expressway interchange, were found less than 0.2 kilometers from subjects' homes. The largest distance was a hospital 23.94 kilometers from the home of the subject who named it. Thus, the logarithmic scale for kilometer distances ranged from –4.60517 to 3.13549. In the present study, values on this transformed scale are referred to as log kilometers.

After the interviewer obtained the name and location for an example of each of the ten facility types, the subject was presented with a series of distance judgments diagrams on which he marked lines to indicate distances. An important difference in Experiment II was that only the shorter judgment form, Form *G*, was used. As in the previous use, the subjects made judgments of the ratio of two lines, and the responses were analyzed as ratios: the length of the line the subject marked divided by the length of the standard line. The facilities were paired in a conventional paired-comparisons format for one presentation of each pair. This type of format produced a symmetric matrix of ratio judgments. Empty cells in the matrix were filled with the inverses of the values in the diagonally opposite cells. Values in the main diagonal were set at 1.00.

Although the responses were in the form of ratios, the range and distribution of the ratios made it reasonable to transform them to natural logarithms and to use their logs as the basis for analysis. For a set of nine values plus the arbitrary 1.00, all having one facility in common, a least squares estimate of the log of the judged distance is the mean of the logs of the ten values (see fig. 18.4 above). These means were calculated, resulting in one such value for each facility for each person. These ten values for each subject are called *subjective scale values* (Torgerson, 1958, pp. 108–111).

For all ten facility types, the judgment ratios most commonly occurring were between 0.01 and 3.05, giving logarithms of about –4.1 to + 1.1. A few subjects tended to make extreme judgments, and a few ratios were as high as 14.8 ($\log_e 14.8 = 2.70136$). For both log kilometer and subjective scales, the extremely high values were rare. In summary, the important principles of the physical distance and behavioral measuring procedures were:

1. The interview judgments consisted of a set of judgment ratios which yielded a subjective scale of ten values for each subject.
2. The physical distance measurements yielded a log kilometer scale with range similar to that of the behavioral measurements.

As a second part of the interview in which the facility locations and the distance judgments were recorded, the respondent was asked several questions about his household conditions. The hypothesis which suggested these questions was, as mentioned earlier, that spatial behavior varied according to the experience and household conditions of the subject. This rather general hypothesis suggests that extensive data analysis could be made. Limited analysis of facility differences and socioeconomic variables is reported here.

B. ANALYSIS OF FACILITY DIFFERENCES

Some analysis of the physical distance data itself is useful in order to understand the nature of the urban space related to the judgments subjects made. The places named were facilities with which the subjects were familiar or which they had recently visited. Except for this restriction, subjects were allowed extensive freedom in specifying an example for each type of facility. The 138 subjects named 82 schools and 86 parking lots or garages. There were 39 parks, 40 expressway interchanges, 30 libraries, and 31 post offices. In other words, some similarities appeared in the numbers of different examples of facilities that were named.

These similarities raised two questions. First, were there similarities in the distances that people traveled for the facilities that were named with equal frequency? Second, were there similarities in the way the distances to these facilities were judged?

In order to answer the first question, average distances traveled to each facility type were compared with frequencies. In the case of two pairs of facility types which were named with about the same frequency, the average distance traveled to each was substantially different. Subjects traveled about 1.7 kilometers further for expressway interchanges and garages than they did for parks and schools. In the case of a third pair, libraries and post offices, the distances traveled were about the same (see table 18.5).

In order to answer the second question, the scale values for judgments,

TABLE 18.5. *Number of facilities of each type named by all subjects and geometric mean of distances from homes.*

Type	*Number*	*Kilometers*[a]
Shopping Center	64	1.64
Bus Stop	109	0.57
Library	30	1.15
Terminal	9	5.28
School	82	0.96
Park	39	1.79
Parking Lot or Garage	86	2.72
Expressway Interchange	40	3.49
Post Office	31	1.17
Hospital	21	2.54
Total	511	21.31

[a] This is the geometric mean of the distances to facilities named from the homes of 138 subjects. In other words, the distances from each subject's home to the specific facility named were measured and the sum of their logarithms was divided by 138.

as well as physical distances, were used. Each subject interviewed had been measured on each of ten facilities and each facility was connected to a physical distance from the subject's home to it. Thus, there were physical distance and behavior scales consisting of ten values each for each subject. Means of these values were used to evaluate both a *psychophysical function* and *error scores* for all subjects. Using logarithms in a regression formula produces a function of the form,

$$\log Y = a \log X + b,$$

where a represents the regression coefficient or slope and b represents the scale constant or Y-axis concept. The linear function reflects an exponential curve relationship of the form,

$$Y = b\, X^{a} + e$$

in the original data, where e represents a residual (Luce, Bush, and Galanter, 1963, pp. 273–293).

The apparent linearity of the graph of this function (see fig. 18.11) suggested that subjects had judged the facilities in a similar manner, presumably according to a common subjective rule. The range of the exponents (regression coefficients) was from 0.067 for bus stops to 0.396 for parks. Although this range is relatively broad, the variability associated with each coefficient is also high. Since no statistical tests were conducted, however, it is not entirely justified to conclude that the coefficients are not significantly different. This possibility is discussed further in D below.

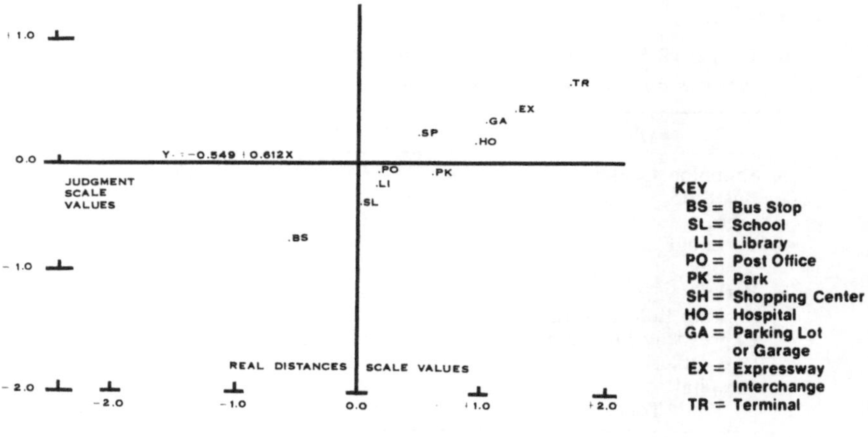

FIGURE 18.11. *Means of scale values for ten facilities (each point represents 138 subjects).*

C. ANALYSIS OF SOCIOECONOMIC VARIABLES

Data were obtained in the interviews not only for distance judgments but also for a series of socioeconomic or household condition variables. Two of these were dichotomous: driving status and sex, and were analyzed by *t* tests. Nine others were treated as continuous variables using product moment correlation formulas.

The two dichotomous household variables showed strong relations to several psychophysical variables. In addition, some of the variables which did not show significant *t* values have significant interpretations. Specifically, no differences appeared in real distances traveled or standard deviations of real distances traveled. Drivers, nondrivers, men, and women all traveled about the same mean number of kilometers to facilities, and they all showed about the same variability in the distances they traveled.

For drivers and nondrivers, the regression coefficient ($p < 0.005$) and the regression line errors ($p < 0.025$) were both significantly different, rather strong evidence of a completely different way of thinking about distances (see table 18.6). For drivers, the mean regression coefficient and the mean regression line errors were both larger than they were for nondrivers. There is no simple way to translate these results into statements abóut behavior, but there are some tentative interpretations. Drivers have a significantly different exponent for their distance judgment function, and they make sloppy or careless mistakes. Or, drivers' log kilometer distances resemble more closely a linear function of log kilometer distances, but there is also more error in their judgments. Nondrivers have a lower exponent for their distance function, but their judgments conform more closely to the derived function or judgment rule. How far away a person thinks a place is may or may not be related to where it really is, but nondrivers are less variable about where they think it is than are drivers.

Men and women are alike on their psychophysical functions but differ on regression line errors ($p < 0.005$). Both sexes have a general idea of the

TABLE 18.6. *Results of T tests*

	Male	*Female*	*Drivers*	*Nondrivers*
Number of Subjects	45	93	63	75
Regression Coefficient	0.4780	0.49551	0.5568	0.3834
	$T = -0.3239$	$p < 0.40$	$T = -0.31925$	$p < 0.005$
$Y-$ Intercept	-0.1718	-0.2510	-0.3034	-0.1632
	$T = -0.9799$	$p < 0.20$	$T = -2.4431$	$p < 0.01$
$Y-$ Axis Errors (Residuals)	0.4251	0.3111	0.3845	0.3178
	$T = -2.7316$	$p < 0.005$	$T = -1.9814$	$p < 0.025$

All values are means of logarithms.

ordering or interrelation of distances but men are always more confused than women. When the subjects are divided into drivers and nondrivers and sex differences are tested in each subgroup, these differences tend to fade. The driver/nondriver distinction is undoubtedly the more important of the two two-valued variables. In looking at the actual values, female nondrivers had the lowest regression line errors of all four groups, while male nondrivers had the highest. Male drivers had the highest regression coefficient and male nondrivers the lowest.

Since the driving status variable shows the major strong effect on distance judgments, a possible conclusion is that whether a person is a driver also influences the distances he travels to facilities. It should be noted, however, that the mean for 138 subjects of all log kilometer scales is 0.4982, and the standard deviation of the scale means is 0.9729, suggesting wide variability among individuals. As noted above, furthermore, there was no difference between groups in physical distances.

Each of nine other household variables for each subject was paired with physical distance and regression parameters. The nine household variables measured age, education, and family composition. Of about 90 correlation coefficients calculated, none exceeded 0.30. For the small number of variables available, multiple regression procedures were not considered appropriate.

D. DISCUSSION OF CONSTANT EXPONENTS AND MOTIVATIONAL BIAS

Although the roughly linear plot of logarithms in figure 18.10 is interpreted to mean a single psychophysical function exists for urban distance concepts, the analyses presented later for separate facility types raise some doubts. These analyses showed a moderate range of exponents for different types of facilities. Although no statistical tests were conducted on these exponents, the possibility that significant differences may exist is suggested by experiments in psychophysics.

Investigations of exponents for various sensory modalities have noted the dispersion of these exponents, and have suggested that a given exponent is characteristic of, and reasonably constant for, a given modality such as taste, touch, or hearing (Stevens and Galanter, 1957; Luce, Bush, and Galanter, 1963, p. 277). The work on specifying the modality exponents has taken place in a larger context of investigation of, and controversy about, the psychophysical law. This larger context will be ignored here, but it is important to consider the implications of the notion of modality constants.

The range of these constants is from less than 0.1 for smell up to 3.5 for electric shock. Variability of the constant for any given modality has been noted, but no theory has sought to incorporate such variability, which is sometimes attributed to experimental conditions. Similarly, individual subject departures from the function determined for a group have been attributed to temporary conditions. These temporary conditions are sometimes called motivational bias, but are rarely considered as constant factors in perception, since constant factors are presumably all reflected in the exponent.

Experiments on emotional involvement and distance to cities (Bratfisch, 1966) have suggested that some variability in exponents characterizes the relation between subjective and objective distances. Using distances among European and Asian cities, he found exponents ranging from 0.58 to 1.07. Without stating any specific hypotheses, he suggests that further investigation deal with concepts such as emotional involvement, estimated importance, and interest. These concepts are seen as influencing a person's concepts of subjective distance, but their relation to objective or physical distance remains obscure.

Thus, these studies of subjective distance conclude that a link between subjective or judged distance and other subjective conditions exists, but that there is no such link between subjective distance and physical distance. It may be that if distances within one city had been used as stimuli, an association among all three types of concepts would have been found. In any event, variability in the exponent of the psychophysical function suggests that substantial differences in subjective judgment rules exist, whether because of past experience and other motivational factors or because of some discontinuous feature in the stimuli.

Two points are important for further work on this tangle of problems. One point is that the total variability in exponents, while substantial enough to need explanation, is not substantial enough to submit conveniently to experimental investigation. In other words, there really is a lot of variability due to experimental effects. A diverse set of variables for investigation of the associations among the three phenomena is needed.

Second, it is eminently desirable that connections between distance judgments and socioeconomic characteristics be established without intervening variables, such as interest or emotional involvement, intriguing as these may be. For those who are faced with development and investment decisions, connecting distances likely to be traveled with available socioeconomic features is much more useful than knowing that emotional involvement is an important but unknown factor.

E. DISCUSSION OF NEIGHBORHOOD CHOICES

The second discussion point deals with the relation between distance concepts and the geographic concept of neighborhoods. Although census tracts and transportation districts have been invented, the neighborhood as a unit of urban space continues to be popular, perhaps because, as noted earlier in this report, the term continues to show up in interviews and popular writing about cities.

One of the more disturbing features of the concept is that it seems to have very concrete meanings to people who use it as an everyday, household term. Commonly heard phrases such as: "You can play outside, but don't go out of the neighborhood," "Well, there goes the neighborhood," and "Those vandals were from outside the neighborhood" have continued to frustrate social scientists who wish to understand people's concepts of distances, boundaries, and space. Analyses of neighborhood changes accompanying renewal or transportation projects have provided some insight into the meaning of neighborhood, but only in relation to these disruptive events.

Studies of the type conducted by Lynch (1960) have suffered from the lack of measurement techniques to apply to spatial behavior, although recent developments have been encouraging (Lee, chap. 5; Garling, 1969). Although the present study contributes little to neighborhood theory, it does provide some suggestions as to how further neighborhood content or area research might be directed.

19

Urban Cognitive Distance

RONALD BRIGGS

Introduction

Geographers have long recognized that the environment may be described by one set of variables relating to the attributes of places and a second set derived from the relative spatial locations of these places (Berry, 1964). Differentiation between these sets is readily observable in most models of the spatial behavior of individuals in which the positive or negative valences of place attributes are balanced against their relative spatial locations (Nystuen, 1967). Recent urban research has stressed that spatial behavior is an outcome of the interaction between these two sets of environmental variables *and* the cognitive processes of the individual. Increasingly, explicit attention has been paid to the processes of cognition (e.g. Harvey, 1969b; Saarinen, 1969) and learning (Golledge, 1967, 1969). This work has stressed particularly the cognition of the attributes of places, with only a limited consideration of the cognition of spatial relations, a conclusion which can be readily drawn from Saarinen's (1969) review.

The purpose of this study is to consider certain aspects of the cognition of spatial relations between places. This examination is a step toward more comprehensive models of the urban system utilizing measures of spatial location and separation which are meaningful in terms of individual behavior. The need for such measures has been pointed to by several authors. Marble and Bowlby (1968, p. 73) conclude a consideration of shopping behavior with the observation that "questions of the individual perception and scaling of distance in the physical space of the city remain of central geographic interest." Olsson (1965, p. 57), in a review of the relation between distance and human interaction, noted that the use of straight-line distance measures has been a consequence of operational rather than theoretical considerations and argued for "a more flexible way of measuring distances and a partial abandonment of purely physical concepts."

A major element of many geographic models is the movement of individuals between spatial locations. An hypothesis common to all these models is that the spatial separation of places itself has an influence upon behavior, and, as a consequence, the models include a variable measuring this separation. This variable frequently assumes paramount importance not only in an attempt to explain movements, but also as the basic explanatory construct of the model. As examples, we might cite the model of consumer movements implicit within central place theory (Berry and Pred, 1965; Golledge, Rushton and Clark, 1966; Clark and Rushton, 1970) as well as the more recent, explicit models of consumer shopping behavior (Baumol and Ide, 1956; Huff, 1960, 1961, 1963; Bucklin, 1967; Golledge and Brown, 1967; Marble, 1967; Nystuen, 1967; Rushton, 1969a: Berry, 1967, gives a simple review of central place theory as well as some of these studies concerned specifically with consumer spatial behavior). The models of urban land use (Wingo, 1961; Alonso, 1964; Casetti, 1967; Muth, 1969), of residential site selection (Chapin, 1968; Brown and Moore, 1970), and of social contact (Cox, 1969) are other examples. In general, these models consider decisions by individuals among a set of alternatives which have specific and separate locations within the environment (Demko and Briggs, 1970). Since these alternatives are not necessarily within the individual's direct perceptual field (that is, when the decision is made, the individual cannot directly see, touch, hear, or smell the alternatives), it is his cognition of their spatial location which is the relevant spatial variable influencing choice. If these models are to have a maximum predictive and explanatory power, a deeper understanding of the individual's cognition of spatial location is required so that meaningful measures of spatial separation may be incorporated within them. In this paper we will: (1) Present a conceptualization of the factors influencing the conception of spatial location within an urban environment; (2) Undertake an empirical examination of some of these factors.

The Conception of Spatial Location

REVIEW OF THE LITERATURE

Studies relating to an individual's conception of spatial location exist in psychology, geography, regional science, sociology, planning, and urban design. These have been based upon both laboratory experiments and observed behavior in the real world, and have involved the testing of separate hypotheses which, in general, have not been related to any overall model of the cognition of spatial location. We will consider studies in psychology concerned with laboratory experiments in perception, studies of behavior within an urban environment, and finally, the existing models of an individual's cognition of spatial relations.

Psychophysical and perceptual studies have examined extensively the perception of distance. These studies have been primarily of visual perception where estimates of distances are obtained to or between objects either in, or immediately following removal from, the field of vision of the observer. Forgus lists five factors influencing these perceived distances (Forgus, 1966): (1) straight line physical distance; (2) assumed horizon; (3) distance cues; (4) attitudinal and organismic factors; (5) frame of reference. Although these studies are useful for generating hypotheses concerning the factors influencing cognitive distance, the determinants of cognition and perception cannot be assumed to be identical.

Of greater relevance are studies which emphasize and examine the dependence of perception upon learning, thinking, and other cognitive processes (Piaget and Inhelder, 1967; Wohlwill, 1960; Gibson, 1969). These works examine developmental changes in perception, considering such development largely completed at adulthood. Their concern is with visually perceived space rather than with the cognition of spatial location under conditions in which all objects cannot be perceived simultaneously. However, they do emphasize the critical influence of learning and experience upon perception, a circumstance we may hypothesize to exist for cognition.

Relevant studies of behavior in an urban environment can be classified into four groups. One group emphasizes the image derived from an entire environmental form such as the city. The classic work by Lynch (1960), although it does not formally expound a model of the cognition of spatial location, contains most of the significant ideas with particular emphasis on cognitive and cognized elements. De Jonge (1962), Gulick (1963), and Appleyard (1970), among others, have applied the ideas of Lynch to additional urban areas and different cultural contexts. Similar studies of New York City are reported by Milgram (1970). In addition, he examined the influence upon behavior of the size, density and heterogeneity of the population of large cities. Observed differences in behavior in large cities, as opposed to small towns or villages, was attributed to the adaptation of a system to conditions of overload brought about by the excessive number of perceptual inputs from a large city requiring processing.

A second group of studies emphasizes the delimitation by individuals of such concepts as their cognitive neighborhood and the cognized Central Business District (Fried and Gleicher, 1961; Lee, 1968; Zannaras, 1968; Saarinen, 1969; Sanoff, 1970; Ladd, 1970: Keller, 1968, reviews the neighborhood concept).

A third group of studies emphasizes movement patterns within the city and has its genesis in the studies initiated by Chapin of activity patterns within the urban area (Chapin, 1965; Chapin and Hightower, 1966; Chapin, 1968; Chapin and Brail, 1969). They have centered around the idea of an individual's action space (awareness space, behavior space, contact space): that area with which the individual has contact and within which his activities take place (Wolpert, 1965, p. 163; see also Webber, 1964a, 1964b). The individual's action space is seen as the primary determinant of his cognition of the urban environment. This cognition in turn influences future overt behavior (Horton and Reynolds, 1969; Brown and Moore, 1970). The significant aspect of the action space concept is that it emphasizes the spatial biases existing in an individual's cognition of, and behavior in, the urban environment. This bias is such that knowledge of the urban environment is concentrated around the individual's home location and around the paths he commonly uses in his everyday life. Thus, cognition of the urban environment conforms to the distance decay principle observed in relation to many phenomena: namely, that as distance from some nodal point increases the magnitude of the phenomena decreases (Zipf, 1949; Olsson, 1965). The bias also appears to be a function of the individual's socioeconomic status and the location of his residence relative to the city as a whole (Moore and Brown, 1970; Brown and Holmes, 1970; Horton and Reynolds, 1970). These biases serve to affect future behavior (Adams, 1969).

A final group of studies emphasized the cognition of distance between points in space. A study in Columbus, Ohio, by Golledge, Briggs and Demko (1969) suggested that learning, the general orientation of routes in the city, direction of the location in relation to downtown, and the real distance might be factors influencing the cognition of distance. The overestimation of distances to points situated toward downtown observed in this study was confirmed by Lee (1970) within a different environmental context (Cambridge, England) using much more adequate sampling and statistical analysis techniques. Both studies used a common set of points within the urban environment for all subjects and obtained direct estimates of distances in miles between pairs of points. Lowrey (see chap. 18) allowed each subject to choose his own points within a set of ten facility categories and obtained scale values for distances using a paired comparison technique based upon ratio estimation. By allowing subjects to specify their own points, greater equality in the subjects' familiarity with the points is obtained. While noting the marked variability between subjects, Lowrey concludes that, in spite of possible environmental influences such as urban topographic barriers, neighborhood effects, and travel time, the judgments were closely related to geographic (straight line, Euclidean) distances. However, he does suggest that car ownership and the facility category influence distance cognition. Thompson (1963) found that for equidistant discount and department stores cognitive distance was greater to the former than the latter. It was also greater for people who did not patronize a given store than for those who did. All cognized mileages were greater than their objective measure. His interpretation of these results was that negative attributes of locations lead to greater cognized spatial separation.

Studies at an intercity level have suggested the influence of other factors upon subjective distance. Stea (1969) indicated the relative desirability of the trip ends, the familiarity of the subjects with the end points, and the attractiveness of the interconnecting paths as influents. Tobler (1961, p. 112) hypothesized that the function relating linear distance to cognitive distance would be concave downward; however, his sample design was inadequate for a definite conclusion. Bratfisch (1969), following earlier work by Ekman and Bratfisch (1965), showed that subjectively perceived geographical distance (x) was very consistently related to emotional involvement (y) by the power function $y = b\,x^{-1/2}$, providing that cognized importance, interest, and knowledge are held constant (see chap. 17). While the major concern of these authors was not to relate subjective and geographical distance, they do note the poor fit and variability in the exponent of a power function relating these two variables. Work concerned with relating cognized and physical distance must account for the influence of subjectively cognized importance, interest, and knowledge.

All the above studies are primarily empirical. Only three general models of the cognition of spatial relations have been constructed. Hudson (1969) developed an hierarchical structure of spatial relations that is isomorphic with central place structures. Spatial relations are defined as a piece of locational information of the form "point *B* is a distance *p* in the direction θ from a point *A*." These distances are not necessarily Euclidean, but occur in a "transformed socioeconomic space useful to the individual." However, no suggestions are made for obtaining this transformation nor of factors influencing its formation. The model is nonoperational until these problems can be solved.

Stea (1969) suggested seven factors which influence an individual's cognition of distance and spatial relations: (1) the relative attractiveness of points viewed as origins and goals; (2) noncommutative barriers; (3) kind and number of barriers separating points; (4) familiarity with certain trips; (5) magnitude of the geographical distance; (6) attractiveness of the connecting path; and (7) familiarity with certain areas. Although this study identifies the factors likely to influence the cognition of spatial location and provides substantiating empirical evidence, it does not stress the cognitive developmental process. This process is being investigated by examining the development of the spatial cognitions of children (see chaps. 3 and 12). The model presented below, which draws heavily upon the ideas of Tolman (1932), deals with the learning of cognitive spatial relations.

A MODEL OF SPATIAL COGNITION

Initially, we must distinguish between *objective physical space* and *individual space*. The characteristics of objective physical space are defined on the two-dimensional Euclidean plane. "Rational man's" knowledge includes: (1) equal and total familiarity with the attributes of the infinity of points on this plane; (2) equal and total familiarity with the spatial relationships between every pair of points.

Individual space may be understood as follows: Assume a neonate with zero knowledge of physical space. His world consists of a "random set of lights, noises, touches, tastes and so on, without any connection or any known cause" (Vernon, 1962, p. 17). Through the process of perception as it modifies, and in turn is modified by, learning and thinking, the organism obtains information about physical space. This information generates both an image of a particular environment with which the individual has had direct experience and a set of generalized cognitive categories against which perceptual inputs from new environments can be compared and thus identified, organized, and given meaning. The physical space of a city is learned over time in three ways: (1) through behavior in which information is gained directly from the city structure through the visual, auditory, olfactory, and kinaesthetic sense modalities; (2) from symbolic representations of the city using visual media such as maps, photographs, and written words; and auditory media such as spoken words or recordings; (3) from ideas about parts of the city which have not been gained through behavior or symbolic representations, but are inferred from experiences in other spatial locations (see chaps. 4, 6, and 15).

Knowledge so learned is spatially discontinuous compared with the objective physical space of the Euclidean plane. This knowledge may be organized into a hierarchy in which knowledge of higher levels is partly dependent upon, and in turn is influenced by, knowledge of the lower levels. We recognize four levels and begin with the most fundamental.

1. Knowledge of points. Locations derived from the Euclidean plane become differentiated with regard to their attributes and meaning to the individual, meaning constituting "what behaviors are relevant to a location." Appleyard (1969b, 1970) has pointed to the several methods used by individuals to select and identify nodes, and has attempted to assess their relative importance. Kaplan discusses "meaning" (see chap. 4). We refer to locations about which there is greater than zero knowledge as *nodes*.

Nodes may be of three types: *End nodes* form the origin or destination of movements within the city and are the nodes at which the movement's purpose is realized; *Link nodes* are utilized to reach an end node and are cues for navigation through the environment; *Potential nodes* have functioned neither as link nor end nodes in previous movement. While all end nodes may function as link nodes in movements for other purposes, all link nodes are not end nodes. For example, many street intersections are exclusively link nodes.

The frequency of use of end nodes or link nodes will vary. In part this will determine the degree of knowledge about that node. Nodes used frequently will be designated as *primary nodes*. Nodes primary for only one or a few individuals (such as the homesite or the work place) are *personal nodes*, whereas *landmark nodes* (such as the CBD, outstanding buildings, etc.) are primary for the majority of the population.

2. Knowledge about the "closeness" of nodes. This is a one-dimensional relationship involving knowledge of the relative *proximity* of pairs of nodes. It may be cognized in terms of physical distance (i.e., miles, blocks, etc.), time, or monetary outlays. A *path* is a means of traversing space between two nodes. A *route* is a sequential set of node-path-node connections linking

two end nodes. Whether an individual's cognition of distance is in part a function of route availability and character, is testable (see chap. 5).

3. Knowledge of spatial positioning and relative location. This is a two-dimensional spatial relationship involving knowledge of the bearings or *directional* relations between nodes. In objective space distance and bearings are mutually dependent since the complete specification of either, for all pairs of points, predetermines the other. Cognitive distance and bearings are not necessarily mutually dependent although they may be inter-related. The two-dimensional Euclidean plane can serve as a true model of cognized space only if: (1) both cognitive and objective distances are linearly related, and cognitive bearings are identical to their objective equivalents; or (2) cognized space is Euclidean but with distance and directional relations between certain nodes being distorted to compensate for distortions of their relations between others; or (3) a function can be specified which transforms cognized space into a Euclidean representation (see Tobler, 1961). Failing this, additional dimensions or alternate metrics must be used.

4. Knowledge of sets of nodes and their interlinking paths. Such knowledge defines some region within the city on the basis of common attributes assigned to the nodes and/or paths constituting the region, together with their spatial contiguity. Paths may act as boundaries, or nodes may be defined as "in" or "not in" the region without the existence of any line boundary.

FACTORS AFFECTING THE COGNITION OF SPATIAL LOCATION

The above suggests the kinds of information and types of knowledge that people obtain about the urban environment. We now identify some of the factors influencing the cognition of spatial location, given this framework.

Geographic interest is centered upon the prediction of observable behavior in the environment. Vis à vis the physical environment, behavior has two aspects. First, it occurs within a physical environment, thus the node and path *opportunities* offered by the environment must influence behavior. Second, it requires the *selection* from this opportunity set of particular routes entailing movements from some initiating end node via a series of interconnecting paths through a set of link nodes to a terminal end node. The use of a particular route implies the preference of this route over all other routes available within the environment.

We may consider in more detail four assumed characteristics of behavior. First, it is "purposeful." The individual has some desire to interact with the terminal end node; it is thus a goal object for him. The attractiveness of a goal object is a direct result of its attributes. The particular attributes relevant for attractiveness have been discussed by several authors, the most comprehensive being that by Huff (1960) with particular reference to shopping behavior. They appear to vary with the purpose of behavior and the characteristics of the individual.

Second, behavior utilizes link nodes and interconnecting paths. Because the urban environment is not an isotropic surface, the following will vary between routes: the number and attributes of the link nodes; the ease with which path segments can be traversed (this will depend upon the type of highway, speed limit, and traffic congestion); the orientation and linearity of the total path.

Third, there is great variation in the frequency with which both complete routes and sections of routes are used. Certain paths and link nodes may be common for differing end nodes, both initial and terminal. Different paths and link nodes may be followed from common initial and/or terminal end nodes. Variations in frequency may result in variations in knowledge concerning the attributes of end nodes, the number and attributes of link nodes, and the ease with which paths can be traversed.

Fourth, behavior is systematically spatially biased. That is, for an individual there is not an equal probability of behavior occurring in all sections of the environment. There is a spatial bias for behavior close to his residence owing to least effort considerations.

Given the above conceptualization, we may hypothesize that the individual's cognition of the spatial location of nodes is a function of: (1) the behavior by which he obtained information concerning the environment; and (2) the characteristics of the environment.

Four characteristics were recognized concerning behavior and its relationship to the urban environment. Any or all of these may influence the cognition of spatial location. The following experiment examines the influence of two factors: the type of route; and the orientation of locations, relative to the city as a whole, on the cognition of distance.

The Experiment

SUBJECTS

The sample consisted of 248 students, all enrolled in introductory courses in geography during the Summer Quarter of 1970 at the Ohio State University. The experiment was conducted in class groups of approximately 45 students each. The use of students has advantages in addition to convenience. Most lived within the immediate neighborhood of the university. Thus, since similar activity patterns can be postulated, similarity should exist between the individual subjects' patterns of knowledge about the urban environment. This is particularly important given the existence of spatial biases in knowledge, and its dependence upon behavior.

STIMULI

Twenty points (locations) within the city of Columbus, Ohio, were chosen such that each could be assigned to one of four groups determined on the basis of two binary variables: away from downtown (groups 1 and 2) versus toward downtown (groups 3 and 4); along a major north-south artery (High Street) (groups 2 and 4) versus off this artery (groups 1 and 3). These relations are specified relative to a twenty-first point, the main entrance to the University, which lies on the north-south artery approximately 2½ miles north of downtown (figure 19.1). This served as a common origin and the experiment examined the cognition of distance from this point to each of the twenty locations.

The locations were chosen on the basis of three criteria: first, distances measured within any one of the four groups should be similar to those in the other groups; second, the locations should be familiar to the subjects; and third, all locations should be of the same node type. Major road intersections, which should all function as link nodes, were the only locations likely to meet the three criteria and be sufficient in number. However, the

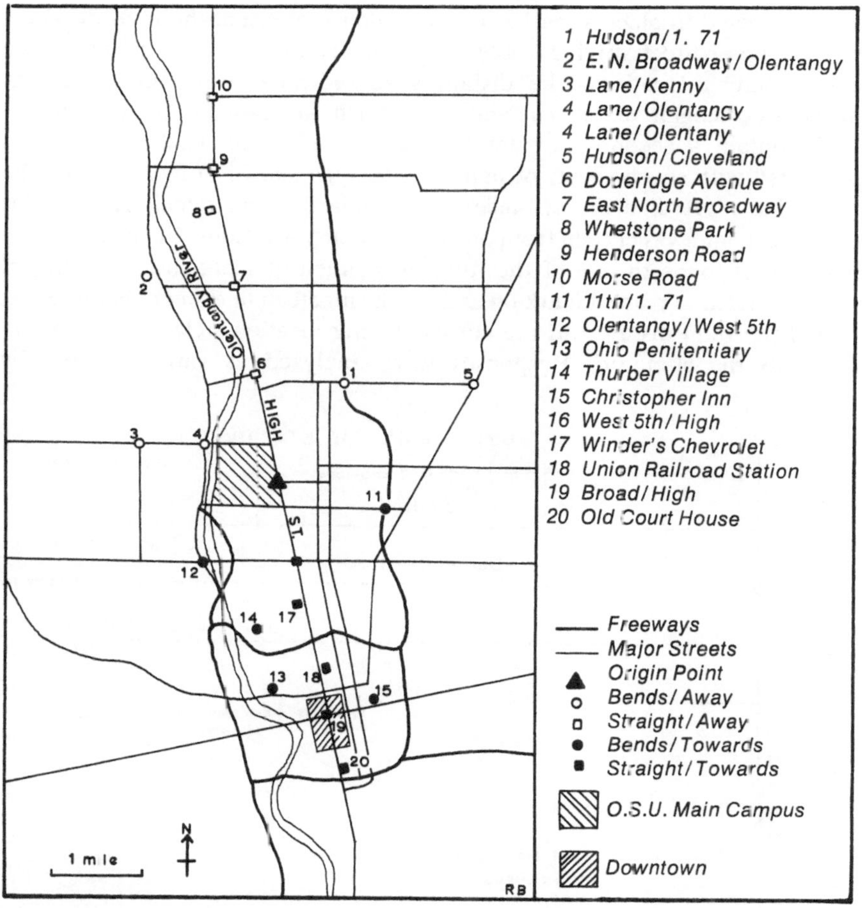

FIGURE 19.1. *The locations, in Columbus, Ohio, used in the study.*

street-naming system—which in part uses numbers—restricted the locations available since extensive use of numbered streets would provide crucial information on distance relations. Therefore, although the locations finally chosen were primarily street intersections (none of which were based on two numbered streets) a few landmarks were of necessity included (figure 19.1).

DATA

Measures of each subject's cognition of the distance from the common origin point (the main entrance to the university) to each of the twenty stimulus points (locations) were obtained in two ways. First every subject was given twenty line scales, one for each location. The scales were 50 centimeters in length and divided in one-mile increments up to seven miles (the longest objective distance being a little over four miles). The subject was instructed to place a mark on each scale to represent the road distance, by the most direct driving route, from the common origin point to the corresponding location. The distances so indicated were interpreted as measures of cognitive distance and, henceforth, are called *mile estimates.*

To obtain a second measure of cognitive distance, each subject was presented with seven sheets of paper of the type shown in figure 19.2. The short line represents the distance to a standard point, which differed for each sheet, and was chosen from one of the twenty locations. Each of the long lines corresponds to one of the remaining nineteen locations. The subject was instructed to place a mark on each of the nineteen lines such that it represented the *road* distance to the corresponding location, relative to the distance to the standard. Responses were analyzed as ratios formed by

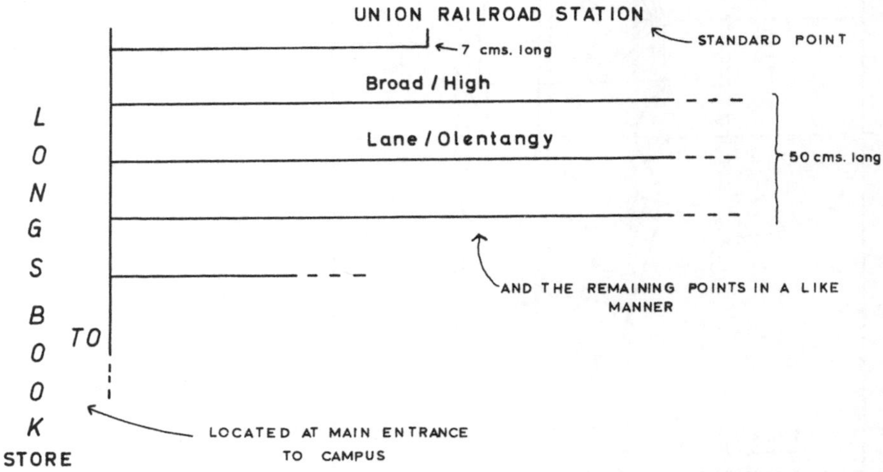

FIGURE 19.2. *The type of questionnaire sheet used to obtain ratio estimates.*

dividing the length of the standard line (7 cms. in all cases) by the length of the line marked by the subjects. These ratios are assumed to represent estimates of the ratio of the road distances from the common origin point to the corresponding two locations. Each location in turn acted as the standard point, and the remaining nineteen points were separately and randomly ordered, this order of *locations* being the same over all subjects for any given standard. This methodology follows that of Lowrey (see chap. 18).

To reduce the large number of estimates necessary from each subject (which would have been 380 if he were presented with all stimuli as standards), each subject used only seven locations as standards. Thus, each location was used as a standard by one-third of the sample, giving a minimum of eighty sample sheets for each of the twenty locations, with 133 estimates requested of each subject. Since subjects were instructed to avoid making distance estimates between locations with which they were unfamiliar, the actual number of ratio estimates available for analysis from a subject was somewhat less than 133. The order in which individual sheets were presented to subjects was varied to control for learning and order effects.

Additional data collected from each subject included the familiarity of the subject with each location (indicated on a four-point scale), and demographic information such as age, sex, marital status, grade-point average, year in school, possession of a car, and places of present and past residence.

DATA ANALYSIS

The ratio estimates were analyzed in the following way to obtain a measure of the cognitive distance from the common origin point to each of the twenty locations. There were a maximum of nineteen ratios estimated relative to each of seven locations for a given subject, and each of twenty locations for the sample as a whole. For each of these 380 ratios (19 x 20), the arithmetic mean over the subjects who estimated that ratio was calculated, the number of observations for a given mean varying from 32 to 74. For each of the twenty locations, a least square estimate of its cognitive distance from the common origin point is given by the geometric mean (or arithmetic mean of the log values) of its associated nineteen mean ratio estimates (Torgerson, 1958, pp. 104–112; Lowrey, chap. 18). These geometric means will be referred to as scale values to differentiate these measures of cognitive distance from the mile estimate measures.

By averaging over subjects, individual differences are ignored. This is legitimate since interest is directed not toward subject differences, but toward differences between groups of locations. In a future study, differences between groups of subjects will be examined by forming ratio averages over some subset of the sample (e.g., car owners, long time residents of the city, etc.).

Objective distance was measured from street plans using driving routes.

For the locations lying off High Street (fig. 19.1) several routes were possible, but because of the grid pattern street layout, distances were similar to within a tenth of a mile. In two cases a slightly more direct route was available than that through the general grid pattern of streets, and the average of this route and the grid pattern route was used. This was legitimate since ratio estimates had been averaged over subjects who were likely to use either route. To allow direct comparison between these objective measures and the scale value measures of cognitive distance, the same transformation applied to the subjective data was applied to these objective measures; that is, ratios were formed and the geometric mean calculated.

Two of the simplest functions relating objective distance to cognitive distance were investigated: (1) a linear function,

$$Y = a + bX,$$

where $Y =$ cognitive distance, $X =$ objective distance, and a, $b =$ constants; and (2) a power function,

$$Y = aX^b$$

Coefficients were estimated by linear regression, using a natural logarithmic transformation in the case of the power function. Standard tests were used to examine the goodness of fit of the data to the linear and power functions. The testing of two research hypotheses was based upon differences between regression lines calculated for selected subsets of the locations. The coincidence of two regression lines formed upon independent samples is tested by comparing the sum of the residual sum of squares for each line to the sum of squared deviations from a regression line formed from a pooling of the two samples (Smillie, 1966, pp. 72–75). A test for differences in the slope, or b coefficients, of two regression lines is given by Hald (1952, pp. 571–579).

RESEARCH HYPOTHESES

Two hypotheses were tested: (1) that for equivalent objective distances, cognized distance is greater for locations toward the downtown than for locations away from downtown; and (2) that for equivalent objective distances, cognized distance is greater for routes involving turns than for straight routes.

Hypothesis 1 is derived from the model on the assumption that: first, the existence of a larger number of link nodes connecting two end nodes increases cognized distance; and second, that cognized distance is directly related to the time required to traverse a path. Downtown, the density of population and buildings increases (Berry, Simmons and Tennant, 1963); the buildings become more varied as regards size and facade as commercial, retailing, and political functions replace residential; and the number of noticeable road intersections increases as traffic lights and stop signs become more frequent. An equally important factor is the increase in time required to traverse the paths toward downtown as a result of greater traffic volume and the increased number of major intersections. This basis for Hypothesis 1 differs from that given by Lee (1970) for a similar hypothesis. His argument was based upon the spatial elongation, or bias, toward downtown of an individual's sociospatial schema. Hypothesis 2 has a similar derivation in that a turn involves identifying the important link node of the intersection and executing the decision to turn, thus increasing cognized distance. There is also laboratory evidence that nonlinear lines are *perceived* longer than straight lines (Lee, 1970), although a similar relation cannot be assumed to hold for cognition.

RESULTS

Goodness-of-fit of Regression Lines. Figures 19.3 through 19.8 are plots of the objective distance (*X* axis) against the two measures of cognitive distance (*Y* axis), scale values (figs. 19.3, 19.5, 19.7) and mile estimates (figs. 19.4, 19.6, 19.8). The least squares regression lines are plotted for

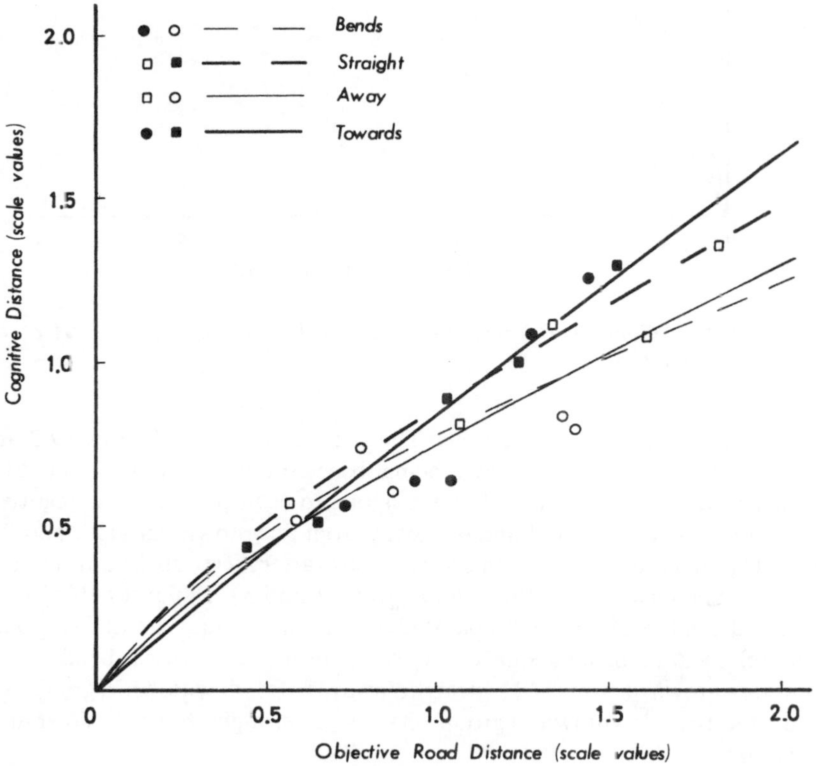

FIGURE 19.3. *"Combined" regression lines (each based on ten points) of the power fit for scale values.*

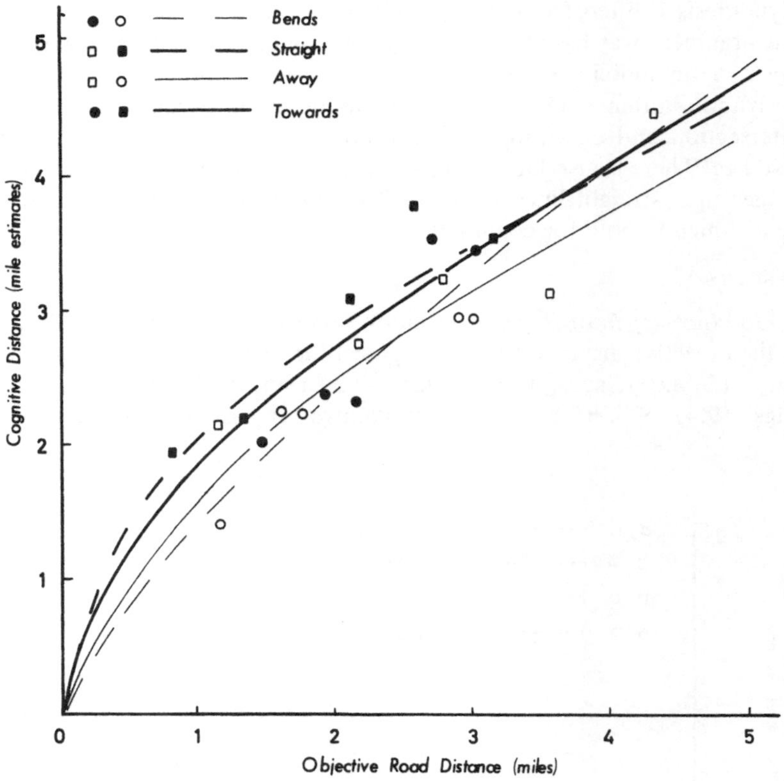

FIGURE 19.4. *"Combined" regression lines (each based on ten points) of the power fit for mile estimates.*

both the linear and power functions. The regression lines in figures 19.3 and 19.4 are "combined" lines each being based upon ten points, the total number of locations which could be assigned to each of the four categories: toward downtown (group 3 and 4); away from downtown (group 1 and 2); straight, that is locations on the north-south artery (group 2 and 4); and bends, that is, locations off this artery (group 1 and 3). In figures 19.5, 19.6, 19.7, and 19.8, each line is "separate," based on a unique set of five points with each location being assigned to only one of four categories: bends/away from downtown (group 1); straight/away from downtown (group 2); bends/toward downtown (group 3); and straight/toward downtown (group 4).

Both the linear and power functions described the data well. Error about

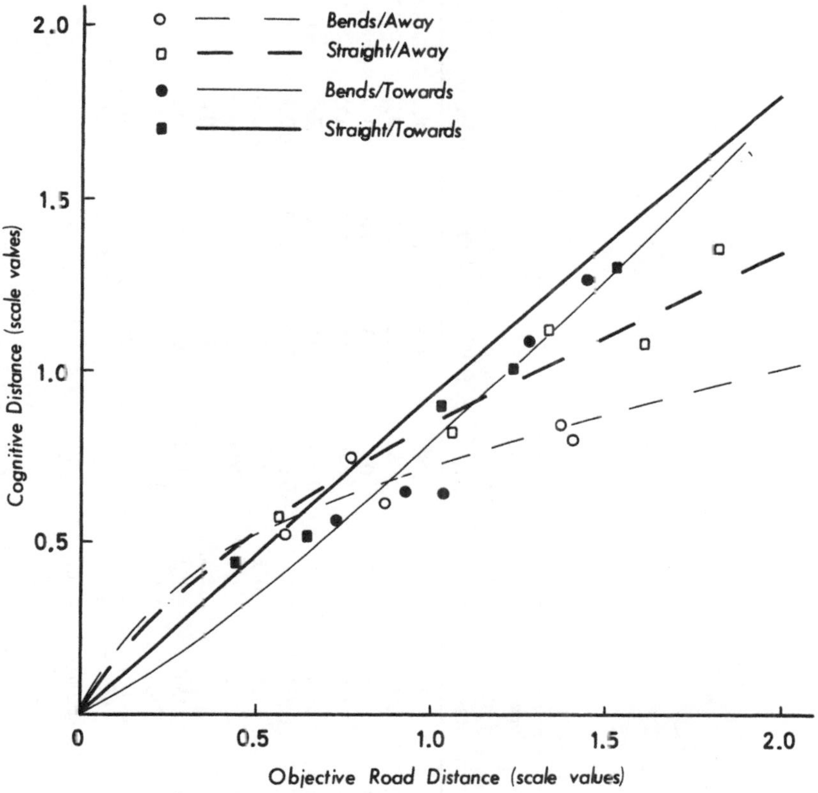

FIGURE 19.5. *"Single" regression lines (each based on five points) of the power fit for scale values.*

the separate regression lines varied from a maximum of 27% to a minimum of 2% of total variance (table 19.1). Tests on the *b* coefficients indicated that a highly significant regression existed in all cases (table 19.1). However, the relatively large deviations of some points from their regression lines suggests that factors specific to these points may influence cognitive distance. If this is the case, a correlation should exist between errors about the regression lines for the mile estimate and scale value measures of cognitive distance since the factors should operate irrespective of the measurement method. A significant relationship did exist for a linear fit, $r = 0.61$, $t = 3.27$, $d.f. = 18$, $p < .01$, and for a power fit, $r = 0.67$, $t = 3.83$, $d.f. = 18$, $p < .01$.

Factors which were not completely controlled include the type of node

TABLE 19.1. *Summary of regressions for objective distance against cognitive distance*

Scale Values (*No. of Locations*)		*All Locations* (20)	*Straight Away* (5)	*Bends Away* (5)	*Bends Towards* (5)	*Straight Towards* (5)	*Straight* (10)	*Bends* (10)	*Away* (10)	*Towards* (10)
Coefficient of	*Linear*	.776	.918	.731	.884	.982	.869	.622	.833	.897
determination (r^2)	*Power*	.817	.947	.753	.846	.962	.922	.672	.837	.871
Standard error of	*Linear*	.140	.090	.075	.110	.054	.123	.149	.111	.107
estimate	*Power*	.150	.081	.107	.143	.095	.114	.164	.127	.145
Regression intercept	*Linear*	.164	.290	.390	−0.290	−0.011	.215	.153	.249	−0.078
	Power	−0.211	−0.167	−0.325	−0.254	−0.090	−0.152	−0.270	−0.238	−0.168
Regression	*Linear*	.645	.543	.325	1.072	.929	.638	.614	.529	.933
coefficient (*b*)	*Power*	.786	.678	.473	1.219	.976	.797	.736	.659	.974
Standard error of	*Linear*	.082	.093	.114	.224	.074	.088	.169	.084	.112
reg. coeff.	*Power*	.088	.093	.156	.302	.108	.082	.182	.103	.133
t *value*	*Linear*	7.89	5.85	2.86	4.78	12.63	7.26	3.63	6.31	8.32
(*significance level*)		($p < .01$)	($p < .01$)	($p < .05$)	($p < .01$)	($p < .01$)	($p < .01$)	($p < .01$)	($p < .01$)	($p < .01$)
	Power	8.95	7.26	3.03	4.03	9.03	9.74	4.05	6.42	7.34
		($p < .01$)	($p < .01$)	($p < .05$)	($p < .02$)	($p < .01$)	($p < .01$)	($p < .01$)	($p < .01$)	($p < .01$)

Mile Estimates (No. of Locations)		*All Locations* (20)	*Straight Away* (5)	*Bends Away* (5)	*Bends Towards* (5)	*Straight Towards* (5)	*Straight* (10)	*Bends* (10)	*Away* (10)	*Towards* (10)
Coefficient of determination	*Linear*	.738	.867	.878	.865	.889	.833	.810	.863	.787
	Power	.760	.868	.876	.856	.927	.878	.825	.824	.757
Standard error of estimate	*Linear*	.348	.320	.222	.268	.272	.322	.288	.304	.331
	Power	.135	.099	.107	.099	.080	.093	.106	.129	.127
Regression intercept	*Linear*	1.098	1.337	.881	.306	1.252	1.451	.669	.975	1.012
	Power	.563	.672	.363	.323	.709	.708	.351	.480	.625
Regression coefficient (b)	*Linear*	.738	.651	.703	1.077	.821	.657	.863	.728	.848
	Power	.570	.481	.682	.839	.539	.481	.756	.612	.554
Standard error of reg. coeff.	*Linear*	.091	.148	.151	.246	.167	.103	.147	.102	.156
	Power	.075	.108	.148	.199	.087	.064	.122	.100	.111
t *value (significance level)*	*Linear*	8.07 ($p < .01$)	4.41 ($p < .02$)	4.64 ($p < .01$)	4.37 ($p < .02$)	4.92 ($p < .01$)	6.34 ($p < .01$)	5.85 ($p < .01$)	7.12 ($p < .01$)	5.42 ($p < .01$)
	Power	7.57 ($p < .01$)	4.46 ($p < .02$)	4.61 ($p < .01$)	4.22 ($p < .02$)	6.23 ($p < .01$)	7.58 ($p < .01$)	6.77 ($p < .01$)	6.14 ($p < .01$)	4.99 ($p < .01$)

$$t = \frac{b}{\sqrt{\dfrac{s^2}{\Sigma (x - \bar{x})^2}}}$$

where: s = standard error of estimate,
x = objective distance.

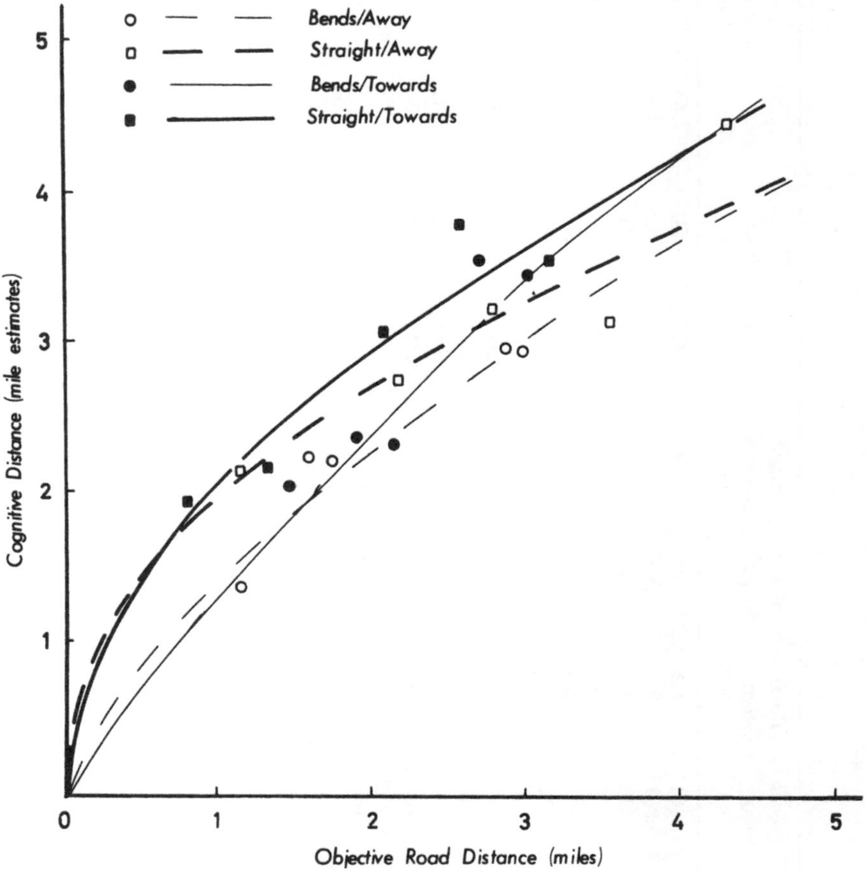

FIGURE 19.6. *"Single" regression lines (each based on five points) of the power fit for mile estimates.*

each location represented to the individual (link node, end node, or landmark node), and the level of familiarity with the location—a function not only of the node type, but of the subject's behavior or activity pattern and the imageability of the location. Although the locations which could function as landmark or end nodes did not appear to be particularly deviant from their regression lines, familiarity may have been influential.

Subjects were instructed to avoid making distance estimates to unfamiliar points: Thus the number of mile estimates made for a given point provides a measure of its familiarity. A similar measure for scale values is not possible since that would have involved estimating ratios of distances to two loca-

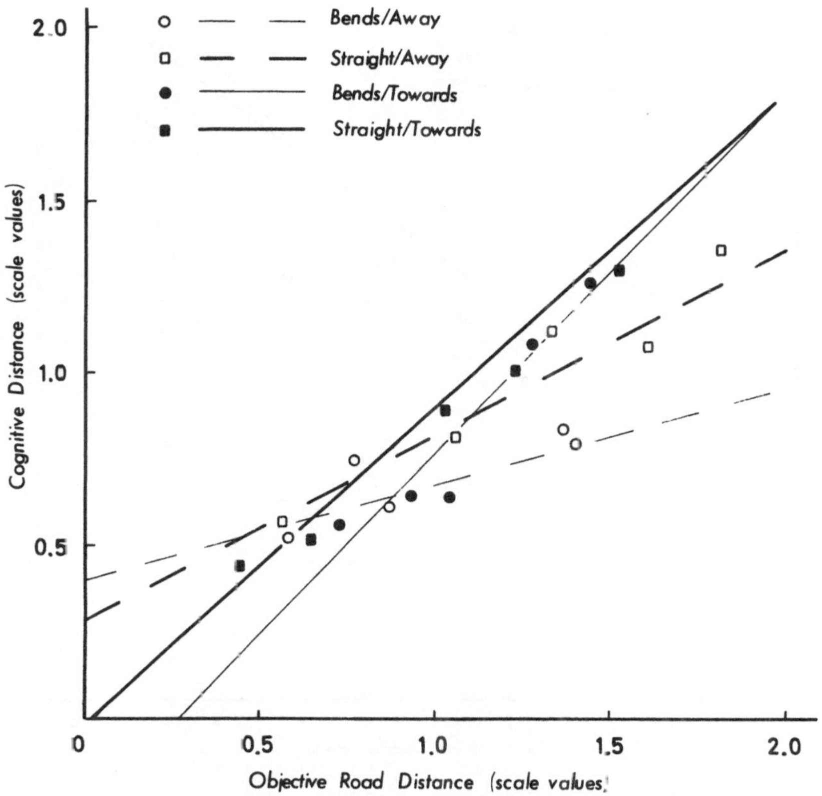

FIGURE 19.7. *"Single" regression lines (each based on five points) of the linear fit for scale values.*

tions, either of which could have been unfamiliar if the estimate was omitted. A very weak correlation did exist between the measure of familiarity and errors about the separate regression lines for mile estimates: for a linear fit, $r = 0.37$, $t = 1.68$, $d.f. = 18$, $p < .10$, and for a power fit, $r = 0.32$, $t = 1.43$, $d.f. = 18$, $p < .10$. This relationship was positive, indicating that less familiar locations are cognized closer than those more familiar. This result would also follow from two independent influences: the existence of a distance decay phenomenon in urban cognition and a tendency for cognized distance to increase at a decreasing rate with objective distance. However, for the locations used in this study, no correlation existed between objective distance from the common origin point and familiarity; thus familiarity itself may have an effect upon cognitive distance. Another influence may be the attributes of the paths to the nodes.

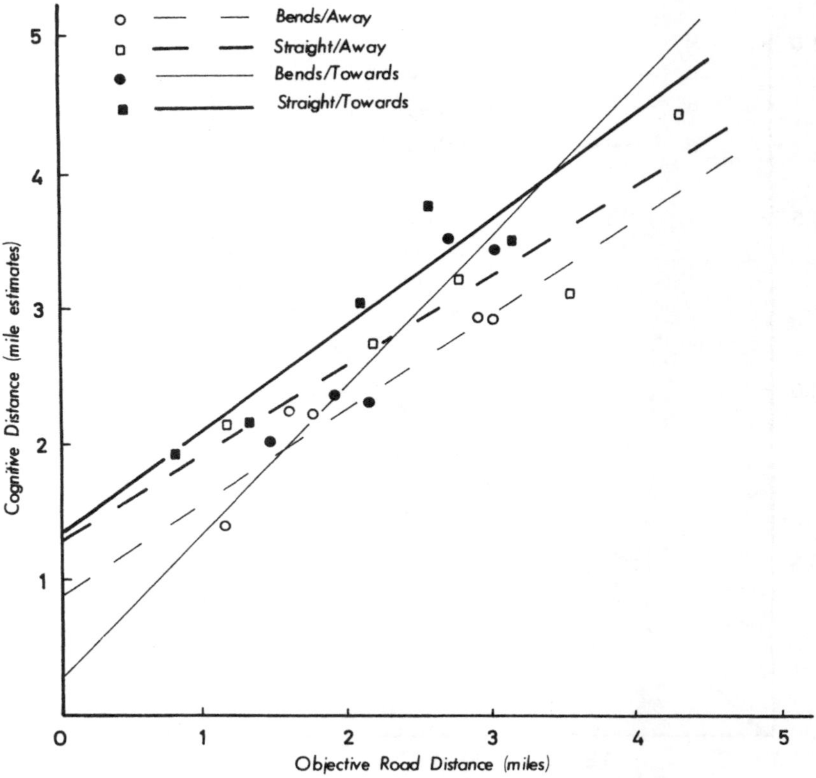

FIGURE 19.8. *"Single" regression lines (each based on five points) of the linear fit for mile estimates.*

Better fits to regression lines were obtained for the straight categories (groups 2 and 4) where all points lie along a common path, than for the bend categories (groups 1 and 3).

The efficacy of the power versus the linear fit to describe the data was tested (table 19.2). In *no* instance did one function give a statistically better fit at even a 10% significance level. This result may be attributable to the small number of points available for the fit within any one category.

Hypotheses Tests. Since regression analysis simultaneously determines values for the two parameters which provide a best fit for the given data, attention should be directed to the relative heights of the regression lines rather than to the relative values of individual parameters, and to their heights only within the range of points actually measured. The relative heights of the regression lines are generally very similar for both scale

TABLE 19.2: *Significance tests upon power versus linear fit*

		Scale Values			*Mile Estimates*		
	Degrees of Freedom	*Residual Sum of Squares* *Linear*	*Power*	*F Statistic*[a]	*Residual Sum of Squares* *Linear*	*Power*	*F Statistic*
All locations	1, 18	.351	.368	.868	2.179	2.360	1.493
Straight/Away	1, 3	.024	.028	.514	.307	.500	1.880
Bends/Away	1, 3	.017	.020	.499	.147	.167	.412
Bends/Towards	1, 3	.036	.046	.800	.216	.283	.929
Straight/Towards	1, 3	.009	.012	1.050	.222	.238	.222
Straight	1, 8	.120	.122	.113	.829	.899	.671
Bends	1, 8	.177	.203	1.176	.666	.721	.666
Away	1, 8	.099	.119	1.595	.737	.909	1.862
Towards	1, 8	.092	.117	2.158	.877	1.086	1.905

[a] Snedecor, 1956, pp. 454–455.
$F = 4.41$, $d.f. = 1, 18$ ($p < .05$); $F = 10.13$, $d.f. = 1, 3$ ($p < .05$); $F = 5.32$, $d.f. = 1, 8$ ($p < .05$).

values (figs. 19.3, 19.5, 19.7) and for mile estimates (figs. 19.4, 19.6, 19.8), thus lending reliability to the experiment since alternative methods of measuring cognitive distance yield similar results (see Nunnally, 1967, pp. 206–235 for discussion of reliability).

The implications for Hypothesis 1 of a visual inspection of the graphs, together with statistical tests on the coincidence of the regression lines, are summarized in table 19.3. More detailed information on the test statistic is given in table 19.4. To uphold Hypothesis 1, the regression line for locations toward downtown should lie above that for locations away from downtown (fig. 19.9). This is the case except for extreme low values in the data range. Results are least conclusive for the comparison of routes involving bends. For scale values all regression lines were statistically noncoincidental but for mile estimates this could not be demonstrated. This difference can be attributed to the general lower variability of the points about their respective regression lines for scale values, a consequence of each of these measures being based upon the mean of 19 ratio estimates rather than upon one estimate as is the case with mile estimate measures.

Hypothesis 2 must be rejected. Without exception the regression lines for straight routes lie above those for routes with bends, this difference being statistically significant in half of the cases (table 19.4). This result may be attributable to the grid pattern street network which causes *road* distances to points lying off the main north/south artery to be considerably longer than the airline distance. Cognitive road distance is apparently influenced by the airline distance. This interpretation is upheld by the fact that, with one exception (scales values, away from downtown), the re-

TABLE 19.3. *Summary of visual inspection of graphs and statistical tests for hypothesis 1*

Mile Estimates	*Estimating Equation*	*Figure #*	*Visual Inspection of Graph*	*Significant difference in coincidence of regression lines? (level of significance)*
Combined	Power	4	Hypothesis upheld over whole range of data	No
Straight Held Constant	Linear	8	,,	No
	Power	6	,,	No
Bends Held Constant	Linear	8	Hypothesis upheld only in upper half of data values	No
	Power	6	Hypothesis upheld except for extreme lower values of data	No
Scale Values				
Combined	Power	3	Hypothesis upheld except for extreme low values of data	Yes ($p < .10$)
Straight Held Constant	Linear	7	,,	Yes ($p < .025$)
	Power	5	,,	Yes ($p < .10$)
Bends Held Constant	Linear	7	Hypothesis upheld only in upper values of data range	Yes ($p < .25$)
	Power	5	,,	Yes ($p < .10$)

gression lines for straight as against bends converge. Convergence occurs because, given the locations used in the study, the relative difference between airline and road distance decreases as distance from the common origin increases.

Coefficients. Additional information can be gained by examining the values, estimated by regression, for the a and b coefficients of the two equations (linear and power) used to describe the data. Attention will be directed toward the a and b coefficients of the power equation, $Y = aX^b$ (where Y is cognitive distance and X objective distance), rather than those of the linear equation, $Y = a + bX$. Since it appears legitimate to assume that zero objective distance will be cognized as zero, it is difficult to give a valid interpretation to the constant term (a coefficient) in the linear

TABLE 19.4. *Significance tests on coincidence and slopes of regression lines.*

Scale Values	*Linear Fit*		*Power Fit*	
Straight Away	b = .543	a = .290	b = .678	a = .846
vs.				
Straight Towards	b = .929	a = −.011	b = .976	a = .914
	$F = 7.909$	$t = 3.413$	$F = 3.639$	$t = 2.337$
	$(p < .025)$	$(p < .01)$	$(p < .10)$	$(p < .05)$
Bends Away	b = .325	a = .389	b = .473	a = .723
vs.				
Bends Towards	b = 1.07	a = −.289	b = 1.219	a = .776
	$F = 7.019$	$t = 3.488$	$F = 3.718$	$t = 2.577$
	$(p < .05)$	$(p < .01)$	$(p < .10)$	$(p < .02)$
Away Straight	b = .543	a = .290	b = .678	a = .846
vs.				
Away Bends	b = .325	a = .389	b = .473	a = .722
	$F = 4.244$	$t = 1.605$	$F = 4.337$	$t = 1.305$
	$(p < .10)$	$(p < .10)$	$(p < .10)$	(NS)
Towards Straight	b = .929	a = −.011	b = .9763	a = .914
vs.				
Towards Bends	b = 1.071	a = −.290	b = 1.218	a = .776
	$F = 3.134$	$t = .750$	$F = 2.727$	$t = .910$
	(NS)	(NS)	(NS)	(NS)
Away	b = .529	a = .249	b = .659	a = .788
vs.				
Towards	b = .933	a = −.078	b = .974	a = .845
	$F = 6.702$	$t = 3.035$	$F = 2.909$	$t = 1.996$
	$(p < .01)$	$(p < .005)$	$(p < .10)$	$(p < .05)$
Bends	b = .614	a = .153	b = .736	a = .763
vs.				
Straight	b = .638	a = .215	b = .797	a = .859
	$F = 1.455$	$t = .142$	$F = 2.157$	$t = .349$
	(NS)	(NS)	(NS)	(NS)

(Continued on p. 384)

TABLE 19.4 (*Continued*)

Mile Estimates	*Linear Fit*		*Power Fit*	
Straight Away	b = .651	a = 1.337	b = .481	a = 1.958
vs.				
Straight Towards	b = .821	a = 1.252	b = .539	a = 2.032
	F = 1.702	t = .833	F = 1.313	t = .262
	(NS)	(NS)	(NS)	(NS)
Bends Away	b = .702	a = .881	b = .682	a = 1.438
vs.				
Bends Towards	b = 1.076	a = .306	b = .839	a = 1.381
	F = 2.504	t = 1.484	F = 1.238	t = .697
	(NS)	($p < .10$)	(NS)	(NS)
Away Straight	b = .651	a = 1.337	b = .481	a = 1.958
vs.				
Away Bends	b = .703	a = .881	b = .682	a = 1.438
	F = 1.870	t = .254	F = 3.381	t = 1.240
	(NS)	(NS)	(NS)	(NS)
Towards Straight	b = .821	a = 1.252	b = .539	a = 2.032
vs.				
Towards Bends	b = 1.076	a = .306	b = .839	a = 1.381
	F = 3.008	t = .957	F = 5.063	t = 1.629
	(NS)	(NS)	($p < .10$)	($p < .10$)
Away	b = .728	a = .975	b = .612	a = 1.616
vs.				
Towards	b = .848	a = 1.011	b = .554	a = 1.868
	F = 2.801	t = .687	F = 2.008	t = .411
	($p < .10$)	(NS)	(NS)	(NS)
Bends	b = .863	a = .669	b = .756	a = 1.421
vs.				
Straight	b = .657	a = 1.451	b = .481	a = 2.030
	F = 3.660	t = 1.134	F = 8.658	t = 2.059
	($p < .05$)	(NS)	($p < .01$)	($p < .05$)

The numbers in parentheses following each test statistic refer to its level of significance (NS: only significant at some level less than $p = .10$).

Test on coincidence of regression lines (Smillie, 1966, pp. 72–75):

$$F = \frac{(E_0 - E_1 - E_2)/(p + 1)}{(E_1 + E_2)/(n_1 + n_2 - 2p - 2)}$$

where: E_i = residual sum of squares for locations in category i (i = 1, 2),
E_0 = residual sum of squares for the two categories pooled,
n_i = number of observations in category i,
p = number of independent variables.

Test on slopes (Hald, 1952, pp. 571–575):

$$t = \frac{b_1 - b_2}{S\sqrt{\dfrac{1}{\Sigma\,(X_1 - \bar{X}_1)^2} + \dfrac{1}{\Sigma\,(X_2 - \bar{X}_2)^2}}}$$

where: $S = [(n_2 - 2)S_1{}^2 + (n_2 - 2)S_2{}^2]/(n_1 + n_2 - 4)$,
S_i = standard error of estimate for category i (i = 1, 2),
X_i = values of independent variable for category i.

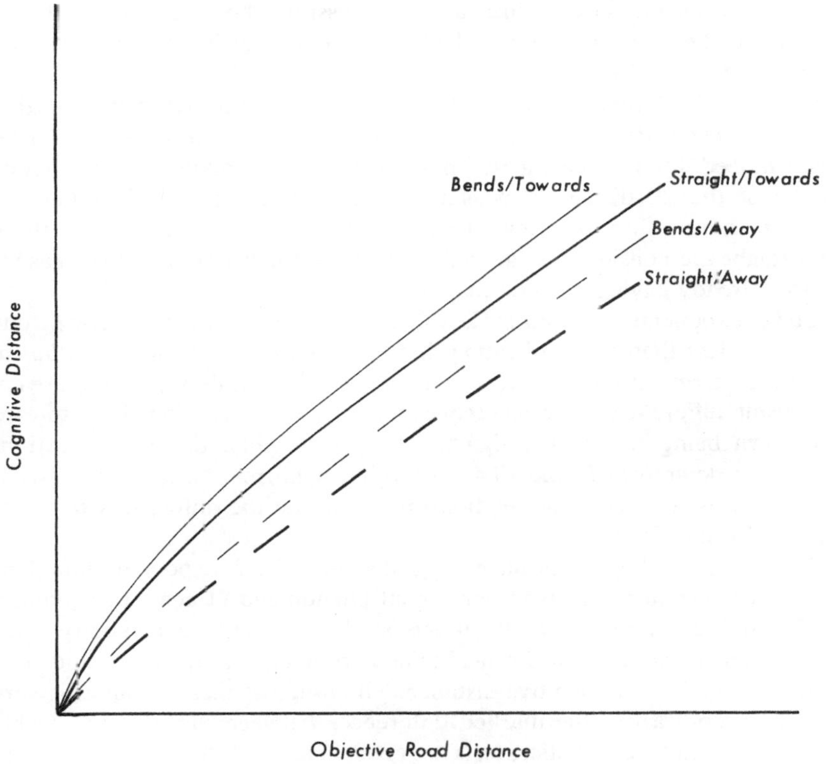

FIGURE 19.9. *The relative heights of the "single" regression lines as hypothesized in the two Research Hypotheses.*

equation, unless it approximates zero. This was not generally the case (table 19.1), thus it may be concluded that a linear function is inappropriate for relating objective and cognized distance.

For the power fit, the *a* coefficient is a scale factor which, for mile estimates, relates an objectively measured mile unit to an individual's cognition of a mile unit. The *b* coefficient reflects the rate of change in a subject's cognition of a mile unit as distance increases. For the scale values, the *a* coefficient measures the rate of change of cognized with objective distance and indicates a general tendency either to underestimate ($a < 1$) or overestimate distance ($a > 1$). The *b* value indicates any change in the rate of change of cognized with objective distance.

For all the scale value power fits the *a* coefficient is less than one, thus indicating a general tendency for *cognized distance to be less than objective distance* (table 19.2). Differences between these coefficients support the hypothesis that cognized distance toward downtown is greater than away from downtown since the *a* values are greater for the former than the latter. Statistical tests on these values are not possible, however, unless the *b* values are the same (Williams, 1959, p. 133), which was not the case (table 19.4).

For the mile estimates, all *a* values are greater than one indicating that *a mile unit is cognized to be considerably greater than its objective measurement.* Indeed, the mean estimated mile distance was greater than the actual for 19 of the 20 distances measured. Although there is little difference between the *a* values for toward as against away from downtown, those for straight are considerably greater than those for bends, possible reasons for this already having been discussed.

The *b* exponents for both scale values and mile estimates are, with one exception, less than one, indicating that *cognized distance increases, but at a decreasing rate, relative to objective distance.* For scale values there were significant differences between the exponents, those for locations toward downtown being statistically significantly greater than those for locations away from downtown (table 19.4). This gives confirmation to Hypothesis 1. For mile estimates the same relationships held, but the differences were not statistically significant.

Studies in psychophysics have suggested that the *b* exponent should be relatively constant for a given perceptual phenomena (Luce and Galanter, 1963). If the exponents found in this study have any real validity, then similar values should have been obtained from both scale value and mile estimate measures of cognitive distance. The range of these *b* values is very similar, in contrast to the marked differences between the *a* values (table 19.4). The results of statistical tests are given in table 19.5. Highly significant differences were observed in only two cases. The values also compare well with those found in other studies of cognized distance. Bratfisch (1969; see

TABLE 19.5: *Tests for significant differences between exponents estimated from scale values and from miles estimates*

	Degrees of Freedom	t[a]	*Significance Level*	*Exponent for Scale Values*	*Exponent for Mile Estimates*
Straight Away	6	1.538	NS	.6777	.4805
Bends Away	6	1.091	NS	.4726	.6821
Bends Towards	6	1.185	NS	1.2185	.8388
Straight Towards	6	3.529	$p < .02$	.9763	.5392
Straight	16	3.238	$p < .01$	.7973	.4811
Bends	16	0.095	NS	.7356	.7559
Away	16	0.349	NS	.6589	.6118
Towards	16	2.570	$p < .05$	.9743	.5537
All Locations	36	1.917	$p < .10$	.7856	.5702

[a] See Table 19.4 for equation.

also chapter 17) found values ranging from 0.58 to 1.08, a similar range to those observed in this study, despite the much larger distances studied. The values observed by Lowrey were much lower (0.06 to 0.40) but these were obtained by individually scaling each subject rather than the averages used both here and by Bratfisch.

Conclusion

The hypothesis was made that cognitive distance is a function of both city structure and the behavior by which this structure is learned. The empirical research, although not conclusive, suggested a relationship between certain elements of city structure and cognitive distance. The behavioral basis for this relationship, developed in the model of spatial cognition, was demonstrated by implication only. The question as to whether cognitive distance is influenced by differences in behaviors resulting from the physical structure of the city, or differences in behaviors by which this structure is learned, remains unanswered.

Future studies should involve the direct measurement or a rigorous control of all variables thought to influence the cognition of distance. Unfortunately, in the laboratory of the real world, this is exceedingly difficult to achieve. There are three primary requisites. First, locations chosen for distance estimates all should be as similar as possible, and certainly of the same node type. A larger number of locations than that used in this study would also be desirable. Second, the actual routes, as well as the number and characteristics of link nodes utilized to reach these locations, should be ascertained. This was not attempted in the present study. Third, the activity patterns of subjects must be examined in order to isolate spatial biases in the behaviors by which city structure is learnt. Any grouping of subjects should be on the basis of similar activity patterns in order to insure homogeneity in their knowledge of the elements of the urban environment. This study *assumed* a similarity of activity patterns among students without investigating its basis.

The research hypotheses tested could not be accepted with certainty since the levels of significance achieved in the statistical tests were far from satisfactory. However, this is in part a result of the method of data analysis. Rather than using the means of the ratio estimates as suggested by Torgerson (1958), scaling could be undertaken separately for each individual. Provided that the individual scale values for any one location were approximately normally distributed and homoscedastic, the coefficients of the estimating equations would be very similar to those found by the present method. More degrees of freedom would be available and, if variability did not increase too markedly, results at a higher significance level might be obtained.

Finally, we might ask: For what purpose is cognitive distance being examined? If our purpose is to understand the phenomenon per se, then it is critical to isolate independently the factors affecting it. Examining distances toward versus away from downtown is insufficient since the critical influents may be travel time, building intensity and variability, bias in the number of trips towards downtown, or any combination of these. Designing experiments and finding real world situations where these variables can be separately examined is an exciting and demanding challenge which may prove impossible to meet directly. On the other hand, if our purpose is to find more realistic measures of spatial separation to use in predictive models of, for example, consumer movement, then our task is simpler. There is no need to determine the separate influence of factors which always covary spatially. As a geographer, my purpose is the latter, but my measures of cognitive distance would rest upon a firmer foundation if the former could be achieved.

Epilogue

We are fully cognizant of the criticisms made of progress in the field named "behavioral geography," especially with regard to the question of "relevance." We are justifiably called upon to flesh out the skeleton, so to speak: to prove, or at least to demonstrate, that the Emperor does in fact have some clothes. We would like to *feel* that he does, and have searched many of the academic kingdom's wardrobes. While not entirely bare, they have yielded an odd assortment of ill-fitting, time or shop-worn clothing unfitting for a royal parade, but perhaps suitable for a less pretentious occasion—clearing up after the parade is over, for example.

The question we, the editors, must ask ourselves is: Does the world seem to be a better place for this book having appeared? If this means "what does the world need right now," we might have plunged into problem-solving up to our necks, ignoring issues of theory formation, generalization, and learning. But would this have enabled us to better solve the next problem for having imperfectly solved the current one? On the other hand, if we were not at all swayed by the mood of the moment, we might have believed that "things will turn out all right in the end" *because* we have provided a world ignorant of questions with a set of definitive answers.

At this point in time, neither of the paths caricatured seems feasible or desirable. We are *not* trying to attack the problems of the moment or the targets of opportunity or the issues of convenience, nor to provide timeless verities, but rather to indicate approaches to an area. These approaches are perhaps better defined than the area as a whole. It is entirely possible that they are inappropriate but they may be of heuristic use in the near future at least.

We have been, hopefully with justification, somewhat less concerned with problems of application than with trying to give coherence to some heretofore diverse ideas, concepts, and findings. We felt that by doing so we might be able to point the way to some future directions of research and toward some possible applications of findings that might emerge.

Some of the contributors to this volume have already suggested such applications. Implementation must await both improvement in the predictive capacity of the research models used, and the elaboration of more coherent theory. But now it is time to go to press. New ideas, new concepts, new findings are already emerging, too late for inclusion in this volume. For those of our readers through whose veins nostalgia courses, we can only quote the closing line of every old radio serial: "Don't miss the next episode."

Bibliography

Aangeenbrug, R. T. (1968) "Regional perception and its effects on industrial location," *Kansas Business Review*, 21, 3–6.

Abderhalden, E. (1919) "Beobachtunger zur Frage der morphologischen und functionellen Asymmetrie des menschlichen Körpers," *Pflugers Archiv ful die Gesamte fur die Physiologie des Menschen und der Tierre,* 177, 213–216.

Abler, R., Adams, J.S., and Gould, P. (1971) *Spatial Organization: The Geographer's View of the World.* Englewood Cliffs, N.J.: Prentice-Hall.

Abu-Lugod, J. (n.d.) "The city is dead—long live the city: some thoughts on Urbanity," unpublished paper, Center for Planning and Development Research, University of California at Berkeley.

Adams, J. S. (1969) "Directional bias in intra-urban migration," *Economic Geography*, 45, 302–323.

Aginsky, B. W. (1952) "The fragmentation of the American community," *Journal of Educational Sociology*, 26, 125–133.

Aiken, R. (1968) "Content analysis: the technique and its application to information flows," unpublished seminar paper, Department of Geography, The Pennsylvania State University.

Alexander, C. (1965) "A city is not a tree," *Architectural Forum,* 122, (1), 58–62 and (2), 58–61.

———, (1966) "The pattern of streets," *Journal of the American Institute of Planners*, 32, 273–278.

———, (1967) "The city as a mechanism for sustaining human contact," in Ewald, W. (ed.), *Environment for Man.* Bloomington: Indiana University Press, 60–102.

Allport, F. H. (1955) *Theories of Perception and the Concept of Structure.* New York: Wiley.

Allport, G. W., and Postman, L. J. (1945) "The Basic Psychology of Rumor," *Transactions of the New York Academy of Sciences*, 8, 61–81.

Almy, M. (1967) "The psychologist looks at spatial concept formation: children's concepts of space and time," in *Research Needs in Geographic Education.* National Council on Geographic Education, Normal, Illinois: Illinois State University, 23–40.

Alonso, W. (1964) *Location and Land Use*. Cambridge, Mass.: Harvard University Press.

Ames, L. B., and Learned, J. (1948) "The development of verbalized space in the young child," *Journal of Genetic Psychology*, 72, 63–84.

Anderson, J., and Tindal, M. (1970) "The concept of home range: new data for the study of territorial behavior," unpublished paper, Department of Geography, Clark University.

Angyal, A. (1930) "Über die Raumlage vorgestellter Örter," *Archiv fur die gesamte psychologie*, 78, 47–94.

———, (1965) Unpublished lecture at the Massachusetts Institute of Technology.

Appleyard, D. (1969a) "City designers and the pluralistic city," in Rodwin, L. et al. (eds.), *Planning, Urban Growth, and Regional Development: The Experience of the Guayana Program of Venezuela*. Cambridge, Mass.: MIT Press, 422–452.

———, (1969b) "Why buildings are known," *Environment and Behavior*, 1, 131–156.

———, (1969c) "Signs on urban highways; messages, audiences, and media," *Dot Zero*, 26–31.

———, (1970) "Styles and methods of structuring a city," *Environment and Behavior*, 2, 100–118.

Appleyard, D., Lynch, K., and Myer, J. (1964) *The View from the Road* Cambridge, Mass.: MIT Press.

Ardrey, R. (1961) *African Genesis*. New York: Atheneum.

———, (1966) *The Territorial Imperative*. New York: Atheneum.

Arnheim, R. (1966) *Toward a Psychology of Art*. Berkeley: University of California Press.

———, (1967) *Art and Visual Perception: A Psychology of the Creative Eye*. Berkeley: University of California Press.

———, (1969) *Visual Thinking*. Berkeley: University of California Press.

Ashby, W. R. (1952) *Design for a Brain*. New York: Wiley.

Attneave, F. H. (1950) "Dimensions of similarity," *American Journal of Psychology*, 63, 516–556.

———, (1954) "Some informative aspects of visual perception," *Psychological Review*, 61, 183–193.

———, (1957) "Transfer of experience with a class-schema to identification-learning of patterns and shapes," *Journal of Experimental Psychology*, 54, 81–88.

Augustine, Saint (1949) *The Confessions of St. Augustine*. Translated by E. P. Pusey. New York: The Modern Library.

Ausubel, D. P. (1963) *The Psychology of Meaningful Verbal Learning*. New York: Gruen and Stratton.

Bachelard, G. (1969) *The Poetics of Space*. Boston: Beacon.

Baldwin, A. L. (1967) *Theories of Child Development*. New York: Wiley.

Banfield, E. (1970) *The Unheavenly City*. Boston: Little, Brown.

Banham, R. (1971) *Los Angeles: The Architecture of Four Ecologies*. London: Penguin Press.

Barker, R. G., ed. (1963) *The Stream of Behavior*. New York: Appleton-Century-Crofts.

———, (1968) *Ecological Psychology*. Stanford, California: Stanford University Press.

Barker, R. G., and Gump, P. V. (1964) *Big School, Small School*. Stanford, California: Stanford University Press.

Barlow, J. (1964) "Inertial navigation as a basis for animal navigation," *Journal of Theoretical Biology*, 6, 76–117.

Bartlett, F. C. (1932) *Remembering*. Cambridge, England: Cambridge University Press.

Bartz, B. S. (1970) "Experimental use of the search task in an analysis of type legibility in cartography," *The Cartographic Journal*, 7, 103–112.

Baumgarten, F. (1927) "Die Orientierungstäuschugen," *Zeitschrift fur Psychologie und Physiologie der Sinnesorgane*, 103, 111–122.

Baumol, W. J., and Ide, E. A. (1956) "Variety in retailing," *Management Science*, 3, 93–101.

Beck, R. J. (1964) *A Comparative Study of Spatial Meaning*, unpublished M. A. thesis, University of Chicago.

———, (1967) "Spatial meaning and the properties of the environment," in Lowenthal, D. (ed.) *Environmental Perception and Behavior* Department of Geography, University of Chicago, Research Paper #109, 18–41.

Bell, C. (1928) *The People of Tibet*. Oxford: Clarendon Press.

Bell, W. (1953) "The social areas of the San Francisco Bay region," *American Sociological Review*, 18, 39–47.

———, (1968) "The city, the suburb, and a theory of social choice," in Greer, S., McElrath, D. C., Miner, M. W., and Orleans, P. (eds.), *The New Urbanization*. New York: St. Martin's Press, 132–168.

Berkhoff, G. (1940) *Lattice Theory*. American Mathematical Society, Colloquium Publication, Volume 25.

Berle, A. A., Jr. (1968) "What the GNP doesn't tell us," *Saturday Review*, August 31, 10.

Berry, B. J. L. (1964) "Approaches to regional analysis: a synthesis," *Annals of the Association of American Geographers*, 54, 2–11.

———, (1967) *Geography of Market Centers and Retail Distribution*. Englewood Cliffs, N. J.,: Prentice-Hall.

Berry, B. J. L., and Pred, A. (1965) *Central Place Studies: A Bibliography of Theory and Applications*. Philadelphia: Regional Science Research Institute, Bibliography Series #1.

Berry, B. J. L., Simmons, J. W., and Tennant, R. J. (1963) "Urban population densities: structure and change," *Geographical Review*, 53, 389–405.

Best, E. (1924) *The Maori*. Wellington: H. H. Tombs.

Bijou, S. W., and Baer, D. M. (1961) *Child Development: A Systematic and Empirical Theory*. New York: Appleton-Century-Crofts.

Binet, M. A. (1894) "Reverse illusions of orientation," *Psychological Review*, 1, 337–350.

Birch, H. G., and Korn, S. J. (1958) "Place-learning, cognitive maps, and parsimony," *Journal of General Psychology*, 58, 17–35.

Blackith, R. E. (1965) "Morphometrics," in T. H. Waterman, and H. Morowitz (eds.), *Theoretical and Mathematical Biology*. New York: Blaisdell Publishing, Chapter 9.

Blaut, J. M. (1961) "Space and process," *The Professional Geographer*, 13 (4), 1–7.

———, (1962) "Object and relationship," *The Professional Geographer*, 14, 1–7.

———, (1969) *Studies in Developmental Geography*, Graduate School of Geography, Clark University, Place Perception Research Report #1.

Blaut, J. M., McCleary, G. F., and Blaut, A. S. (1970) "Environmental mapping in young children," *Environment and Behavior*, 2, 335–349.

Blaut, J. M., and Stea, D. (1969) *Place Learning*. Graduate School of Geography, Clark University, Place Perception Research Report #4.

———, (1971) "Studies of geographic learning," *Annals of the Association of American Geographers,* 61, 387–393.

Blumenthal, A. (1928) "Über den einflus der kopf haltung auf die gehrichtung," *Beitr. Anat. etc., Ohr.,* 26, 390–422.

Bogen, J. E., Marsh, J. F., and Tenhouten, W. D. (1970) "A theory of cognitive functioning and social stratification," unpublished paper, U.C.L.A.

Bogoraz-Tan, V. G. (1940) "The Chukchee," *Memoirs of the American Museum of Natural History.* New York: G. E. Stechert (1907, 1909).

Bordessa, R. (1969) "Perception research in geography: an appraisal and contribution to urban perception," University of Newcastle-upon-Tyne, Department of Geography Seminar Paper #8.

Boulding, K. E. (1965) *The Image.* Ann Arbor: University of Michigan Press.

———, (1959) "National images and international systems," *Journal of Conflict Resolution,* 3, 120–131.

———, (1970) "The end is in sight for galloping science," *The Washington Post,* September 6.

Bowden, M. J. (1969) "The perception of the Western interior of the United States, 1800–1870: a problem in historical geography," *Proceedings of the Association of American Geographers,* 1, 16–21.

Bower, G. H. (1970) "Analysis of a mnemonic device," *American Scientist,* 58, 496–510.

Bower, T. G. R. (1966) "The visual world of infants," *Scientific American,* 215, 80–92.

Bracey, H. E. (1964) *Neighbours: Subdivision Life in England and the United States.* Baton Rouge, Louisiana: Louisiana State University Press.

Brain, W. R. (1941) "Visual disorientation with special reference to lesions of the right cerebral hemisphere," *Brain,* 64, 244–272.

Bratfisch, O. (1966) "A further study of the relation between subjective distance and emotional involvement," *Reports from the Psychological Laboratories,* The University of Stockholm, #208.

———, (1969) "A further study of the relation between subjective distance and emotional involvement," *Acta Psychologica,* 29, 244–255.

Bratfisch, O., Ekman, G., Lundberg, U., and Krüger, K. (1970) "Subjective temporal distance and emotional involvement," *Reports from the Psychological Laboratories,* The University of Stockholm, #299.

Brennan, T. (1948) *Midland City.* London: Dobson.

Bridgen, R. J. (1935) "The dynamics of spiral movement in man," *Journal of Comparative Psychology,* 20, 59–74.

Brody, E. B., ed. (1970) *Behavior in New Environments: Adaptation of Migrant Populations.* Beverly Hills, California: Sage Publications.

Brookfield, H. C. (1969) "On the environment as perceived," in Board, C., Chorley, R. J., Haggett, P., and Stoddart, D. R., eds., *Progress in Geography,* Volume 1. London: Edward Arnold, 51–80.

Brower, S. (1965) "The signs we learn to read," *Landscape,* 15, 9–12.

Brown, B. R., and Evans, S. H. (1969) "Perceptual learning in pattern discrimination tasks with two and three schema categories," *Psychonomic Science,* 15, 101–103.

Brown, L. A., and Holmes, J. (1970) "Search behavior in an intra-urban migration context," paper presented at Population Association of America meetings, Atlanta, Georgia. Reprinted in Brown, L. A., ed., *Population and Migration in an Urban Context: A Spatial Perspective.* Ohio State University, Department of Geography Discussion Paper 13.

Brown, L. A., and Moore, E. G. (1970) "The intra-urban migration process: a perspective," *Geografiska Annaler*, 52B.

Brown, L. B. (1969) "The 3D reconstruction of a 2D visual display," *Journal of Genetic Psychology*, 115, 257–262.

Brown, W. (1932) "Spatial integrations in a human maze," *University of California Publications in Psychology*, 5, 123–134.

Bruner, J. S. (1957a) "On going beyond the information given," *Contemporary Approaches to Cognition*. Cambridge, Mass.: Harvard University Press.

———, (1957b) "On perceptual readiness," *Psychological Review*, 64, 123–152.

———, (1964) "The course of cognitive growth," *American Psychologist*, 19, 1–15.

———, (1966) "On cognitive growth," in Bruner, J. S., Olver, R. R., and Greenfield, P. M., eds., *Studies in Cognitive Growth*. New York: Wiley, 1–67.

Bruner, J. S., Goodnow, J., and Austin, G. (1962) *Study of Thinking*. New York: Science Editions.

Bruner, J. S., and Kenny, J. H. (1966) "On multiple ordering," in Bruner, J. S., Olver, R. R., and Greenfield, P. M., eds., *Studies in Cognitive Growth*. New York: Wiley, 154–167.

Bruner, J. S., Olver, R. R., and Greenfield, P. M., eds. (1966) *Studies in Cognitive Growth*. New York: Wiley.

Bruner, J. S., and Postman, L. (1949) "On the perception of incongruity: a paradigm," *Journal of Personality*, 18, 206–223.

Brunswik, E. (1943) "Organismic achievement and environmental probability," *Psychological Review*, 50, 255–272.

Buck, R. C. (1963) "Reflexive prediction," *Philosophy of Science*, 30, 359–369.

Bucklin, L. P. (1967) *Shopping Patterns in Urban Areas*. Berkeley, Calif.: University of California Institute of Business and Economic Research.

Budd, R., and Thorpe, R. (1963) *An Introduction to Content Analysis*. Iowa City: State University of Iowa Press.

Bugelski, B. R. (1970) "Words and things and images," *American Psychologist*, 25, 1002–1012.

Bunge, W. (1962) *Theoretical Geography*. Lund: C. W. K. Gleerup, Lund Studies in Geography, Series C. General and Mathematical Geography #1.

———, (1969) *The First Years of the Detroit Geographical Expedition: A Personal Report*. Detroit: Detroit Geographical Expedition, Discussion Paper #1.

Burrill, M. F. (1968) "The language of geography," *Annals of the Association of American Geographers*, 58, 1–11.

Burton, I., and Kates, R. W. (1964) "The perception of natural hazards in resource management," *Natural Resources Journal*, 3, 412–441.

Bushurova, V. V. (1956) "On the initial formation of functional asymmetry of the hands in connection with the differentiation of direction in space," *Izvestiya Akademii Pedagogicheskikh Naut RSFSR*, #86. Cited in Shemyakin, F. N., "Orientation in Space," in Ananyev, B. G. et al., eds., *Psychological Science in the U.S.S.R.*, 1. Washington: Office of Technical Services, 1962, 186–225.

Buttimer, A. (1969) "Social space in interdisciplinary perceptive," *Geographical Review*, 59, 417–426.

Campbell, R. D. (1968) "Personality as an element of regional geography," *Annals of the Association of American Geographers*, 58, 748–759.

Cannitello, N. (1970) "Route cognition: a part of environmental cognition," unpublished seminar paper, Department of Psychology, Clark University.

Caplow, T. (1955) "The definition and measurement of ambiences," *Social Forces*, 34, 28–33.
Carmichael, L. (1933) "Origin and prenatal growth of behavior," in Murchison, C.A., ed., *Handbook of Child Psychology*. Worcester, Mass.: Clark University Press. Cited in Werner, H., *Comparative Psychology of Mental Development*. Rev. ed. New York: International Universities Press, 1948.
Carmichael, L., Hogan, H. P., and Walter, A. A. (1932) "An experimental study of the effect of language on the reproduction of visually perceived forms," *Journal of Experimental Psychology*, 15, 73–86.
Carr, S. (1970) "The city of the mind," in Proshansky, H. M., Ittelson, W. H., and Rivlin, L. G., eds., *Environmental Psychology: Man and His Physical Setting*. New York: Holt, Rinehart, and Winston, 518–533.
Carr, S., and Lynch, K. (1968) "Where learning happens," *Daedalus: Journal of the American Academy of Arts and Sciences*, 97, 1277–1291.
Carr, S., and Schissler, D. (1969) "The city as a trip: perceptual selection and memory in the view from the road," *Environment and Behavior*, 1, 7–36.
Carson, D. (1967) "Comments on 'The Pattern of Streets,'" *Journal of the American Institute of Planners*, 33, 409–412.
Casamorata, C. (1944) "I canti di Firenze," *L'Universo*, 25, #3.
Casetti, E. (1967) "Urban population densities: an alternative explanation," *Canadian Geographer*, 11, 96–100.
Cassirer, E. (1944) *An Essay on Man: An Introduction of the Philosophy of Human Culture*. New Haven: Yale University Press.
———, (1953) *The Philosophy of Symbolic Forms*. Volume 1, *Language*. New Haven: Yale University Press.
———, (1955) *The Philosophy of Symbolic Forms*. Volume 2, *Mythical Thought*. New Haven: Yale University Press.
———, (1957) *The Philosophy of Symbolic Forms*. Volume 3, *The Phenomenonology of Knowledge*. New Haven: Yale University Press.
Castner, H. W., and Robinson, A. H. (1969) *Dot Area Symbols in Cartography: the Influence of Pattern on their Perception*. Washington: American Congress on Surveying and Mapping, Technical Monograph No. CA–4.
Cattell, R. (1965a) "Factor analysis: an introduction to essentials. I. The purpose and underlying models," *Biometrics*, 21, 190–215.
———, (1965b) "Factor analysis: an introduction to essentials. II. The role of factor analysis in research," *Biometrics*, 21, 405–435.
Chapin, F. S. (1965) *Urban Land Use Planning*. 2nd ed. Urbana, Ill.: University of Illinois Press.
———, (1968) "Activity systems as sources of inputs for land use models," in Hemmens, G. C., ed., *Urban Development Models*. Washington: Highway Research Board Special Report #97.
Chapin, F. S., and Brail, R. K. (1969) "Human activity systems in the Metropolitan United States," *Environment and Behavior*, 1, 107–130.
Chapin, F. S., and Hightower, H. C. (1966) *Household Activity Systems: A Pilot Investigation*. Chapel Hill, North Carolina: Center for Urban and Regional Studies, University of North Carolina.
Chein, I. (1954) "The environment as a determinant of behavior," *Journal of Social Psychology*, 39, 115–127.
Chorley, R. J., and Haggett, P. (1965) "Trend surface mapping in geographical research," *Transactions and Papers of the Institute of British Geographers*, 37, 47–67.
Claparède, E. (1903) "La faculté d'orientation lointaine (sens de direction, sens de retour.)," *Archive de Psychologie, Gèneve*, 2, 133–180.

———, (1924) "Note sur la localisation du moi," *Archive de Psychologie*, Gèneve, 19, 172–182.

———, (1943) "L'orientation lointaine," *Nouveau Traité de Psychologie*, 8, 3. Presse Universitaire, Paris.

Clark, W. A. V., and Rushton, G. (1970) "Models of intra-urban consumer behavior and their implications for central place theory," *Economic Geography*, 46, 486–497.

Cobb, E. (1959) "The ecology of imagination in childhood," *Daedalus*, 88, 537–548. Reprinted in Shepard, P., and McKinley, D., eds., *The Subversive Science: Essays Towards an Ecology of Man.* Boston: Houghton-Mifflin, 1969, 122–132.

Cochran, T. C. (1965) "The social impact of the railway," in Mazlish, B., ed., *The Railway and the Space Program: An Exploration in Historical Analogy.* Cambridge, Mass.: MIT Press.

Coghill, G. E. (1929) *Anatomy and the Problem of Behavior.* Cambridge, England: Cambridge University Press. Cited in Werner, H., *Comparative Psychology of Mental Development.* Rev. ed. New York: International Universities Press, 1948.

Cole, J. P., Falconer, M., Brandwood, G., Simmons, P., and Young, J. (1968) "Notes on perception in geography," *Bulletin of Quantitative Data for Geographers*, University of Nottingham, #18.

Cole, J. P., and Whysall, P. (1968) "Places in the news, a study of geographical information," *Bulletin of Quantitative Data for Geographers*, University of Nottingham, #7.

Colucci, C. (1902) "Sui disturbi dell'orientamento topografico," *Annali di Neurologia*, 20, 555–596.

Colyard, Y., ed. (1971) *Field Notes: the Geography of the Children of Detroit.* Detroit: Detroit Geographical Expedition, Discussion Paper #3.

Cooley, C. H. (1902) *Human Nature and the Social Order.* New York: Charles Scribner's Sons.

Cornetz, V. (1913) "Le cas elémentaire du sens de la direction chez l'homme," *Bulletin de la Societé de Géographie d'Alger*, 18, 742.

Cox, H. (1965) *The Secular City: A Celebration of its Liberties and an Invitation to its Discipline.* New York: MacMillan.

Cox, K. R. (1969) "The genesis of acquaintance field spatial structures: a conceptual model and empirical tests," in Cox, K. R., and Golledge, R. G., eds., *Behavioral Problems in Geography: A Symposium.* Northwestern University, Department of Geography, Studies in Geography, #17, 146–168.

Cox, K. R., and Golledge, R. G., eds. (1969) *Behavioral Problems in Geography: A Symposium.* Evanston, Ill.: Northwestern University Press.

Craik, K. M. (1970a) "Environmental psychology," in Newcomb, T. M., ed., *New Directions in Psychology*, Vol. 4. New York: Holt, Rinehart, and Winston, 1–121.

———, (1970b) "Environmental disposition(s) and preferences," in Archea, J., and Eastman, C. M., eds., *EDRA Two: Proceedings of the 2nd Annual Environmental Design Research Association Conference.* Pittsburgh: Carnegie-Mellon University, 309–339.

Crane, D. A. (1961) "Lynch: The Image of the City—Review," *Journal of the American Institute of Planners*, 27, 152–155.

Critchley, M. (1953) *The Parietal Lobes.* London: Edward Arnold.

Dahir, J. (1947) *The Neighborhood Unit Plan: A Selected Bibliography.* New York: Russell Sage Foundation.

Darwin, C. (1959) *On the Origin of Species.* Philadelphia: University of Pennsylvania Press.

Dateson, F. W., ed. (1964) *Selected Poems of William Blake.* London: Heinemann, Ltd.

David, H. A. (1963) *The Method of Paired Comparisons.* New York: Hafner Publishing.

Davis, F. C. (1933) "Effect of maze rotation upon subjects reporting different methods of learning and retention," *Publications of the University of California, Los Angeles, in Education, Philosophy, and Psychology*, 1, 47–64.

Davis, W. M. (1954) *Geographical Essays.* New York: Dover.

Deese, J., and Hulse, S. H. (1967) *The Psychology of Learning.* 3rd ed. New York: McGraw-Hill.

De Groot, J. J. M. (1912) *Religion in China.* New York: G. P. Putnam's.

De Jonge, D. (1962) "Images of urban areas, their structures and psychological foundations," *Journal of the American Institute of Planners*, 28, 266–276.

de Lawe, P. H. C. (1965a) "Field and case studies," in Hauser, P. M., ed., *Handbook for Social Research in Urban Areas.* Paris: UNESCO, 55–72.

________, (1965b) "Social organization in an urban milieu," in Hauser, P. M., ed., *Handbook for Social Research in Urban Areas.* Paris: UNESCO, 140–158.

Demko, D., and Briggs, R. (1970) "An initial conceptualization and operationalization of spatial choice behavior: a migration example using multidimensional unfolding," *Proceedings of the Canadian Association of Geographers*, 1.

Dent, B. D. (1970) *Perceptual Organization and Thematic Map Communication.* Graduate School of Geography, Clark University, Place Perception Research Report #5.

Department of City Planning. (1971) *The Visual Environment of Los Angeles.* Los Angeles: Department of City Planning.

De Santillana, G. (1960) "Perspectives," *American Scientist*, 48, 264–269.

De Silva, H. R. (1931) "A case of a boy possessing an automatic directional orientation," *Science*, 75, 393–394.

Deutsch, K. W., and Isard, W. (1961) "A note on a generalized concept of effective distance," *Behavioral Science*, 6, 308–311.

Doherty, J. M. (1968a) "Residential preferences for urban environments in the United States," *London School of Economics, School of Geography Discussion Papers*, 29, 1–71.

________, (1968b) *A Multivariate Analysis of Residential Preferences for the Cities of the United States,* unpublished M.A. thesis, The Pennsylvania State University.

Donaldson, B. A. (1938) *The Wild Rue: A Study of Muhammadan Magic and Folklore in Iran.* London: Lirzac.

Dornbach, J. E. (1967) *An Analysis of the Maps as an Information Display System*, unpublished Ph.D. dissertation, Clark University.

Dornič, S. (1967) "Subjective distance and emotional involvement: a verification of the exponent invariance," *Reports from the Psychological Laboratories*, Stockholm, 237, 1–7.

Downs, R. M. (1967) "Approaches to, and problems in, the measurement of geographic space perception," *University of Bristol, Department of Geography Seminar Paper Series A*, 9, 1–17.

________, (1968) "The role of perception in modern geography," *University of Bristol, Department of Geography Seminar Paper Series A*, 11, 1–20.

———, (1970a) "The cognitive structure of an urban shopping center," *Environment and Behavior*, 2, 13–39.

———, (1970b) "Geographic space perception: past approaches and future prospects," in Board, C., Chorley, R. J., Haggett, P., and Stoddart, D. R., eds., *Progress in Geography, Volume 2*. London: Edward Arnold, 65–108.

Drever, J. (1955) "Early learning and the perception of space," *American Journal of Psychology*, 68, 605–614.

Drumheller, S. J. (1968) "Conjure up a map—a crucial but neglected skill," *Journal of Geography*, 68, 140–146.

Du Mas, F., and Worchel, P. (1956) "The influence of the spatial context on the relearning of a rotated perceptual motor task," *Journal of Genetic Psychology*, 54, 65–80.

Duncan, B. K. (1934) "A comparative study of finger maze learning by blind and sighted subjects," *Journal of Genetic Psychology*, 44, 69–95.

Dutton, G. H. (1970) "Macroscopic aspects of metropolitan evolution," unpublished seminar paper, Institute for Computer Graphics, Harvard University.

Edmonds, E. M., and Edmonds, S. C. (1969) "Schema mediation in categorization learning," *Psychonomic Science*, 14, 196.

Eisler, H. (1960) "Similarity in the continuum of heaviness with some methodological and theoretical considerations," *Scandinavian Journal of Psychology*, 1, 69–81.

Eisler, H., and Ekman, G. (1959) "A mechanism of subjective similarity," *Acta Psychologica*, 16, 1–10.

Ekman, G. (1958) "Two generalized ratio scaling methods," *Journal of Psychology*, 45, 287–295.

———, (1961) "Some aspects of psychophysical research," in Rosenblith, W.A., ed., *Sensory Communication*. New York: Wiley, 35–47.

———, (1962) "Measurement of moral judgment: a comparison of scaling methods," *Perceptual and Motor Skills*, 15, 3–9.

———, (1963) "A direct method for multidimensional ratio scaling," *Psychometrika*, 28, 33–41.

———, (1965) "Two methods for the analysis of perceptual dimensionality," *Perceptual and Motor Skills*, 20, 557–572.

———, (1970a) "Quantitative approaches to psychological problems," in Lindblom, P., ed., *Theory and Methods in Behavioral Sciences*. Stockholm: Norsteds, 53–72.

———, (1970b) "Comparative studies on multidimensional scaling and related techniques," in Pawlik, K., ed., *Perspectives in Multivariate Psychological Research*. Bern: Huber.

Ekman, G., and Åkesson, C. (1965) "Saltness, sweetness, and preference; a study of quantitative relations in individual subjects," *Scandinavian Journal of Psychology*, 6, 241–253.

Ekman, G., and Bratfisch, O. (1965) "Subjective distance and emotional involvement; a psychological mechanism," *Acta Psychologica*, 24, 446–453.

Ekman, G., Engen, T., Künnapas, T., and Lindman, R. (1964) "A quantitative principle of qualitative similarity," *Journal of Experimental Psychology*, 68, 530–536.

Ekman, G., Goude, G., and Waern, Y. (1961) "Subjective similarity in two perceptual continua," *Journal of Experimental Psychology*, 61, 222–227.

Ekman, G., Hosman, B., Lindman, R., Ljungberg, L., and Åkesson, C. (1968) "Interindividual differences in scaling performance," *Perceptual and Motor Skills*, 26, 815–823.

Ekman, G., Hosman, J., and Lindström, B. (1965) "Roughness, smoothness, and preference; a study of quantitative relations in individual subjects," *Journal of Experimental Psychology*, 70, 18–26.

Ekman, G., Lindman, R., and William-Olsson, W. (1961) "A psychophysical study of cartographic symbols," *Perceptual and Motor Skills*, 13, 355–368.

Ekman, G., and Lundberg, U. (1970) "Emotional reaction to past and future events as a function of temporal distance," *Reports from the Psychological Laboratories*, The University of Stockholm, #302.

Ekman, G., and Sjöberg, L. (1965) "Scaling," *Annual Review of Psychology*, 16, 451–474.

Eliade, M. (1959) *The Sacred and the Profane*. New York: Harcourt, Brace, and World.

Eliot, J. (1970) "Children's spatial visualization," in National Council for the Social Sciences, 40th Yearbook, *Focus on Geography*. Washington: National Council for the Social Sciences, 263–290.

Elliott, H. W. (1886) *Our Arctic Province*. New York: Scribners.

English, H. B., and English, A. C. (1958) *A Comprehensive Dictionary of Psychological and Psychoanalytical Terms*. New York: Longmans, Green & Co.

Erikson, E. K. (1963) "Toys and reasons," in *Childhood and Society*. 2nd ed. New York: Norton, 209–246.

Esser, A. H., ed. (1971) *Behavior and Environment: The Use of Space by Animals and Men*. New York: Plenum Press.

Estes, W. K. (1954) "Individual behavior in uncertain situations: an interpretation in terms of statistical association theory," in Thrall, R. M., Coombs, C.H., and Davis, R. L., eds., *Decision Processes*. New York: Wiley.

Eyles, J. D. (1968) "The inhabitant's image of Highgate Village (London): an example of perception measurement technique." *London School of Economics, Graduate School of Geography Discussion Paper*, 15, 1–12.

Fernald, M. R. (1913) "The mental imagery of two blind subjects," *Psychological Bulletin*, 10, 62–63.

Ferraro, A. (1921) "Ricerche sul valore della prova 'P. Marie-Béhague', diretta a svelare i disturbi dell'orientamento fin," *Riv. Pat., nerv., ment.*, 26, 74–78 (Abs.).

Festinger, L., Schacter, S., and Back, K. (1950) *Social Pressures in Informal Groups*. New York: Harper Row.

Finsch, O. (1888) "Ethnologische erfahrungen und belegstucke aus der Südsee," *Vienna, Naturhistorisches Hofmuseum, Annalen*, 3, 83–160 and 293–364. (See also (1891), 6, 13–36 and 37–130—(1893), 8, 1–106, 119–275, and 295–437.)

Firey, W. (1945) "'Sentiment' and 'symbolism' as ecological variables," *American Sociological Review*, 10, 140–148.

Firey, W. (1960) *Man, Mind, and Land*. Glencoe, Ill.: The Free Press.

Firth, R. (1936) *We, the Tikopia*. London: Allen and Unwin Ltd.

Fishbein, M. (1967a) "A consideration of beliefs and their role in attitude measurement," in Fishbein, M., ed., *Readings in Attitude Theory and Measurement*. New York: Wiley, 257–266.

———, (1967b) "Attitude and the prediction of behavior," in Fishbein, M., ed., *Readings in Attitude Theory and Measurement*. New York: Wiley, 477–492.

Flannery, J. J., (1956) *The Graduated Circle: A Description, Analysis and Evaluation of a Quantitative Map Symbol*, unpublished Ph.D. dissertation, University of Wisconsin.

Flavell, J. H. (1963) *The Developmental Psychology of Jean Piaget*. Princeton: Van Nostrand.

———, (1970) "Concept development," in Mussen, P.H., ed., *Manual of Child Psychology*, Vol. 1. 3rd ed. New York: Wiley, 903–1059.

Flavell, J .H., and Draguns, J. A. (1957) "A microgenetic approach to perception and thought," *Psychological Bulletin*, 59, 197–217.

Flavell, J. H., and Wohlwill, J. F. (1969) "Functional and formal aspects of cognitive development," in Elkind, D., and Flavell, J. H. eds., *Studies in Cognitive Development: Essays in Honor of Jean Piaget*. New York: Oxford University Press, 67–120.

Foley, D. J. (1950) "The use of local facilities in a metropolis," *American Journal of Sociology*, 3, 238–246.

Follini, M. B. (1966) *The Construction of Behavioral Space: A Microgenetic Investigation of Orientation in an Unfamiliar Locality*, unpublished M.A. thesis, Clark University.

Forgus, R. M. (1966) *Perception: The Basic Process in Cognitive Development*. New York: McGraw-Hill.

Forrester, J. W. (1971) *World Dynamics*. Cambridge, Mass.: Wright-Allen Press.

Forster, E. M. (1949) *A Passage to India*. New York: Harcourt.

Freedman, J., Klevansky, S., and Ehrlich, P. (1971) "The effect of crowding on human task performance," *Journal of Applied Social Psychology*, 1, 7–25.

Freeman, F. N. (1916) "Geography: extension of experience through imagination," in *The Psychology of Common Branches*. Boston: Houghton-Mifflin, 161–178.

Fried, M., and Gleicher, P. (1961) "Some sources of residential satisfaction in an urban slum," *Journal of the American Institute of Planners*, 29, 179–198.

Fuller, G. (1966) "Western New York; a cultural hearth?" unpublished seminar paper, Department of Geography, The Pennsylvania State University.

Galpin, C. J. (1915) *The Social Anatomy of an Agricultural Village*. Madison: Research Bulletin 34, Agricultural Experiment Station, University of Wisconsin.

Garling, T. (1969) "Studies in visual perception of architectural spaces and rooms. II. Judgments of open and closed space by category rating and magnitude estimation," *Scandinavian Journal of Psychology*, 10, 257–268.

Gatty, H. (1958) *Nature is your Guide*. New York: E. P. Dutton.

Gautier, E. F. (1908) *Missions au Sahara*. Paris: Librairie A. Colin.

Gay, J. (1922) *Trivia, or, the Art of Walking the Streets of London*. London: D. O'Connor.

Geier, F. M., Levin, M., and Tolman, E. C. (1941) "Individual differences in emotionality, hypothesis formation, vicarious trial and error and visual discrimination learning in rats," *Comparative Psychology Monographs*, 17, No. 3.

Geoghegan, R. H. (1944) *The Aleut Language*. Washington, D.C.: U.S. Department of Interior.

George, F. H., and McIntosh, S. B. (1960) "Experimental disorientation and conceptual confusion," *Quarterly Journal of Experimental Psychology*, 12, 141–148.

Gibson, E. (1969) *Principles of Perceptual Learning and Development*. New York: Appleton-Century-Crofts.

———, (1970) "The development of perception as an adaptive process," *American Scientist*, 58, 98–107.

Gibson, J. J. (1966) *The Senses Considered as Perceptual Systems.* Boston: Houghton-Mifflin.

Gill, E. (1941) *Autobiography.* New York: Devin-Adair.

Ginsburg, N. S. (1968) "On the Chinese perception of a world order," in Tang, T., ed., *China's Policies in Asia and America's Alternatives.* Vol 2. Chicago: University of Chicago Press, 73–91.

Gittins, J. S. (1969) *Forming Impressions of an Unfamiliar City: A Comparative Study of Aesthetic and Scientific Knowing,* unpublished M.A. thesis Clark University.

Glacken, C. C. (1951) *The Idea of the Habitable World,* unpublished Ph.D. thesis, The Johns Hopkins University.

Goldberg, T. (1969) "The automobile: a social institution for adolescents," *Environment and Behavior,* 1, 157–186.

Golding, W. (1966) *The Hot Gates.* New York: Harcourt, Brace.

Goldstein, K., and Scheerer, M. (1941) "Abstract and concrete behavior: an experimental study with special tests," *Psychological Monographs,* 53, Whole #239.

Golledge, R. G. (1967) "Conceptualizing the market decision process," *Journal of Regional Science,* 7 (No. 2, Supplemental), 239–258.

———, (1969) "The geographical relevance of some learning theories," in Cox, K. R., and Golledge, R. G., eds., *Behavioral Problems in Geography.* Evanston, Ill.: Northwestern University, Department of Geography, Studies in Geography, #17, 101–145.

Golledge, R. G., Briggs, R., and Demko, D. (1969) "The configuration of distances in intra-urban space," *Proceedings of the Association of American Geographers,* 1, 60–65.

Golledge, R. G., and Brown, L. A. (1967) "Search, learning and the market decision process," *Geografiska Annaler,* 49B, 116–124.

Golledge, R. G., Rushton, G., and Clark, W. A. V. (1966) "Some spatial characteristics of Iowa's dispersed farm population and their implications for the groupings of central place functions," *Economic Geography,* 42, 261–272.

Goodey, B. (1968a) "A pilot study of the geographic perception of North Dakota students," *University of North Dakota Geographical Research Report,* 1, 1–70.

———, (1968b) "Environmental, extra-environmental and preferential perception in geography," paper presented at the 1968 Meetings of the North Dakota Academy of Sciences.

———, (1970) "Perception, gaming, and Delphi: experimental approaches to environmental education," paper presented at the Architectural Psychology Conference, Kingston-upon-Thames, England.

Gould, P. (1960) *The Development of the Transportation Pattern in Ghana.* Evanston, Ill.: Northwestern University Department of Geography.

———, (1967a) "On the geographic interpretation of eigenvalues," *Transactions of the Institute of British Geographers,* 42, 53–86.

———, (1967b) "Structuring information on spacio-temporal preferences," *Journal of Regional Science,* 7, 2–16.

———, (1969) "Problems of space preference measures and relationships," *Geographical Analysis,* 1, 31–44.

———, (1970a) "Tanzania 1920–63: the spatial impress of the modernization process," *World Politics.*

———, (1970b) "The structure of spatial preferences in Tanzania," *Area,* 4, 29–35.

Gould, P., and Ola, D. (1970) "The perception of residential desirability in the Western region of Nigeria," *Environment and Planning,* 2, 73–87.

Gould, P., and White, R. R. (1968) "The mental maps of British school leavers," *Regional Studies*, 2, 161–182.

Gregg, F. M. (1939) "Are motor accompaniments necessary to orientational perception," *Journal of Psychology*, 8, 63–87.

———, (1940) "Overcoming geographic disorientation," *Journal of Consulting and Clinical Psychology*, 4, 66–68.

Grove, D. (1963) *Population Patterns: Their Impact on Regional Planning*. Kumasi: The Kwame University of Science and Technology.

Gruber, H. E., Terrell, G., and Wertheimer, M., eds. (1962) *Contemporary Approaches to Creative Thinking*. New York: Atherton Press.

Guilford, J. P. (1939) "A Study of psychodynamics," *Psychometrika*, 4, 1–23.

———, (1954) *Psychometric Methods*. New York: McGraw-Hill.

Guldberg, F. O. (1897) "Die Circularbewegung als thierische Grundbewegung, ihre Ursache, Phänomenalitat und Bedeutung," *Zeitschrift fur Biologie*, 17, 419–458.

Gulick, J. (1963) "Images of an Arab City," *Journal of the American Institute of Planners*, 29, 179–198.

Gulliksen, H. (1964a) "The structure of individual differences in optimality judgments," in Shelley, M. W., and Bryan, G., eds., *Human Judgments and Optimality*. New York: Wiley.

———, (1964b) "Intercultural studies of attitudes," in Gulliksen, H., ed., *Contributions to Mathematical Psychology*. New York: Holt, Rinehart and Winston.

Gulliver, F. P. (1908) "Orientation of maps," *Journal of Geography*, 7, 55–58.

Guthrie, E. R. (1935) *The Psychology of Learning*. New York: Harper.

Gutman, R. (1966) "Site planning and social behavior," *Journal of Social Issues*, 22, 103–115.

Guttman, L. (1966) "The Nonmetric Breakthrough for the Behavioral Sciences," invited address to the Automatic Data Processing Conference of the Information Processing Association of Israel.

———, (1968) "A general nonmetric technique for finding the smallest coordinate space for a configuration of points," *Psychometrika*, 33, 469–506.

Haddon, J. (1960) "A view of foreign lands," *Geography*, 65, 286–289.

Halbwachs, M. (1950) *La Mémoire Collective*. Paris: Presses Universitaires de France.

Hald, A. (1952) *Statistical Theory with Engineering Applications*. New York: Wiley.

Hall, E. T. (1966) *The Hidden Dimension*. New York: Doubleday.

Hallowell, A. I. (1942) "Some psychological aspects of measurement among the Salteaux," *American Anthropologist*, 44, 62–77.

———, (1955) *Culture and Experience*. Philadelphia: University of Pennsylvania Press.

Haney, D. G., Anderson, J. L., Katz, R. C., and Peterson, G. D. (1964) *The Value of Time for Passenger Cars: Further Theory and Small Scale Behavioral Studies*. Menlo Park: Stanford Research Institute.

Happold, F. C. (1966) *Religious Faith and Twentieth Century Man*. Harmondsworth: Penguin Books, Ltd.

Harman, H. H. (1960) *Modern Factor Analysis*. Chicago: Universty of Chicago Press.

Harris, C. (1954) "The market as a factor in the localization of industry in the United States," *Annals of the Association of American Geographers*, 44, 315–348.

Harris, D. B., ed. (1957) *The Concept of Development*. Minneapolis: University of Minnesota Press.

Hart, R. A. (1971) *Aerial Geography: An Experiment in Elementary Educaiton,* unpublished M. A. thesis, Clark University.

Hartman, C. (1964) "The housing of relocated families," *Journal of the American Institute of Planners*, 4, 266–286.

Hartman, F. (1902) *Die Orientierung*. Leipzig: Vogel.

Harvard Educational Review. (1969) Vol. 39 (2), entire issue.

Harvey, D. W. (1969a) *Explanation in Geography,* London: Edward Arnold; New York: St. Martin's Press.

———, (1969b) "Conceptual and measurement problems in the cognitive-behavioral approach to location theory," in Cox, K. R., and Golledge, R. G., eds., *Behavioral Problems in Geography: A Symposium.* Evanston, Ill.: Northwestern University, Department of Geography, Studies in Geography, #17, 35–68.

———, (1970a) "Social processes, spatial form and the redistribution of real income in an urban system," in Chisholm, M., Frey, A. E., and Haggett, P., eds., *Regional Forecasting*. London: Butterworths, 267–300.

———, (1970) "Social processes and spatial form: an analysis of the conceptual problems of urban planning," *Papers, Regional Science Association,* 25, 47–69.

———, (1973), *Social Justice and the City,* London: Edward Arnold; Baltimore: Johns Hopkins University Press, 120-52.

Hasler, A. D., and Wisby, W. J. (1958) "Return of displaced large-mouth bass and green sunfish to a home area," *Ecology*, 39, 289–293.

Hauty, G. T., and Adams, T. (1965) *Phase Shifts of the Human Circadian System and Performance Deficit During the Periods of Transition. Part I: East-West Flight. Part II: West-East Flight. Part III: North-South Flight.* Washington, D.C.: Federal Aviation Agency, Office of Aviation Medicine.

Head, H. (1920) *Studies in Neurology*. Oxford: Oxford University Press.

———, (1926) *Aphasia and Kindred Disorders of Speech.* Cambridge: Cambridge University Press.

Hebb, D. O. (1949) *The Organization of Behavior*. New York: Wiley.

———, (1963) "The semiautonomous process: its nature and nurture," *American Psychologist*, 18, 16–27.

———, (1966) *A Textbook of Psychology*. Philadelphia: Saunders.

Hefner, R., Levy, S. G., and Warner, H. L. (1967) "A survey of internationally relevant attitudes and behavior," *Peace Research Society (International) Papers*, 7, 139–150.

Held, R., and Rekosh, J. (1963) "Motor-sensory feedback and the geometry of visual space," *Science*, 141, 722–723.

Helson, H. (1959) "Adaptation level theory," in Koch, S., ed., *Psychology: A Study of a Science*, Vol. 1. New York: McGraw-Hill.

Hendel, C. W. (1953) "Introduction: the philosophy of form in Kant," in Cassirer, E., *The Philosophy of Symbolic Forms*. Vol. 1, *Language*. New Haven: Yale University Press, 1–65.

Herman, T. (1959) "Group values towards the national space: the case of China," *The Geographical Review*, 49, 164–182.

Hershberger, R. G. (1970) "Architecture and meaning," *Journal of Aesthetic Education*, 4, 37–55.

Higginson, G. D. (1936) "Human learning with a rotated maze," *Journal of Psychology*, 1, 277–294.

Hilgard, E. R., and Bower, G. H. (1966) *Theories of Learning*. 3rd ed. New York: Appleton-Century-Crofts.

Hillery, G. A., Jr. (1955) "Definitions of community: areas of agreement," *Rural Sociology*, 20, 111–123.

Hochberg, J. E. (1964) *Perception*. Englewood Cliffs, N.J.: Prentice-Hall.

Hochberg, J. E., and McAlister, E. (1953) "A quantitative approach to figural 'goodness,'" *Journal of Experimental Psychology*, 46, 361–364.

Holloway, G. E. T. (1967a) *An Introduction to the Child's Conception of Space*. London: Routledge and Kegan Paul.

———, (1967b) *An Introduction to the Child's Conception of Geometry*. London: Routledge and Kegan Paul.

Holloway, J. L. (1958) "Smoothing and filtering of time series and space fields," *Advances in Geophysics*, 4, 386–387.

Holt, R. (1964) "Imagery: the return of the ostracized," *American Psychologist*, 19, 254–264.

Homo, L. (1951) *Rome Impériale et L'Urbanisme dans L'Antiquité*. Paris: Michel.

Horowitz, M. J., Duff, D. F., and Stratton, L. O. (1964) "Personal space and the Body-Buffer zone," *Archives of General Psychiatry*, 11, 651–656. Reprinted in Proshansky, H. M., Ittelson, W. H., and Rivlin, L. G., eds., *Environmental Psychology*. New York: Holt, Rinehart, and Winston, 1970, 214–220.

Horst, P. (1965) *Factor Analysis of Data Matrices*. New York: Holt, Rinehart, and Winston.

Horton, F. E., and Reynolds, D. R. (1969) "An investigation of individual action spaces: a progress report," *Proceedings of the Association of American Geographers*, 1, 70–75.

———, (1970) "Intra-urban migration and the perception of residential quality," paper presented at Population Association of America meetings, Atlanta, Ga. Reprinted in Brown, L. A., ed., *Population and Migration in an Urban Context: A Spatial Perspective*. Ohio State University, Department of Geography, Discussion Paper #13.

Howard, I. P., and Templeton, W. B. (1966) *Human Spatial Orientation*. New York: Wiley.

Howe, G. F. (1931) "A study of children's knowledge of directions," *Journal of Geography*, 30, 298–304.

———, (1932) "The teaching of directions in space," *Journal of Geography*, 31, 207–210.

Howell, F. C. (1965) *Early Man*. New York: Time-Life Books.

Hoyt, H. (1939) *Structure and Growth of Residential Neighborhoods in American Cities*. Washington, D.C.: Federal Housing Administration.

Hudson, J. (1969) "A model of spatial relations," *Geographical Analysis*, 1, 260–271.

Hudson, W. H. (1922) "On the sense of direction," *Century Magazine*, 104, 693–701.

Hudson, W. (1960) "Pictorial depth perception in sub-cultural groups in Africa," *Journal of Social Psychology*, 52, 183–208.

Huff, D. L. (1960) "A topographic model of consumer space preferences," *Papers and Proceedings of the Regional Science Association*, 6, 159–164.

———, (1961) "Ecological characteristics of consumer behavior" *Papers and Proceedings of the Regional Science Association*, 7, 19–28.

———, (1963) "A probabilistic analysis of shopping center trade areas," *Land Economics*, 39, 81–90.

Hull, C. (1952) *A Behavior System*. New Haven: Yale University Press.

Hunt, E. B. (1962) *Concept Learning: An Information Processing Problem*. New York: Wiley.

Hunt, J. M. (1965) "Intrinsic motivation and its role in psychological development," *Nebraska Symposium on Motivation*, 13, 189–282.

Hyman, H. H. (1942) "The psychology of status," *Archives of Psychology*, 38, No. 269.

Inhelder, B. (1965) "Operation thought and mental imagery," *Monographs, Society for Research in Child Development*, 30, 4–18.

Isaacs, R. (1965) *Differential Games*. New York: Wiley.

Ittelson, W. H. (1951) "The constancies in perceptual theory," *Psychological Review*, 58, 285–294. Reprinted in Proshansky, H. M., Ittelson, W. H., and Rivlin, L. G., eds., *Environmental Psychology*. New York: Holt, Rinehart, and Winston, 1970, 112–119.

Jaccard, P. (1926) "Une enquête sur la désorientation en montagne," *Bulletin de la Société Vaudoise des Science Naturelles*, 56, 151–159.

________, (1932) *Le Sens de Direction et L'Orientation Lointaine chez L'Homme*. Paris: Payot.

Jackson, J. B. (1956–1957) "Other-directed houses," *Landscape*, 6, 29–35.

Jackson, L. L. (1943) "V.T.E. on an elevated maze," *Journal of Comparative Psychology*, 36, 99–107.

Jackson, P. (1970) *Myth and Reality: Environmental Perception of the Mormons, 1840–1865, an Historical Geosophy*, unpublished Ph.D. dissertation, Clark University.

James, W. (1892) *Psychology: The Briefer Course*. (1962 edition). New York: Harper Torchbooks.

Janowitz, M. (1952) *The Community Press in an Urban Setting*. Glencoe: The Free Press.

Jensen, A. (1969) "How much can we boost IQ and scholastic achievement?" *Harvard Education Review*, 39, 1–123.

Johanssen, G. (1950) *Configurations in Event Perception*. Uppsala: Almquist and Wiksell.

John, E. R. (1967) *Mechanisms of Memory*. New York: Academic Press.

Johnston, R. J. (1968) "Social status and residential desirability: a pilot study of residential location decisions in Christchurch, New Zealand," mimeo paper, Department of Geography, University of Canterbury, New Zealand.

________, (1969) "The residential preferences of New Zealand students: some tests of economic and ecologic concepts," mimeo paper, Department of Geography, University of Canterbury, New Zealand.

________, (1970) "Latent migration potential and the gravity model: a New Zealand study," *Geographical Analysis*, 2, 387–397.

Kabonova-Meller, E. N. (1956) "Formation of geographical representations in students of the fifth to seventh classes," *Izvestiya Akademi Pedagoficheskikh Nauk RSFSR*, 86. Cited in Shemyakin, F. N., "Orientation in Space," in Ananyev, B.G. et al (eds.), *Psychological Science in the U.S.S.R.*, Vol. 1, Washington: Office of Technical Services, 1962, 186–255.

Kagan, J., and Kogan, N. (1969) "Individual variation in cognitive processes," in Mussen, P. H., ed., *Manual of Child Psychology*, Vol. 1. 3rd ed. New York: Wiley, 1273–1365.

Kant, I. (1902) *Critique of Pure Reason*. 2nd Rev. Ed. New York: Macmillan.

Kaplan, B. (1966a) "The study of language in psychiatry: the comparative developmental approach and its applications to symbolization and language in psychopathology," in Arieti, S., ed., *American Handbook of Psychiatry*, Vol. 3. New York: Basic Books, 659–688.

________, (1966b) "The 'latent content' of Heinz Werner's comparative developmental approach," in Wapner, S., and Kaplan, B., eds., *Heinz Werner 1890–1964; Papers in Memorium*. Worcester, Mass.: Clark University Press.

———, (1967) "Meditations on genesis," *Human Development*, 10, 65–87.

———, (1972) "Strife of systems: the tension between organismic and developmental points of view," in Gray, W., and Rizzo, N. D., eds., *Unity Through Diversity: A Festschrift in Honor of Ludwig von Bertalanffy*. 4 Volumes. New York: Gordon and Breach.

Kaplan, S. (1970) "Perception and thought: getting along in a difficult and complex world," University of Michigan: unpublished paper.

Kates, R. W. (1962) *Hazard and Choice Perception in Flood Plain Management*. Chicago: University of Chicago, Department of Geography Research Paper #78.

———, (1966) "Stimulus and symbol: the view from the bridge," *Journal of Social Issues*, 22, 21–28.

———, (1967) "The perception of storm hazard on the shores of megalopolis," in Lowenthal, D., ed., *Environmental Perception and Behavior*. Chicago: University of Chicago, Department of Geography Research Paper #109, 60–74.

———, (1970), "Human perceptions of the environment," *International Social Science Journal*, 22, 648–660.

Kates, R. W., and Wohlwill, J. F. (1966) "Man's response to the physical environment: introduction," *Journal of Social Issues*, 22, 15–20.

Katz, R. D. (1967) "Urban housing and site design," in Schnore, L., and Fagin, H., eds., *Urban Research and Policy Planning*. Beverly Hills: Sage Publishing.

Kawaguchi, E. (1909) *Three Years in Tibet*. Adyar, Madras: The Theosophist Office.

Keller, F. S. (1954) *Learning: Reinforcement Theory*. New York: Random House.

Keller, S. (1965) *Neighbors, Neighboring and Neighborhoods in Sociological Perspective*. Athens: Athens Technological Institute.

———, (1968) *The Urban Neighborhood: A Sociological Perspective*. New York: Random House.

Kelley, T. (1935) *Essential Traits of Mental Life*. Cambridge, Mass.: Harvard University Press.

Kelly, G. A. (1955) *The Psychology of Personal Constructs*. New York: W. W. Norton and Company.

———, (1970) "A brief introduction to personal construct theory," in Bannister, D., ed., *Perspectives in Personal Construct Theory*. New York: Academic Press, 1–23.

Kelman, H. C., ed. (1965) *International Behavior: A Social Psychological Analysis*. New York: Holt, Rinehart, and Winston.

Kennedy, R. (1943) "Premarital residents propinquity," *American Journal of Sociology*, 48, 580–584.

Kepes, G. (1956) *The New Landscape*. Chicago: P. Theobald.

Kershner, J. R. (1970) "Children's spatial representations and horizontal directionality," *Journal of Genetic Psychology*, 116, 177–189.

Khopreninova, N. G. (1956) "Orientation of locality by sightless persons," *Uchenyye Zapiski Chkalovskogo Gos. Pedagogicheskikh In-ta* (Chkalov State Pedagogical Institute), #8. Cited in Shemyakin, F. N., "Orientation in space," in Ananyev, B. G. et al (eds.), *Psychological Science in the U.S.S.R.*, Vol. 1. Washington: Office of Technical Services, 1962, 186–225.

Kinsman, B. (1965) *Wind Waves: Their Generation and Propagation on the Ocean Surface*. Englewood Cliffs, N.J.: Prentice-Hall.

Kirk, W. (1951) "Historical geography and the concept of the behavioral environment," in Kuriyan, G., ed., *Indian Geographical Journal, Silver Jubilee Edition*. Madras: Indian Geographical Society, 152–160.

Kirk, W. (1963) "Problems in Geography," *Geography*, 47, 357–371.

Klein, D. B. (1970) *A History of Scientific Psychology*. New York: Basic Books.

Klein, H. J. (1967) "The delimitation of the town centre in the image of its citizens," in University of Amsterdam, Sociographical Department, eds., *Urban Core and Inner City*. Leiden: E. J. Brill.

Klineberg, O. (1965) *The Human Dimension in International Relations*. New York: Holt, Rinehart and Winston.

Kloven, E. J. (1968) "Selection of target areas by factor analysis," *Western Miner*, 41, 44–54.

Knotts, J. R., and Miles, W. R. (1929) "The maze learning ability of blind compared with sighted children," *Journal of Genetic Psychology*, 36, 21–50.

Koch, H. L., and Ufkess, J. (1926) "A comparative study of stylus maze learning by blind and seeing subjects," *Journal of Experimental Psychology*, 9, 118–131.

Koffka, K. (1935) *Principles of Gestalt Psychology*. New York: Harcourt-Brace.

Kolodnaya, A. Y. (1954) "Disturbance of the differentiation of 'right' and 'left' and the role of the cutaneous analyzer in its restoration," *Izvestiya Akademii Pedagogicheskikh Nauk RSFSR*, #53. Cited in Shemyakin, F. N., "Orientation in Space," in Ananyev, B. G. et al. (eds.) *Psychological Science in the U.S.S.R.*, Vol. 1, Washington: Office of Technical Services, 1962, 186–225.

Kolodnaya, A. Y. (1956) "Development of the differentiation of the 'right' and 'left' directions of children of preschool age," *Izvestiya Akademii Pedagogicheskikh Nauk RSFSR*, #86. Cited in Shemyakin, F. N., "Orientation in Space," in Ananyev, B. G. et al. (eds.). *Psychological Science in the U.S.S.R.*, Vol. 1, Washington: Office of Technical Services, 1962, 186–225.

Kovach, L. D. (1960) "Life can be so nonlinear," *American Scientist*, 48, 218–225.

Kron, G. (1957) "Tearing, tensors and topological models," *American Scientist*, 45, 401–413.

Kruskal, J. B. (1964a) "Multidimensional scaling by optimizing goodness of fit to a non-metric hypothesis," *Psychometrika*, 29, 1–27.

——— (1964b) "Non-metric multidimensional scaling," *Psychometrika*, 29, 115–130.

Ladd, F. C. (1970) "Black youths view their environment: neighborhood maps," *Environment and Behavior*, 2, 64–79.

Langdon, F. J. (1966) "The social and physical environment: a social scientist's view," *Journal of the Royal Institute of British Architects*, 73, 460–464.

Langer, J. (1969) *Theories of Development*. New York: Holt, Rinehart and Winston.

———, (1970) "Werner's comparative organismic theory," in Mussen, P.H., ed., *Manual of Child Psychology*. Vol. 1, 3rd ed. New York: Wiley, 733–773.

Langhorne, M. C. (1948) "The effects of maze rotation on learning," *Journal of Genetic Psychology*, 38, 191–205.

Lashley, K. S. (1951) "The problem of serial order in behavior," in Leffress, L. H. (ed.), *Cerebral Mechanisms in Behavior: the Hixom Symposium*. New York: Wiley.

Laughlin, W. S. (1968) "Hunting: an integrating biobehavior system and its evolutionary importance," in Lee, R. B., and Devore, I., eds., *Man the Hunter*. Chicago: Aldine.

Laurendau, M., and Pinard, A. (1970) *The Development of the Concept of Space in the Child*. New York: International Universities Press.

Lee, T. R. (1954) *A Study of Urban Neighborhood*, unpublished Ph.D. dissertation, University of Cambridge.

———, (1957) "On the relation between the school journey and social and emotional adjustment in rural infant children," *British Journal of Educational Psychology*, 27, 101–114.

———, (1961) "A test of the hypothesis that school reorganization is a cause of rural depopulation," *Durham Research Review*, 3, 64.

———, (1962) " "Brennan's Law" of shopping behavior," *Psychological Report*, 11, 662.

———, (1963) "The optimum provision and siting of social clubs," *Durham Research Review*, 4, 53–61.

———, (1968) "Urban neighborhood as a socio-spatial schema," *Human Relations*, 21, 241–267.

———, (1969) "The psychology of spatial orientation," *Architectural Association Quarterly*, 1, 11–15.

———, (1970) "Perceived distance as a function of direction in the city," *Environment and Behavior*, 2, 40–51.

Lenkner, H. (1934) *Die Psychologischen Grundlagen der Fortbewegung des Menschen in der Zweidimensionalen unter Besonderer Berucksichtigung der Verkehrstechnik (das Drall-Problem)*. Wurzburg: Trilsch.

Levi-Strauss, C. (1949) "History and anthropology," *Revue de Metaphysique et de Morale*, 54, 363–391. Reprinted in *Structural Anthropology*. New York: Doubleday Anchor, 1967, 1–28.

Levy, L. H. (1970) *Conceptions of Personality: Theories and Research*. New York: Random House.

Lewin, K. (1936) *Principles of Topological Psychology*. New York: McGraw-Hill.

Lewis, C. S. (1957) "The Shoddy Lands," in *The Best from Fantasy and Science Fiction*. New York: Doubleday.

Liebig, F. G. (1933) "Uber unsere Orientierung im Raume bei Ausschluss der Augen," (On orientation in space without sight), *Zeitschrift für Sinnesphysiology*, 64, 251–282. Cited in Howard, I. P., and Templeton, W. B., *Human Spatial Orientation*. New York: Wiley, 1966.

Lord, F. E. (1941) "A study of spatial orientation in children," *Journal of Educational Research*, 34, 481–505.

Lorenz, K. (1966) *On Aggression*. New York: Harcourt, Brace, and World.

Lowenthal, D. (1961) "Geography, experience, and imagination: towards a geographical epistemology," *Annals of the Association of American Geographers*, 51, 241–260.

Lowenthal, D., and Prince, H. (1965) "English landscape tastes," *The Geographical Review*, 186–222

Lowrey, R. A. (1970a) "Distance concepts of urban residents," *Environment and Behavior*, 2, 52–73.

———, (1970b) "Distance concepts of urban residents," in Archea, J., and Eastman, C. M., eds., *EDRA Two: Proceedings of the 2nd Annual Environmental Design Research Association Conference*. Pittsburgh: Carnegie-Mellon University, 135–142.

Lucannas, F. von. (1924) "On the sense of locality in men and animals," *Review of Reviews*, 70, 218.

Lucas, R. C. (1963) "Wilderness perception and use: the example of the Boundary Waters Canoe Area," *Natural Resources Journal*, 3, 394–411.

Luce, R. D., Bush, R., and Galanter, E. (1963) "Psychophysical scaling," in Luce, R. D., and Galanter, E., eds., *Handbook of Mathematical Psychology*. New York: Wiley.

Luce, R. D., and Galanter, E., eds. (1963) *Handbook of Mathematical Psychology*. New York: Wiley.

Lukashok, A. K., and Lynch, K. (1956) "Some childhood memories of the city," *Journal of the American Institute of Planners*, 22, 142–152.

Lund, F. H. (1930) "Physical asymmetries and disorientation," *American Journal of Psychology*, 42, 51–62.

Lundberg, U., and Ekman, G. (1970a) "Emotional involvement while anticipating an examination: a psychophysical study," *Perceptual and Motor Skills*, 31, 603–609.

———, (1970b) "Geographical data as psychophysical stimuli," *Reports from the Psychological Laboratories*, The University of Stockholm, #293.

Lundberg, U., Ekman, G., and Frankenhaeuser, M. (1970) "Anticipation of electric shock. A psychophysical study," *Reports from the Psychological Laboratories*, The University of Stockholm, #308.

Luneburg, R. K. (1947) *The Mathematical Analysis of Binocular Vision*. Princeton, New Jersey: Princeton University Press.

Lyman, S. M., and Scott, M. B. (1967) "Territoriality: a neglected sociological dimension," *Social Problems*, 15, 236–249.

Lynch, K. (1960) *The Image of the City*. Cambridge, Mass.: MIT Press.

———, (1962) *Site Planning*. Cambridge, Mass.: MIT Press.

Lynch, K., and Rivkin, M. (1959) "A walk around the block," *Landscape*, 8, 24–34.

Lyons, H. (1928) "The sailing charts of the Marshall Islanders," *Geographical Journal*, 72, 325–328.

Lyublinskaya, A. A. (1948) "Learning spatial relations by a child of pre-school age," *Anthology Problemy Psikhologii*, Leningrad: Izd-vo LGU Leningrad State University. Cited in Shemyakin, F. N., "Orientation in Space," in Ananyev, B. B. et al. (eds.), *Psychological Science in the U.S.S.R.*, Vol. 1, Washington: Office of Technical Services, 1962, 186–255.

———, (1954) "The role of speech in the development of visual perception in children," *Sb. Vaprosy Detskoy: Obshechey Psikhologii*, Moscow: Izd-vo Academiia Pedagogiehe-skikh Nauk RSFRS. Cited in Shemyakin, F. N. "Orientation in space," in Ananyev, B. G. et al. (eds.), *Psychological. Science in the U.S.S.R.*, Vol. 1, Washington: Office of Technical Services, 1962, 186–255.

———, (1956) "Peculiarities of the assimilation of space by children of preschool age," *Izvestiya Akademii Pedagogicheskikh Nauk RSFSR*, #86. Cited in Shemyakin, F. N., "Orientation in space," in Ananyev, B. G. et al. (eds.), *Psychological Science in the U.S.S.R.*, Vol. 1, Washington: Office of Technical Services, 1962, 186–255.

Mackay, D. M. (1951) "Mindlike behavior in artefacts," *British Journal for the Philosophy of Science*, 2, 105–121.

Mackay, J. R. (1958) "The interactance hypothesis and boundaries in Canada," *Canadian Geographer*, 11, 1–8.

Maegraith, B. G. (1932) "The astronomy of the Aranda and Luritja tribes," Adelaide University Field Anthropology, Central Australia no. 10, taken from *Transactions of the Royal Society of South Australia*, 56.

Malán, M. (1940) "Zur Erblichkeit der Orientierungsfähigkeit im Raum," *Zeitschrift fur Morphologie un Okologie der Tierre*, 39, 1–23.

Malinowski, B. (1922) *Argonauts of the Western Pacific*. London: Routledge.

Mandler, G. (1962) "From association to structure," *Psychological Review*, 69, 415–427.

Manning, P. (1965) "Human consequences of building design decisions," *Architect's Journal*, 142, 1577–1580.

Marble, D. F. (1967) "A theoretical exploration of individual travel behavior," in Garrison, W. L. and Marble, D. F. (eds.), *Quantitative Geography, Part 1*,

Economic and Cultural Topics. Evaston, Ill.: Northwestern University, Department of Geography, Studies in Geography, #13, 33–53.
Marble, D. F., and Bowlby, S. R. (1968) "Shopping alternatives and recurrent travel patterns," in Horton, F., (ed.), *Geographic Studies of Urban Transportation and Network Analysis*. Evanston, Ill.: Northwestern University, Department of Geography, Studies in Geography #16, 42–75.
Marie, P., and Béhague, B. (1919) "Syndrome de désorientation dans l'espace consecutif aux plaies profondes du lobe frontal," *Revue Neurologique*, 26, 1–14.
Mark, L. (1970) "The effect of concrete operations on the mental maps of children," unpublished seminar paper, Department of Psychology, Clark University.
Marshall, D. R. (1968) "Who participates in what? A bibliographic essay on individual participation in urban areas," *Urban Affairs Quarterly*, 4, 201–223.
Matthews, G. V. T. (1951) "Experimental investigation of navigation in homing pigeons," *Journal of Experimental Biology*, 28, 508–536.
———, (1953) "Navigation in the Manx shearwater," *Journal of Experimental Biology*, 30, 370–396.
Mazlish, B., ed. (1965) *The Railway and the Space Program: An Exploration in Historical Analogy*. Cambridge, Mass.: MIT Press.
McCleary, G. F., Jr. (1970) "Beyond simple psychophysics—approaches to the understanding of map perception," *Technical Papers of the Thirtieth Annual Meeting, American Congress on Surveying and Mapping*, 189–209.
McClelland, D. (1961) *The Achieving Society*. New York: Van Nostrand.
McClenahan, B. (1929) *The Changing Urban Neighborhood*. Los Angeles: University of Southern California Press.
McFie, J., Piercy, M. F., and Zangwill, O. L. (1950) "Visual-spatial agnosia associated with lesions of the right cerebral hemisphere," *Brain*, 73, 167–190.
McGee, V. E. (1966) "Multidimensional analysis of 'elastic distances,'" *British Journal of Mathematical and Statistical Psychology*, 19, 181–196.
———, (1968) "Multidimensional scaling of N sets of similarity measures: a nonmetric individual differences approach," *Multivariate Behavioral Research*, 3, 233–248.
McKenzie, R. D. (1921) "The neighborhood: a study of local life in the city of Columbus, Ohio," *American Journal of Sociology*, 27, 145–168, 344–363, 486–513, 588–610, 780–799.
McReynolds, J., and Worchel, P. (1954) "Geographic orientation in the blind," *Journal of Genetic Psychology*, 51, 221–236.
Mead, M. (1932) "An investigation of the thought of primitive children, with special reference to animism," *Journal of the Royal Anthropological Institute*, 62, 173–190. Reprinted in Hunt, R., ed., *Personalities and Cultures: Readings in Psychological Anthropology*. New York: Natural History Press, 1967, 213–237.
Meier, R. L. (1959) "Measuring social and cultural change in urban regions," *Journal of the American Institute of Planners*, 25, 180–190.
———, (1962) *A Communication Theory of Urban Growth*. Cambridge, Mass.: The MIT Press.
Merritt, R. (1962) "Systems and the disintegration of empires," *General Systems Yearbook*, 8, 91–103.
Merton, R. K. (1957) *Social Theory and Social Structure*. Glencoe, Ill.: The Free Press.
Meyer, E. (1935) "La répresentation des relations spatiales chez l'enfant," *Cahiers de Pedagogie Experimental et de Psychologie de l'Enfant* (Geneva), 8.

Meyer, E. (1940) "Comprehension of spatial relationships in preschool children," *Journal of Genetic Psychology*, 57, 119–151.

Meyer, O. (1900) "Ein-und doppelseitige homonyme Hemianopsie mit Orientierungsstörungen," *Mschr. Psychiat. Neurol*, 8, 440–456.

Michelson, W. (1970) *Man and his Urban Environment: A Sociological Approach*. Reading, Mass.: Addison-Wesley.

Milgram, S. (1970) "The experience of living in cities," *Science*, 167, 1461–1468.

Miller, G. A. (1956) "The magical number seven plus or minus two: some limits in our capacity for processing information," *Psychological Review*, 63, 81–97.

———, (1962) *Psychology: the Science of Mental Life*. New York: Harper Row.

Miller, G. A., Galanter, E., and Pribram, K. H. (1960) *Plans and the Structure of Behavior*. New York: Holt.

Miller, J. W. (1967) "Measuring perspective ability," *Journal of Geography*, 66, 167–171.

Milner, P. M. (1957) "The cell assembly: Mark II," *Psychological Review*, 64, 242–252.

Montgomery, R. (1965) "Improving the design process in urban renewal," *Journal of the American Institute of Planners*, 31, 7–20.

Moore, E. G., and Brown, L. A. (1970) *Spatial Properties of Urban Contact Fields: An Empirical Analysis*. Northwestern University, Department of Geography, Discussion Paper #52.

Moore, G. T. (1970) "A comparative analysis of organismic-developmental theory and general systems theory," unpublished paper, Department of Psychology, Clark University.

Moore, T. V. (1915) "The temporal relations of meaning and imagery," *Psychological Review*, 22, 177–225.

Morison, S. E. (1970) *The European Discovery of America*. Oxford: Oxford University Press.

Morrill, R. L., and Pitts, F. R. (1967) "Marriage, migration and the mean information field," *Annals of the Association of American Geographers*, 57, 401–422.

Muchow, M., and Muchow, H. (1935) *Der Lebensraum des Grosstadtkindes. (The Life Space of the Child in the Large City)*, Hamburg: Verlag. Cited in Werner, H., *Comparative Psychology of Mental Development* (rev. ed.), New York: International Universities Press, 1948.

Muenzinger, K. F. (1938) "Vicarious trial and error at a point of choice: I. A general survey of its relation to learning efficiency." *Journal of Genetic Psychology*, 53, 75–86.

Muir, M. E. (1970) "The use of aerial photographs as an aid in teaching map skills in the first grade," *Place Perception Reports*, No. 3. Worcester, Mass.: Clark University.

Muir, M. E., and Blaut, J. M. (1969–1970) "The use of aerial photographs in teaching mapping to children in the first grade: an experimental study," *Minnesota Geographer*, 22 (3), 4–19.

Munn, N. L. (1950) *Handbook of Psychological Research on the Rat*. Boston: Houghton-Mifflin.

Muth, R. F. (1969) *Cities and Housing*. Chicago: University of Chicago Press.

Neisser, U. (1968) "The processes of vision," *Scientific American*, 219, 204–214.

"A New Yorker's idea of The United States of America," (1936), *The Saturday Review of Literature*, 15, no. 5, 4.

Nice, M. (1941) "Territory in bird life," *American Midland Naturalist*, 26, 441–487.

North, O. H., Zaninovich, M. G., and Zinnes, D. (1963) *Content Analysis*. Evanston: Northwestern University Press.

Noyes, H. P. (1957) "The physical description of elementary particles," *American Scientist*, 45, 431–448.

Nunnally, J. C. (1967) *Psychometric Theory*. New York: McGraw-Hill.

Nystuen, J. D. (1967) "A theory and simulation of intra-urban travel," in Garrison, W. L., and Marble, D. F., eds., *Quantitative Geography, Part 1, Economic and Cultural Topics*. Evanston, Ill.: Northwestern University, Department of Geography, Studies in Geography #13, 54–83.

Ola, D. (1968) *The Perception of Geographic Space in Lagos and West States of Nigeria*, unpublished M.A. thesis, Pennsylvania State University.

Oldfield, R. C. (1954) "Memory mechanisms and the theory of schemata," *British Journal of Psychology*, 45, 14–23.

Olds, J. (1969) "The central nervous system and the reinforcement of behavior," *American Psychologist*, 24, 114–132.

Olds, J., and Milner, P. M. (1954) "Positive reinforcement produced by electrical stimulation of septal area and other regions of rat brain," *Journal of Comparative and Physiological Psychology*, 47, 419–427.

Olson, D. R., and Baker, N. E. (1969) "Children's recall of spatial orientation of objects," *Journal of Genetic Psychology*, 114, 273–281.

Olsson, G. (1965) *Distance and Human Interaction*. Philadelphia: Regional Science Research Institute, Bibliography Series #2.

Ore, O. (1962) *Theory of Graphs*, American Mathematical Society Colloquium Publications, Vol. 38.

Park, R. E. (1952) "Sociology, community, and society," in Park, R. E., *Human Communities: The City and Human Ecology*. Glencoe: The Free Press, 178–209.

Parr, A. E. (1963) "Mind and milieu," *Sociological Inquiry*, 107, 273–288.

———, (1967) "The child in the city: urbanity and the urban scene," *Landscape*, 16 (Spring), 1–3.

Pastalan, L. A., and Carson, D. (1970) *Spatial Behavior of Older People*. Ann Arbor, Michigan: Institute of Gerontology, University of Michigan.

Paterson, A., and Zangwill, O. L. (1944) "Recovery of spatial orientation in the post-traumatic confusional state," *Brain*, 67, 54–68.

———, (1945) "A case of topographical disorientation associated with a unilateral cerebral lesion," *Brain*, 68, 188–212.

Paul, I. H. (1959) *Studies in Remembering*. New York: International University Press.

Pauling, L. (1955) "The stochastic method and the structure of proteins," *American Scientist*, 43, 285–297.

Perrin, F. A. C. (1914) "An experimental and introspective study of the human learning process in the maze," *Psychological Monographs*, 16, #70.

Peterson, G. L. (1967) "A model of preference: quantitative analysis of the perception and of the visual appearance of residential neighborhoods," *Journal of Regional Science*, 7, 19–31.

Peterson, J. (1916) "Illusions of direction orientation," *Journal of Philosophy, Psychology, and Scientific Methods*, 13, 225–236.

Pfeiffer, J. E. (1969) *The Emergence of Man*. New York: Harper and Row.

Piaget, J. (1953) "How children form mathematical concepts," *Scientific American*, 189 (5), 74–79.

———, (1954a) *The Child's Construction of Reality*. New York: Basic Books.

———, (1954b) "Perceptual and cognitive (or operational) structures in the

development of the concept of space in the child (Summary)," *Proceedings, 14th International Congress of Psychology*. Amsterdam: North Holland Publishing Co., 41–46. Also in *Acta Psychologica*, Amsterdam, 1955, 11, 41–46.
Piaget, J. (1962) "Beginnings of representative imitation and further development of imitation," in *Plays, Dreams, and Imitation in Childhood*. New York: Norton, 62–86.
———, (1963a) *The Origins of Intelligence in Children*. New York: Norton.
———, (1963b) *The Psychology of Intelligence*. Totowa, New Jersey: Littlefield, Adams.
———, (1963c) *The Child's Conception of the World*. Patterson, N.J.: Littlefield, Adams.
———, (1968) *Six Psychological Studies*. New York: Vintage.
———, (1969) *The Mechanisms of Perception*. New York: Basic Books.
———, (1970) *Structuralism*. New York: Basic Books.
Piaget, J., and Inhelder, B. (1967) *The Child's Conception of Space*. New York: Norton.
———, (1969) *The Psychology of the Child*. New York: Basic Books.
Piaget, J., Inhelder, B., and Szeminska, A. (1960) *The Child's Conception of Geometry*. New York: Basic Books.
Piercy, M. (1957) "Experimental disorientation in the horizontal plane," *Quarterly Journal of Experimental Psychology*, 9, 65–77.
Pinard, A., and Laurendau, M. (1966) "Le caractère topologique des premières représentations spatiales de l'enfant: examen des hypothèses de Piaget. (The topological character of the child's first spatial representations: examination of some of Piaget's hypotheses.)" *Journal International de Psychologie*, 1, 243–255.
———, (1969) " 'Stage' in Piaget's developmental theory: exegesis of a concept," in Elkind, D., and Flavell, J. H., eds., *Studies in Cognitive Development: Essays in Honor of Jean Piaget*. New York: Oxford University Press, 121–170.
Pink, O. M. (1933) "Spirit ancestors in a Northern Arande horde country," *Oceania*, 4, 176–186.
———, (1936) "The landowners in the Northern division of the Aranda tribe," *Oceania*, 6, 275–305.
Pollock, F. (1938) "Zur Pathologie und Klinik der Orientierung (Isolierte Orientierungsstorung im Raum infolge ubergrossen, linksseitigen Stirnhirntumors)," *Schweiz. Arch. Neurol. Psychiat.*, 42, 141–164.
Pomerantz, J. R., Kaplan, S., and Kaplan, R. (1969) "Satiation effects in the perception of single letters," *Perception and Psychophysics*, 6, 129–132.
Porteus, S. D. (1931) *The Psychology of a Primitive People*. New York: Longmans, Green.
Posner, M. I. (1969) "Representational systems for storing information in memory," in Talland, G. A., and Waugh, N., eds., *Psychopathology of Memory*. New York: Academic Press.
———, (1970) "Abstraction and the processes of recognition," in Bower, G. H., and Spence, J. T., eds., *The Psychology of Learning and Motivation: Advances in research and theory*, Vol. 3. New York: Academic Press.
Posner, M. I., and Keele, S. W. (1968) "On the genesis of abstract ideas," *Journal of Experimental Psychology*, 77, 353–363.
Pratolini, V. (1947) *Il Quartiere*. Florence: Valleschi.
Prince, H. C. (1971) "Real, imagined and abstract worlds of the past," in Board, C., Chorley, R., Haggett, P., and Stoddart, D., eds., *Progress in Geography*, Volume 3. London: Edward Arnold, 1–86.

Proshansky, H. M., Ittelson, W., and Rivlin, L. G., eds., (1970) *Environmental Psychology: Man and his Physical Setting*. New York: Holt, Rinehart, and Winston.
Proust, M. (1925) *Albertine Disparue*. Paris: Nouvelle Revue Française.
———, (1954) *Du Côté de chez Swann*. Paris: Gallimard.
Rand, G. (1969) "Some Copernican views of the city," *Architectural Forum*, 132 (9), 77–81.
Rasmussen, K. J. V. (1931) *The Netsilik Eskimos* (Report of the Fifth Thule Expedition, 1921–24, Volume 8, No. 1–2.) Copenhagen: Gyldendal.
Rasmussen, S. E. (1959) *Experiencing Architecture*. Cambridge, Mass.: MIT Press.
Rattray, R. S. (1927) *Religion and Art in the Ashanti*. Oxford: Clarendon Press.
Reed, C. F. (1969) "The solution of hidden-figures: brightness discrimination and eye movement," *Neuropsychologia*, 7, 121–133.
Reichard, G. A. (1950) *Navaho Religion: a Study of Symbolism*. New York: Pantheon.
Restle, F. (1961) *The Psychology of Judgment and Choice*. New York: Wiley.
Reynolds, D., and McNulty, M. L. (1968) "Political boundaries, barrier effects, and space perception," paper presented at the 1968 Annual meeting of the Association of American Geographers.
Richardson, A. (1969) *Mental Imagery*. New York: Springer.
Ridgley, D. (1922) "The teaching of directions in space and on maps," *Journal of Geography*, 21, 66–72.
Robinson, A. R. (1952) *The Look of Maps: An Examination of Cartographic Design and Relative Values*. New Haven: Yale University Map Laboratory.
———, (1967) "Psychological aspects of color in cartography," *International Yearbook of Cartography*, 7, 50–61.
Robinson, J. P., and Hefner, R. (1967) "Multidimensional difference in public and academic perceptions of nations," *Journal of Personality and Social Psychology*, 7, 251–259.
———, (1968) "Perceptual maps of the world," *Public Opinion Quarterly*, 32, 273–280.
Rokeach, M. (1960) *The Open and Closed Mind*. New York: Basic Books.
Romney, A. K., and D'Andrade, R. G., eds. (1964) "Transcultural studies in cognition," *American Anthropologist*, 66.
Roosen-Runge, P. (1967) "Comments on 'The Pattern of Streets,'" *Journal of the American Institute of Planners*, 33, 412–414.
Roskam, E., and Lingoes, J. (1970) "MINISSA -1, a FORTRAN IV (G) program for smallest space analysis of square symmetric matrices," *Behavioral Science*, 15, 204–205.
Rossetti, G. G. (1943) "The Choice," in Cooper, C. W., ed., *Preface to Poetry*. New York: Harcourt, Brace, and Co., 237.
Rostlund, E. (1956) "Twentieth-century magic," *Landscape*, 5, 23–26.
Rudner, R. S. (1966) *Philosophy of Social Science*. Englewood Cliffs, N.J.: Prentice-Hall.
Rummel, R. (1967) "Understanding factor analysis," *Journal of Conflict Resolution*, 11, 445–480.
Rusch, C. W. (1970) "On understanding awareness," *Journal of Aesthetic Education*, 4 (4), 57–79.
Rushton, G. (1969a) "Analysis of spatial behavior by revealed space preference," *Annals of the Association of American Geographers*, 59, 391–400.
———, (1969b) "The scaling of locational preferences," in Cox, K. R., and Golledge. R. G., eds., *Behavioral Problems in Geography: A Symposium.*

Evanston, Ill.: Northwestern University, Department of Geography, Studies in Geography, #17, 197–227.
Russett, B. M. (1966) "Discovering voting groups in the United Nations," *The American Political Science Review*, 60, 327–339.
Ryan, T. A., and Ryan, M. S. (1940) "Geographical orientation," *American Journal of Psychology*, 53, 204–215.
Saarinen, T. (1966) *Perception of the Drought Hazard on the Great Plains*. Chicago: University of Chicago, Department of Geography Research Paper #106.
———, (1968) "Image of the Chicago Loop," University of Chicago: unpublished paper.
———, (1969) "Perception of environment," *Commission on College Geography Resource Paper No. 5*. Washington, D.C.: Association of American Geographers.
Samuel, A. L. (1963) "Some studies of machine learning using the game of checkers," in Feigenbaum, E. A., and Feldman, J., eds., *Computers and Thought*. New York: McGraw-Hill.
Sander, F. (1930) "Structures, totality of experience, and gestalt," in Murchison, C. A., eds., *Psychologies of 1930*. Worcester, Mass.: Clark University Press, 188–204.
Sandstrom, C. (1951) *Orientation in the Present Space*. Stockholm: Almqvist and Wiksell.
Sanoff, H. (1970) "Social perception of the ecological neighborhood," *Ekistics*, 30, 130–132.
Sapir, E. (1912) "Language and environment," *American Anthropologist*, 14, 226–242.
Sato, C. (1960) "Orientation in the perception of space," *Japanese Journal of Psychology*, 31, 153–160.
Schaeffer, A. (1928) "Spiral movement in man," *Journal of Morphology*, 45, 293–398.
Schnore, L. F. (1967) "Community," in Smelser, N. J., ed., *Sociology: an Introduction*. New York: Wiley, 82–150.
Scully, V. (1962) *The Earth, the Temple, and the Gods*. New Haven: Yale University Press.
Searles, H. F. (1960) *The Non-Human Environment in Normal Development and in Schizophrenia*. New York: International Universities Press.
Seminar and Ad Hoc Committee (1959), "Education for research in psychology," *American Psychologist*, 14, 167–179.
Semmes, J., Weinstein, S., Ghent, L., and Teuber, H. L. (1955) "Spatial orientation in man after cerebral injury: I. Analysis by locus of lesion," *Journal of Psychology*, 39, 227–244.
———, (1963) "Correlates of impaired orientation in personal and extra-personal space," *Brain*, 86, 747–772.
Sewell, W. R. D., ed. (1966) *Human Dimensions in Weather Modification*. Chicago: University of Chicago, Department of Geography Research Series #105.
Shelley, M.W., and Bryan, G., eds. (1964) *Human Judgments and Optimality*. New York: Wiley.
Shelton, F. C. (1967) "A note on the world across the street," *Harvard Graduate School of Education Association Bulletin*, 12, 47–48.
Shemyakin, F. N. (1940) "On the psychology of space representations," *Uchenye Zapiski Gos. In-ta Psikhologii* (Moscow), 1, n.p. Cited in Shemyakin, F. N., "Orientation in space," in Ananyev, B. G. et al. (eds.) *Psy-*

chological Science in the U.S.S.R., Vol. 1, Washington: Office of Technical Services, 1962, 186–255.

———, (1962) "Orientation in space," in Ananyev, B. G. et al. (eds.), *Psychological Science in the U.S.S.R.*, Vol. 1, Washington: Office of Technical Services, Report #62–11083, 186–255.

Shen, T–L., and Shen, L. S. C. (1953) *Tibet and the Tibetans*. Stanford: Stanford University Press.

Shepard, P. (1956–57) "Dead cities in the American West," *Landscape*, 6.

Shepard, R. N. (1962a) "The analysis of proximities: multi-dimensional scaling with an unknown distance function, I," *Psychometrika*, 27, 125–140.

———, (1962b) "The analysis of proximities: multi-dimensional scaling with an unknown distance function, II," *Psychometrika*, 27, 219–246.

———, (1968) "*Cognitive Psychology*: a review of the book by U. Neisser," *American Journal of Psychology*, 81, 285–289.

Shepard, R. N., and Chipman, S. (1970) "Second-order isomorphism of internal representations: shapes of states," *Cognitive Psychology*, 1, 1–7.

Shevky, E., and Bell, W. (1955) *Social Area Analysis: Theory, Illustrative Application and Computational Procedures*. Stanford: Stanford University Press.

Silkin, Lord. (1948) *Royal Institute of British Architects Journal*, 55 (#10).

Simon, H. A. (1957) *Models of Man*. New York: Wiley.

———, (1965) *Administrative Behavior*. New York: The Free Press.

———, (1969) *The Sciences of the Artificial*. Cambridge, Mass.: MIT Press.

Sjoberg, L. (1968) "The dimensionality paradox in comparative judgment: a resolution," *Scandinavian Journal of Psychology*, 9, 97–108.

Skinner, B. F. (1938) *The Behavior of Organisms: An Experimental Analysis*. New York: Appleton-Century-Crofts.

———, (1948) *Walden Two*. New York: MacMillan.

———, (1950) "Are theories of learning necessary?" *Psychological Review*, 57, 193–216.

———, (1953) *Science and Human Behavior*. New York: MacMillan.

Smillie, K. W. (1966) *An Introduction to Regression and Correlation*. London: Academic Press.

Smith, W. F. (1933) "Direction orientation in children," *Journal of Genetic Psychology*, 42, 154–166.

Snedecor, G. W. (1956) *Statistical Methods*. 5th Ed. Ames, Iowa: Iowa State College Press.

Sommer, R. (1969) *Personal Space: The Behavioral Basis of Design*. New York: Prentice-Hall.

Sonnenfeld, J. (1966) "Variable values in space landscape: an inquiry into the nature of environmental necessity," *Journal of Social Issues*, 22, 71–82.

———, (1967) "Environmental perception and adaptation level in the Arctic," in Lowenthal, D. (ed.), *Environmental Perception and Behavior*. Chicago: University of Chicago, Department of Geography Research Paper #109, 42–59.

Southworth, M. (1969) "The sonic environment of cities," *Environment and Behavior*, 1, 49–70.

Spearman, C. (1937) *Psychology Down the Ages*. London: MacMillan and Co.

Spencer, B., and Gillen, F. J. (1899) *The Native Tribes of Central Australia*, London: Macmillan.

Sprout, H., and Sprout, M. (1965) *The Ecological Perspective on Human Affairs with Special Reference to International Politics*. Princeton, New Jersey: Princeton University Press.

Stagner, R. (1967) *Psychological Aspects of International Conflict.* Belmont, Calif.: Brooks, Cole Publishing Co.

Stanley, G. (1968) "Emotional involvement and geographic distance," *Journal of Social Psychology*, 75, 165–167.

Stapf, K. -H. (1968) *Untersuchunger zur subjektiven Landkarte,* Dissertation in natural science, Institute of Technology, Carolo-Wilhelmina zu Braunschweig.

Stea, D. (1967) "Reasons for our moving," *Landscape*, 17, 27–28.

———, (1968) "Environmental perception and cognition: toward a model for mental maps," *Student Publication of the School of Design, North Carolina State University*, 18, 64–75.

———, (1969) "The measurement of mental maps: an experimental model for studying conceptual spaces," in Cox, K. R., and Golledge, R. G., eds., *Behavioral Problems in Geography: A Symposium.* Evanston, Ill.: Northwestern University Press, 228–253.

Stea, D., and Bower, G. H. (1963) "A study of schedule preferences," in Interamerican Society of Psychology, *Proceedings of the Seventh Congress, Interamerican Society of Psychology.* México, D.F.: Sociedad Interamericana de Psicologiá.

Stea, D., Douglas, F., Emerson, W., and Hart, R. (1969) "Orientation in the blind: a case study," Clark University: unpublished manuscript.

Stea, D., and Downs, R. M. (1970) "From the outside looking in at the inside looking out," *Environment and Behavior*, 2, 3–12.

Stea, D., and Wood, D. (Forthcoming) *A Cognitive Atlas: The Psychological Geography of Four Mexican Cities.*

Stefánsson, V. (1914) "The Stefánsson-Anderson Arctic expedition of the American Museum; preliminary ethnological report," *Anthropological Papers of the American Museum of Natural History,* 14, Part 1.

Steinitz, C. (1968) "Meaning and the congruence of urban form and activity," *Journal of the American Institute of Planners*, 34, 233–248.

Stern, W. (1930) *Psychology of Early Childhood.* New York: Holt.

Stevens, S. S. (1957) "On the psychophysical law," *Psychological Review*, 64, 153–181.

———, (1960) "The psychophysics of sensory function," *American Scientist*, 48, 226–253.

———, (1966) "A metric for social consensus," *Science*, 151, 530–541.

Stevens, S. S., and Galanter, E. H. (1957) "Ratio scales and category scales for a dozen perceptual continua," *Journal of Experimental Psychology*, 54, 377–411.

Straus, E. (1963) *The Primary World of Senses.* Glencoe, Ill.: Free Press.

———, (1966) *Phenomenalogical Psychology: Selected Papers.* New York: Basic Books.

Strauss, A. (1961) *Images of the American City.* Glencoe: The Free Press.

Strehlow, C. (1907) *Die Aranda und Loritza-stämme in Zentral Australien.* Frankfurt am Main: J. Baer (1907–1920).

Studer, R. G. (1966) "On environmental programming," *Architectural Association Journal*, 81, 290–296.

Studer, R. G., and Stea, D. (1966) "Architectural programming, environmental design, and human behavior," *Journal of Social Issues*, 22, 127–136.

Sverlov, V. S. (1951) *Prostranstevennaya Orientirovka Sepykh,* (Space Orientation by the Blind), Moscow: n.p., cited in Shemyakin, F. N. "Orientation in space," in Ananyev, B. G. et al. (eds.), *Psychological Science in the U.S.S.R.*, Vol. 1, Washington: Office of Technical Sciences, 1962, 186–225.

Sylvester, R. H., (1913) "The mental imagery of the blind," *Psychological Bulletin*, 10, 210–211.

Szymanski, J. S. (1913) "Versuche über den richtungssinn beim menschen," *Pflugers Archiv fur die Gesamte fur die Physiologie des Menschen und der Tiere*, 151, 158–170.

Taaffe, E. J. (1962) "The urban hierarchy: an air passenger definition," *Economic Geography*, 38, 1–14.

Taaffe, E., Morrill, R., and Gould, P. (1963) "Transport expansion in underdeveloped countries," *The Geographical Review*, 53, 502–529.

Tatham, G. (1951) "Environmentalism and possibilism," in Taylor, G., ed., *Geography in the Twentieth Century*. New York: Philosophical Library, 128–162.

Taylor, M., Abraham, J., Mills, D., Jasionowski, L., Kalewski, M.A., Spencer, F., and Caddel, A. (1971) "Mack Avenue and Bloomfield Hills from a child's point of view," in Colyard, Y., ed., *Field Notes: the geography of the children of Detroit*. Detroit: Detroit Geographical Expedition, Discussion Paper #3, 19–21.

Thomlinson, R. (1961) "A model for migrational analysis," *Journal of the American Statistical Association*, 56, 675–689.

Thompson, D. L. (1965) "New concept: subjective distance," *Journal of Retailing*, 39, 1–6.

Thompson, G. G. (1962) "The development of space, distance, and position concepts," *Child Psychology: Growth Trends in Psychological Adjustment*. 2nd ed. Boston: Houghton-Mifflin, 326–330.

Tindal, M. (1971) *The Home Range of Black Elementary School Children: A Study in Personal Geography*, unpublished M.A. thesis, Clark University.

Tobler, W. R. (1961) *Map Transformations of Geographic Space*, unpublished Ph.D. dissertation, University of Washington.

———, (1963) "Geographic area and map projections," *Geographical Review*, 53, 59–78.

Tolman, E. C. (1932) *Purposive Behavior in Animals and Men*. New York: Century.

———, (1951) "A psychological model," in Parsons, T., and Shils, E., eds. *Towards a General Theory of Action*. Cambridge, Mass.: Harvard University Press, 279–361.

———, (1958) *Behavior and Psychological Man: Essays in Motivation and Learning*. Berkeley: University of California Press.

Torgerson, W. (1958) *Theory and Methods of Scaling*. New York: Wiley.

Törnqvist, G. (1962) *Transport Costs as a Location Factor for the Manufacturing Industry*. Lund, Sweden: Studies in Geography, Series C.

———, (1963) *Studier i Industrilokadisering*. Stockholm: Geografiska Institionen vid Stockholms Universitet.

Toulmin, S. (1961) *Foresight and Understanding: An Enquiry into the Aims of Science*. New York: Harper Torchbooks.

Towler, J. O. (1970) "The elementary school child's concept of reference systems," *Journal of Geography*, 69, 89–93.

Towler, J. O., and Nelson, L. D. (1968) "The elementary school child's concept of scale," *Journal of Geography*, 67, 24–28.

Trowbridge, C. C. (1913) "Fundamental methods of orientation and imaginary maps," *Science*, 38, 888–897.

Tuan, Y. (1967) "Attitudes toward environment: themes and approaches," in Lowenthal, D. (ed.), *Environmental Perception and Behavior*. Chicago:

University of Chicago, Department of Geography Research Paper #109, 4–17.
Tucker, L. (1964) "Systematic differences between individuals in perceptual judgments," in Shelley, M. W., and Bryan, G., eds., *Human Judgments and Optimality*. New York: Wiley.
Turnbull, C. (1961) "Some observations regarding the experiences and behavior of the BaMbuti pygmies," *American Journal of Psychology*, 74, 304–308.
Twain, M. (1917) *Life on the Mississippi*. New York: Harper.
Tyler, S. A., ed. (1971) *Cognitive Anthropology*. New York: Holt, Rinehart and Winston.
Veldman, D. J. (1967) *FORTRAN Programming for the Behavioral Sciences*. New York: Holt.
Vernon, M. D. (1962) *The Psychology of Perception*. Baltimore: Pelican Books.
Viguier, C. (1882) "Le sens de l'orientation et ses organs chez les animaux et chez l'homme,' *Revue de Philomathematique*, 14, 1–36.
Von Senden, M. (1960) *Space and Sight*. London: Methuen.
Voss, J. F. (1969) "Associative learning and thought: the nature of an association and its relation to thought," in Voss, J. F., ed., *Approaches to Thought*. Columbus, Ohio: Merrill.
Vygotsky, L. S. (1962) *Thought and Language*. Cambridge, Mass.: MIT Press.
Waddell, L. A. (1895) *The Buddhism of Tibet or Lamaism*. London: W. H. Allen.
Wallace, A. F. C. (1961) *Culture and Personality*. New York: Random House.
Wallace, J. G. (1965) "Mathematical and scientific concepts," *Concept Growth and the Education of the Child: A Survey of Research on Conceptualization*. London: National Foundation for Educational Research, Publication Series #12, 95–122.
Wallach, M., and Kogan, N. (1965) *Modes of Thinking in Young Children*. New York: Holt, Rinehart and Winston.
Wapner, S. (1969) "Organismic-developmental theory: some applications to cognition," in Mussen, P. H., Langer, P. and Covington, M., eds., *Trends and Issues in Developmental Psychology*. New York: Holt, Rinehart and Winston, 38–67.
Wapner, S., Cirillo, L., and Baker, A. H. (1973) "Some aspects of the development of space perception," in Hill, J. P., ed., *Minnesota Symposium on Child Psychology*, Vol. 4., Minneapolis: University of Minnesota Press.
Wapner, S., and Werner, H. (1957) *Perceptual Development*. Worcester, Mass.: Clark University Press.
Warren, H. C. (1908) "Magnetic sense of direction," *Psychological Bulletin*, 5, 376–377.
———, (1921) *A History of the Association Psychology*. New York: Charles Scribner's Sons.
Watanabe, S. (1969) *Knowing and Guessing: A Quantitative Study of Inference and Information*. New York: Wiley.
Watson, J. B., and Lashley, K. S. (1915) *Homing and Related Activities of Birds*. Washington: Carnegie Institute.
Watson, J. D. (1968) *The Double Helix*. New York: Atheneum.
Weaver, W. (1961) "The imperfections of science," *American Scientist*, 49, 99–113.
Webber, M. M. (1963) "Order and diversity: community without propinquity," in Wingo, L., ed., *Cities and Space: The Future Use of Urban Land*. Baltimore: The Johns Hopkins Press, 23–56.
———, (1964a) "The urban place and the non-place urban realm," in Webber, M. M., Dyckham, J. W., Foley, D. L., Guttenberg, A. Z., Wheaton,

W. L. C., and Wurster, C. B., eds., *Explorations in Urban Structure*. Philadelphia: University of Pennsylvania Press.

———, (1964b) "Culture, territoriality and the elastic mile," *Papers and Proceedings of the Regional Science Association*, 13, 59–69.

Weinstein, S., Semmes, J., Ghent, L., and Teuber, H. L. (1956) "Spatial orientation in man after cerebral injury: II. Analysis according to concommitant defects," *Journal of Psychology*, 42, 249–263.

Wells, E. F. (1971) "Imagineering," *TWA Magazine*, 24–29.

Werner, H. (1940) "Perception of spatial relationship in mentally retarded children," *Journal of Genetic Psychology*, 57, 93–100.

———, (1948) *Comparative Psychology of Mental Development*. New York: International Universities Press.

———, (1956) "Microgenesis and aphasia," *Journal of Abnormal and Social Psychology*, 52, 347–353.

———, (1957) "The concept of development from a comparative and organismic point of view," in Harris, D. B., ed., *The Concept of Development*. Minneapolis: University of Minnesota Press, 125–148.

Werner, H., and Kaplan, B. (1963) *Symbol Formation: An Organismic-Developmental Approach to Language and the Expression of Thought*. New York: Wiley.

Werner, H., and Wapner, S. (1949) "Sensory-tonic field theory of perception," *Journal of Personality*, 18, 88–107.

———, (1952) "Experiments on sensori-tonic field theory of perception: IV. Effects of initial position of a rod on apparent verticality," *Journal of Experimental Psychology*, 43, 68–74.

———, (1954) "Studies of physiognomic perception: I. Effect of configurational dynamics and meaning induced sets on the position of the apparent median place," *Journal of Psychology*, 38, 51–65.

———, (1956) "Sensori-tonic field theory of perception: basic concepts and experiments," *Revista di Psicologia*, 50, 315–337.

Wertheimer, M. (1923) "Untersuchugen zur lehre von der gestalt, (principles of perception organization)," *Psychologie Forsche*, 4, 301–350. Cited in

Beardslee, D. C., and Wertheimer, M. (eds.), *Readings in Perception*. New York: Van Nostrand, 115–135.

White, G. F. (1966) "Formation and role of public attitudes," in Jarrett, H. (ed.), *Environmental Quality in a Growing Economy: Essays from the Sixth R.F.F. Forum*. Baltimore: The Johns Hopkins Press, 105–127.

White, R. R. (1967a) *Space Preference and Migration: A Multi-dimensional Analysis of the Spatial Preferences of British School Leavers*, unpublished master's thesis, The Pennsylvania State University.

———, (1967b) "The measurement of spatial perception," *University of Bristol, Department of Geography Seminar Papers Series A*., 8. 1–21.

White, S. H. (1970) "The learning theory tradition and child psychology," in Mussen, P.H., ed., *Manual of Child Psychology*. Vol. 1 3rd ed. New York: Wiley, 657–701.

Whitten, E. H. T. (1965) *A Surface-Fitting Program Suitable for Testing Geological Models which Involves Areally-Distributed Data*. Washington, D.C.: Office of Naval Research, Geography Branch, Technical Report No. 2.

Whyte, W. F. (1943) *Streetcorner Society*. Chicago: University of Chicago Press.

———, (1956) *The Organization Man*. New York: Simon and Schuster.

Wiener, N. (1948) *Cybernetics or Control and Communication in the Animal and the Machine*. Cambridge, Mass.: MIT Press.

Williams, E. J. (1959) *Regression Analysis*. New York: Wiley.

Williams, R. L. (1958) "Map symbols: equal-appearing intervals for printed screens," *Annals of the Association of American Geographers*, 48, 132–139.

Wilson, A. (1966) "The impact of climate on industrial growth: Tucson, Arizona—a case study," in Sewell, W. R. D., ed., *Human Dimensions of Weather Modification*. Chicago: Department of Geography, University of Chicago, Research Paper #105.

Wilson, G., and Wilson, M. (1945) *The Analysis of Social Change: Based on Observations in Central Africa*. Cambridge: Cambridge University Press.

Winfield, G. F. (1948) *China: the Land and the People*. New York: William Sloane Association.

Wingo, L. (1961) *Transportation and Urban Land*. Baltimore: The Johns Hopkins Press.

Wisner, B. (1970) "Protogeography: search for the beginnings," unpublished paper presented to the Annual Conference of the Association of American Geographers, San Francisco.

Witkin, H. A. (1946) "Studies in geographic orientation," *Yearbook of the American Philosophical Society*, 152–155.

———, (1949) "Orientation in space," *Research Reviews*, Office of Naval Research, Washington, D.C., December.

Witkin, H. A., Dyk, R. B., Faterson, H. F., Goodenough, D. R., and Karp, S. A. (1962) *Psychological Differentiation: Studies of Development*. New York: Wiley.

Witkin, H. A., Lewis, H. B., Hertzman, M., Machover, K. Meissner, P. B., and Wapner, S. (1954) *Personality Through Perception: An Experimental and Clinical Study*. New York: Harper.

Wohl, R. R., and Strauss, A. L. (1958) "Symbolic representation and the urban milieu," *American Journal of Sociology*, 63, 523–532.

Wohlwill, J. F. (1960) "Developmental studies of perception," *Psychological Bulletin*, 57, 249–288.

———, (1970) "The emerging discipline of environmental psychology," *American Psychologist*, 25, 303–312.

Wolpert, J. (1964) "The decision process in a spatial context," *Annals of the Association of American Geographers*, 54, 537–558.

——— (1965) "Behavioral aspects of the decision to migrate," *Papers and Proceedings of the Regional Science Association*, 15, 159–169.

Wood, D. (1969) "The image of San Cristobal," *Monadnock* (Clark Geographical Society), 43, 29–45.

———, (1971) *Fleeting Glimpses: Adolescent and Other Images of the Entity called San Cristobal las Casas, Chiapas, Mexico*, unpublished M.A. thesis, Clark University.

Wood, M. (1968) "Visual perception and map design," *The Cartographic Journal*, 5, 54–64.

Woodworth, R. S. (1938) *Experimental Psychology*. New York: Holt.

Worchel, P. (1951) "Space perception and orientation in the blind," *Psychological Monographs*, 65, Whole #332.

———, (1952) "The role of vestibular organs in space orientation," *Journal of Experimental Psychology*, 44, 4–10.

Worchel, P., and Rockett, F. C. (1955) "The frame of reference in perceptual and motor skill: 1. The effect of changing frames of reference," *Perceptual and Motor Skills*, 4, 115–121.

Wright, J. K. (1925) *The Geographical Lore of the Time of the Crusades*. New York: American Geographical Society Research Series #15.

———, (1947) "Terrae incognitae: the place of the imagination in geography," *Annals of the Association of American Geographers*, 37, 1–15.

Young, F. W., and Torgerson, W. S. (1967) "TORSCA, a FORTRAN IV program for Shepard-Kruskal multidimensional scaling analysis," *Behavioral Science*, 12, 498.

Yung, E. (1918) "Le sens de la direction," *Echo des Alpes*, 4, 110.

Zangwill, O. L. (1951) "Discussion on parietal lobe syndromes," *Proceedings of the Royal Society of Medicine*, 44, 343–346.

Zannaras, G. (1968) *An Empirical Analysis of Urban, Neighborhood Perception*, unpublished M.A. thesis, The Ohio State University.

Zemtsova, M. I. (1956) *Puti Kompensatsii Slepoty* (*Ways of Compensating for Blindness*), Moscow: Izd-vo Akedemiya Pedagogisheskikh Nauk RSFSR. Cited in Shemyakin, F. N., "Orientation in space," in Ananyev, G. G. et al. (ed.), *Psychological Science in the U.S.S.R.*, Vol. 1, Washington: Office of Technical Services, 1962, 186–255.

Zerner, C. (1970) "Notes on toy play, toy construction, and play fantasies: a study of the play behavior of the children of St. Vincent within the context of place perception theory," Clark University: Unpublished manuscript.

Zinnes, J. L. (1969) "Scaling," *Annual Review of Psychology*, 20, 447–478.

Zipf, G. K. (1949) *Human Behavior and the Principle of Least Effort*. Cambridge, Mass.: Addison-Wesley.

Indexes

Name Index

Index of Place Names

This index contains references to places mentioned in the text. Only those geographic locations figuring significantly are indexed here; those mentioned only incidentally are not included.

Subject Index